T0192816

Bayesian Inference for Stochastic Processes

Bayesian Inference for Stochastic Processes

Lyle D. Broemeling

CRC Press is an imprint of the
Taylor & Francis Group, an **informa** business
A CHAPMAN & HALL BOOK

CRC Press
Taylor & Francis Group
6000 Broken Sound Parkway NW, Suite 300
Boca Raton, FL 33487-2742

First issued in paperback 2020

ISBN-13: 978-0-367-57243-3 (pbk)
ISBN-13: 978-1-138-19613-1 (hbk)

Visit the Taylor & Francis Web site at
http://www.taylorandfrancis.com

and the CRC Press Web site at
http://www.crcpress.com

I would like to dedicate this book to my wife, Ana Broemeling, who provided the support and love that helped me face the many challenges in this effort.

Contents

Preface

Bayesian methods are being used in many areas of scientific investigation. This is demonstrated by referring to scientific literature, which shows that the Bayesian approach is prevalent in medicine, finance, astronomy, economics, cryptology, engineering, and various branches of biology. For example, in medicine, Bayesian sequential stopping rules are employed in the design and analysis of clinical trials and are used to assess the accuracy of various diagnostic tests.

This book is intended to be a textbook for graduate students in statistics and biostatistics and a reference for consulting statisticians. It will be quite valuable for those involved in using stochastic processes to model various phenomena. The book adopts a unique data analytic approach to making inferences for the unknown parameters of the process. The examples are taken from biology, finance, and sociology; thus, the book should be a good resource for those consulting in those areas of scientific endeavor. A good background in probability and statistical inference is required. Courses in stochastic processes would further aid the student; however, the book does provide a solid background in stochastic processes and in Bayesian inference. Two software packages are used in this book. The first is the R package, which simulates realizations from the relevant stochastic process, while WinBUGS provides the Bayesian analysis for the unknown parameters. The code for R and WinBUGS is available at the author's website: http://www.lbroemeling.com/.

Bayesian Inference for Stochastic Processes is a valuable reference for the consulting statistician and for the Bayesian working in the area of stochastic processes. The book can be thought of as a companion to *Bayesian Analysis of Stochastic Process Models* by Insua, Ruggeri, and Wiper, whose approach is more theoretical than that of *Bayesian Inference for Stochastic Processes*, which takes a more Bayesian data analysis method.

Author

Lyle D. Broemeling, PhD, is the director of Broemeling and Associates, Inc., and is a consulting biostatistician. He has been involved with academic health science centers for about 20 years and has taught and been a consultant at the University of Texas Medical Branch in Galveston, the University of Texas MD Anderson Cancer Center, and the University of Texas School of Public Health. His main interest is developing Bayesian methods for use in medical and biological problems and in authoring textbooks in statistics. His numerous books include *Bayesian Biostatistics and Diagnostic Medicine*, *Bayesian Methods in Epidemiology*, and *Bayesian Methods for Agreement*.

1

Introduction to Bayesian Inference for Stochastic Processes

1.1 Introduction

Chapter 1 presents the overall objectives of the book and a preview of the book. Bayesian inference is becoming the predominant way to make statistical inferences about the parameters of a model, and in this case, the model is a stochastic process. Stochastic processes is a branch of probability theory, where the parameters of the process are assumed to be known, but in this book, based on data about the process, inferences about the parameters will be performed from a Bayesian perspective. Thus, the book requires a good background in both stochastic processes and Bayesian inference. Fortunately, the reader will be able to learn these two subjects, namely, stochastic processes and Bayesian inference. Inference means estimating parameters, testing hypotheses about those parameters, and predicting future observations of the process. Of course, in order to provide Bayesian inferences, essential information about the relevant stochastic process is necessary, and this preview begins with that information.

This book uses R, mainly for generating observations for stochastic processes and for plotting various observations pertinent for stochastic processes. Also, WinBUGS plays an important role for executing a Bayesian analysis for the unknown parameters of the model. It is sometimes employed to generate observations from stochastic processes with known parameters.

1.2 Bayesian Statistical Fundamentals

Chapter 2 is an introduction to Bayesian inference beginning with the Bayes theorem, which is expressed in terms of prior information about unknown parameters; sample information via the likelihood function; and, subsequently, the posterior distribution, as determined by the Bayes theorem.

Our first population to be considered is binomial with one parameter, the probability of success, where the prior distribution is a conjugate beta distribution, resulting in a posterior distribution, which is beta. As a special case, the uniform prior is used if little information is available, also resulting in a beta posterior distribution.

The next population to be considered is normal with three cases: (1) the mean unknown and variance known, (2) the mean known and the precision unknown, and (3) both mean and variance of the normal unknown. In the first case, the prior distribution for the mean is assumed to be normal with known mean and precision, and then applying the Bayes theorem results in a normal distribution for the mean. In the second case, a gamma distribution is assigned to the precision, resulting in a gamma posterior distribution. The third case is the most realistic, where both the mean and precision are unknown, where a normal gamma prior is assigned, resulting in a normal gamma posterior for the mean and precision. The marginal posterior distribution of the mean is a univariate t while that for the precision is a gamma.

Chapter 2 continues with presenting Bayesian inferential techniques based on the posterior distribution for the binomial and normal populations.

For example, for the normal with both mean and precision unknown, a point estimate of the mean is the mean of the posterior distribution of the mean. In all three cases of the normal, the predictive distribution of future observations is derived.

Also, Chapter 2 introduces the fundamental ideas of computing the posterior distribution via simulation methods. In particular, Monte Carlo Markov chain (MCMC) methods are described and illustrated with WinBUGS for the normal and binomial populations.

1.3 Fundamentals of Stochastic Processes

Chapter 3 describes the theory that is necessary to understand the probability ideas of stochastic processes beginning with the definition. Basically, a stochastic process is a collection of random variables whose indices are points of time and where the joint distribution of the collection can be determined for every choice of time points. The general definition is followed by two examples of stochastic processes, the Poisson and Wiener processes. A Poisson process is a counting process in continuous time that counts the number of events occurring over time, and the distribution of the probability of the counts follows a Poisson process. On the other hand, the Wiener process is continuous in time and has a continuous state space, and the joint distribution of a collection of Wiener variables follows a multivariate normal distribution. It needs to be stressed that the Wiener process and the Brownian motion are the same stochastic process.

There are two fundamental ideas when describing a stochastic process, namely, the state space and the time index called the parameter of the process. This use of the word *parameter* is not to be confused when the word is used as the parameter of a distribution.

Next to be presented in Chapter 3 is the classification of the states of a Markov chain, which is a process that can have either a discrete or a continuous index set, but where the state space is discrete. The basic idea in dealing with the classification of the states is that of accessibility.

One state is accessible from another if there is a positive probability of going from this state to the other in a finite number of steps. Once one has the definition of accessibility, one can define a communicating class, where it is possible for each state of the class to be accessible from all the other states. In addition, the definition of accessibility allows one to define the recurrence, periodicity, and transience of a state. Other ideas introduced in Chapter 3 are that of a homogenous chain and that of an embedded Markov chain. Of prime importance to a homogenous and stationary Markov chain is the knowledge of the one-step transition probability matrix and the initial distribution that allows one to determine all the probabilistic properties of the chain.

Continuous Markov chains are those with a continuous index set, and the birth and death process is a good example. For such a process at a given time, there is the positive probability of a birth, the positive probability of a death, and, hence, the probability of neither. These probabilities might depend on the current state of the process, or they may be independent of the current state, and both situations are explained. What happens in the long run for a Markov process? For example, suppose the Markov chain initially is in a given state, then what is the probability that the process will remain in that state in the long run? Therefore, limiting distributions are important to know, and several examples illustrate the limiting distributions of several stochastic process.

As mentioned earlier, the Wiener process (or the Brownian motion) is an example of a Markov process with continuous state and continuous time, which is used to characterize the behavior of tiny microscopic particles called the Brownian motion.

Chapter 3 ends with a discussion of stochastic calculus, where the derivative and integral of Wiener processes are defined and illustrated with many examples. The student will benefit from solving the many exercises at the end of Chapter 3.

1.4 Discrete Markov Chains

Chapter 4 is the first chapter where Bayesian inferences are introduced for a specific class of Markov processes, those with a discrete index set. R and WinBUGS are employed to generate observations from a given chain and to compute posterior distributions of the unknown parameters, the one-step transition probabilities of the chain, etc. A rainy day is an example of a four-state chain. Suppose that whether or not it rains today is a function of what happened the past two days. In particular, suppose that if it rained for the past two days, then it will rain tomorrow with a probability of .7, and if it rained today but not yesterday, then it will rain tomorrow with a probability of .5. In addition, if it rained yesterday but not today, it will rain the following day with a probability of .4, and lastly, if it has not rained yesterday and the day before yesterday, it will rain tomorrow with a probability of .2. This is enough information to define a four-state Markov chain as follows:

- State 1: It rained today and yesterday.
- State 2: It rained today but not yesterday.
- State 3: It rained yesterday but not today.
- State 4: It did not rain yesterday or today.

The resulting Markov chain has a probability transition matrix as follows:

$$P = \begin{pmatrix} .7, .0, .3, .0 \\ .5, .0, .5, .0 \\ .0, .4, .0, .6 \\ .0, .2, .0, .8 \end{pmatrix}. \tag{1.1}$$

For this example, the R language was used to generate multinomial observations for the first row of the one-step transition probability matrix, then assuming a uniform prior for the first-row probabilities, the resulting posterior Dirichlet distribution allows one to execute a Bayesian analysis for estimating these one-step transition probabilities. In addition to estimation, a formal Bayesian test of the null hypothesis is performed. The last step of Bayesian techniques is to derive the predictive distribution for the cell counts of the first row of the process.

It is very important to know the long-term behavior of a stochastic process. In the long run, what are the possible states of a Markov chain? And how do they depend on the initial state? Such problems come under the subject of limiting probabilities. R Code 4.2 (Matrixpower) will be employed to determine the long-term behavior of a Markov chain. An example based on a study by Albert[1] reveals the long-term behavior of the Canadian Forest Fire Weather Index. A five-state transition matrix is based on data taken over 26 years at 15 weather stations. The time unit is a day, and the following matrix, taken from one location in the early summer, gives the probability of daily changes in the fire index (the states of the chain). The fire index has five values: nil, low, moderate, high, and extreme.

$$P = \begin{pmatrix} .575, .118, .172, .109, .026 \\ .453, .243, .148, .123, .033 \\ .104, .343, .367, .167, .019 \\ .015, .066, .381, .505, .096 \\ .000, .060, .149, .567, .224 \end{pmatrix}. \tag{1.2}$$

The limiting probabilities of the five states are determined by computing the matrix powers of the preceding one-step transition matrix. Stationary distributions are those that satisfy the matrix equation

$$\pi = \pi P, \tag{1.3}$$

where P is the one-step transition matrix and π is the vector representing the stationary distribution. It is explained under what conditions the stationary distribution exists and the idea illustrated with the preceding five-state chain.

Chapter 4 continues with a presentation of irreducible chains and Bayesian inferences for transient and recurrent states and the estimation of the period of a state. Also discussed are ergodic chains and time reversibility and finding the probability of recurrence using Bayesian methods of inference. Chapter 4 is concluded with a social mobility example where the stationary distribution is determined by a Bayesian analysis.

1.5 Bayesian Inference for Markov Chains in Biology

Markov chains are often used to explain interesting phenomena in biology. Bayesian inferential techniques will be employed to gain a deeper understanding of the mechanism of various biological phenomena.

Several examples will illustrate the Bayesian approach to making inferences: (1) an example of inbreeding in genetics; (2) the general birth and death process; (3) the logistic growth process; (4) a simple model for an epidemic; (5) the chain binomial model and the Greenwood and the Reed–Frost versions of an epidemic; (6) several genetic models, including the Wright model; and (7) the Ehrenfest model for diffusion through a cell membrane.

Bayesian inferences will include determining the posterior distribution of the relevant parameters, testing hypotheses about those parameters, and determining the Bayesian predictive distribution of future observations.

Such Bayesian procedures will closely follow those presented in Chapter 4 and will be composed of generating simulations from the chain, displaying the associated transition graph for the chain of each example, and employing the appropriate R Code and WinBUGS Code for the analysis. Inheritance depends on the information contained in the chromosomes that are passed down to future generations; humans have two sets of chromosomes, one from the mother and one from the father. Certain locations of the chromosome contain detailed information about the physical characteristics of the individual. Chemicals that make up the chromosome at specific locations (loci) are called genes, and at each locus, the genes are manifested in one of several forms called alleles.

The first process to be considered is the so-called inbreeding problem, which leads to a probability transition matrix with three states.

Suppose there are two forms of alleles for a given gene symbolized by *a* and A; thus, humans could then have one of three types, namely, *aa*, AA, or A*a*, and are called genotypes of the locus. In addition, the two genotypes AA and *aa* are denoted as homozygous, while A*a* is referred to as a heterozygous genotype.

The inbreeding problem has six states: (1) AA × AA, (2) AA × A*a*, (3) A*a* × A*a*, (4) A*a* × *aa*, (5) AA × *aa*, and (6) *aa* × *aa*.

The laws of inheritance imply the following makeup of the next generation: (1) If the parents are both of type AA, the offspring will be AA individuals, so that the crossing of a brother and a sister will be of one type, and $P_{11} = 1$. (2) Now suppose the parents are type 2, namely, AA × A*a*, and the offspring will occur in the following proportions: ½ AA and ½ A*a*; therefore, the crossing of a brother and a sister will be ¼ type {AA × AA}, ½ type {AA × A*a*}, and ¼ type {Aa × A*a*}. (3) Lastly, if the parents are of type A*a* × A*a*, the offspring are in the proportion of ¼ type AA, ½ type A*a*, and ¼ type *aa*; thus, brother-and-sister mating will give 1/16 type {AA × AA}, ¼ type {AA × A*a*}, ¼ type {A*a* × A*a*}, ¼ type { A*a* × *aa*}, 1/8 type {AA × *aa*}, and 1/16 type {*aa* × *aa*}. It can be shown that the transition matrix is

$$P = \begin{pmatrix} 1, .0, .0, .0, .0, .0 \\ 1/4, 1/2, 1/4, 0, 0, .0 \\ 1/16, 1/4, 1/4, 1/4, 1/8, 1/16 \\ .0, .0, 1/4, 1/2, .0, 1/4 \\ .0, .0, 1.0, .0, .0, .0 \\ .0, .0, .0, .0, .0, 1.0 \end{pmatrix}. \quad (1.4)$$

Using R, cell counts corresponding to the third row of Equation 1.4 are generated using a multinomial distribution; then with WinBUGS, the posterior Dirichlet distribution of the transition probabilities is computed. In the same way, Bayesian inferences for the Wright model for genetics are described.

Next to be considered in Chapter 5 is a general birth and death process that is formulated as a discrete-time Markov chain (DTMC). We consider a finite population of maximum size N and a chain $\{X(n), n = 0,1,2,...\}$ with state space $\{0,1,2,\ldots,N\}$, where $X(n)$ is the size of the population at time n. The birth and death probabilities b_i and d_i depend on the size of population where

$$\begin{aligned} P_{ij} &= \Pr\{X(n+1) = j | X(n) = i\}, \\ &= b_i, j = i+1, \\ &= d_i, j = i-1, \\ &= 1-(b_i+d_i), j = 1, \\ &= 0, \text{otherwise}, \end{aligned} \tag{1.5}$$

and $i = 1,2,\ldots$, $P_{00} = 1$, and $P_{0j} = 0$, $j \neq 0$ Also note that $P_{N,N+1} = b_N = 0$; therefore, the $N+1 \times N+1$ transition matrix is given by

$$P = \begin{pmatrix} 1, 0, 0, 0, 0, \ldots\ldots\ldots\ldots\ldots\ldots, 0 \\ d_1, 1-(d_1+b_1), b_1, 0, 0, 0, 0, 0, 0, 0, \ldots,, 0 \\ 0, d_2, 1-(d_2+b_2), b_2, 0, 0, 0, 0, 0, 0, \ldots, 0 \\ . \\ . \\ 0, 0, 0, \ldots\ldots\ldots, 0, d_{N-1}, 1-(d_{N-1}+b_{N-1}), b_{N-1} \\ 0, 0, \ldots\ldots\ldots\ldots\ldots\ldots\ldots, 0, d_N, 1-d_N \end{pmatrix}. \tag{1.6}$$

Thus, the population increases by one, decreases by one, or remains the same. There are two communicating classes, namely, $\{0\}$ and $\{1,2,\ldots,N\}$, where 0 is an absorbing state and the remaining are transient. Based on the one-step transition probability matrix in Equation 1.6, corresponding cell counts are generated with R using a multinomial generator for the rows; then assuming a uniform prior for the transition probabilities, the resulting posterior distribution for a given row has a Dirichlet distribution, from which WinBUGS is employed to determine appropriate Bayesian inferences, including estimates for the stationary distribution for the birth and death process. This is generalized to the logistic growth process, which is a variation of the birth and death process, where similar inferences are carried out.

Chapter 5 continues with using Markov chains to model the evolution of an epidemic and then explores the use of Bayesian techniques to provide inferences for the unknown transition probabilities (which contain the basic parameters that describe the epidemic) of the process. The first to be discussed is an explanation of the basic principles of an epidemic, that is, the biological foundation of an epidemic. Next, a deterministic version of a simple epidemic is presented, which lays rudiments of the stochastic version of an epidemic model. Of primary importance in the study of epidemics is to determine the average duration of the epidemic. Various versions which generalize the simple epidemic are the chain binomial models, which include the Greenwood model and the Reed–Frost model. The relationship between the epidemic model and the previously discussed birth and death process is obvious.

It can be shown for the stochastic version of a simple epidemic that the one-step transition matrix of the number of infected people is

$$P = \begin{pmatrix} 1, 0, \ldots\ldots\ldots\ldots\ldots\ldots\ldots\ldots, 0 \\ b+\gamma, 1-b-\gamma-\lambda_1, \lambda_1, 0, \ldots\ldots\ldots, 0 \\ 0, 2(b+\gamma), 1-2(b+\gamma)-\lambda_2, 0, \ldots\ldots, 0 \\ . \\ . \\ . \\ 0, \ldots\ldots\ldots 0, N(b+\gamma), 1-N(b+\gamma) \end{pmatrix}. \quad (1.7)$$

Bayesian inferences are executed as before, and based on R, multinomial realizations are generated for the cell counts according to the transition probabilities of Equation 1.7. Also considered is another version of the stochastic epidemic called the chain binomial epidemic model.

Chapter 5 is concluded with various molecular modes for human evolution, and finally, the Ehrenfest model of cell diffusion is considered. Note that a one-step probability transition matrix corresponds to each model, allowing one to generate multinomial observations for the rows followed up with Bayesian inferences for the transition probabilities executed with WinBUGS.

1.6 Posterior Distributions for the Parameters of Markov Chains in Continuous Time

The main goal of Chapter 6 is to present Bayesian inferences for processes in continuous time. As in earlier chapters, Bayesian ways to estimate parameters, test hypotheses about those parameters, and predict future observations will be developed. There are many examples of Markov chains in continuous times, and the best-known example is the Poisson process. The Poisson process is a counting process that records many interesting phenomena such as the number of accidents in a given stretch of highway over a selected period, the number of telephone calls at a switchboard, the arrival of customers at a counter, the number of visits to a website, and earthquake occurrences in a particular region.

Chapter 6 begins with the definition of the Poisson process, which is followed by a description of the arrival and interarrival times, and then various generalizations are considered such as nonhomogeneous Poisson processes, which include compound processes and processes that contain covariates. Of course, R is employed to generate realizations from several examples of homogeneous and nonhomogeneous Poisson processes.

The Poisson process is defined as a counting process with stationary independent increments and has one parameter λ, the average number of events per unit time. It is known that the interarrival times between events have independent exponentially distributed with a common mean $1/\lambda$; thus, the waiting times to the occurrence of events are gamma distributions. Bayesian inferences are based on the prior information and sample information expressed by the likelihood function.

Suppose $Y_1, Y_2, \ldots, Y_n$ is a random sample from an exponential population of size n with parameter λ and corresponding observations $y_1, y_2, \ldots, y_n$, then the likelihood function for λ is

$$l(\lambda|\text{data}) \propto \lambda^n \exp\left(-\lambda \sum_{i=1}^{i=n} y_i\right), \quad \lambda > 0 \tag{1.8}$$

Note that $\sum_{i=1}^{i=n} y_i$ is the waiting time to the nth event.

Prior information will be expressed with the gamma distribution with density

$$f(\lambda) \propto \lambda^{\alpha-1} \exp(-\beta\lambda), \quad \lambda > 0. \tag{1.9}$$

Sometimes prior information is expressed with the improper density

$$g(\lambda) \propto 1/\lambda, \quad \lambda > 0. \tag{1.10}$$

In the former case with the gamma prior, the posterior distribution of λ is gamma $(n + \alpha, \beta + \sum_{i=1}^{i=n} y_i)$, while for the improper prior, the posterior distribution is gamma $(n, \sum_{i=1}^{i=n} y_i)$.

Bayesian inferences are illustrated with the following waiting times for a bus problem. A bus station serves three routes labeled 1, 2, and 3. Buses on each route arrive at the bus station according to three independent Poisson processes.

Buses on route 1 arrive at the station on the average every 10 minutes; those on route 2 arrive on the average every 15 minutes; while route 3 buses arrive on the average every 20 minutes. Some interesting problems are as follows:

1. When a person arrives at the station, what is the probability that the first bus to arrive is from route 2?
2. On average, how long will the person wait for some bus to arrive?
3. The person has been waiting for 20 minutes for a bus on route 3 to arrive, and during this time, three route 1 buses arrive at the station. What is the expected additional time the person will have to wait for the arrival of a route 3 bus?

Note that we have three independent Poisson processes with parameters: $\lambda_1 = 10$, $\lambda_2 = 15$, and $\lambda_3 = 20$. Since the parameters are known, it is straightforward to calculate these probabilities. Let Y_1, Y_2, and Y_3 denote the waiting times for buses from routes 1, 2, and 3, respectively, then it is known that $Y_1 \sim \exp(1/10)$, $Y_2 \sim \exp(1/15)$, and $Y_3 \sim \exp(1/20)$.

To answer question 1, note that the desired probability is given by

$$P[\min(Y_1, Y_2, Y_3) = Y_2] = \lambda_2/(\lambda_1 + \lambda_2 + \lambda_3) = (1/15)/(13/60) = .31. \tag{1.11}$$

Taking a statistical approach, interarrival times for the three Poisson processes are generated with R, then assuming the three parameters are unknown, and that the priors are improper, and using as sample information the interarrival times generated by R, WinBUGS is used to execute the Bayesian analysis, which determines the posterior distribution of the probability in Equation 1.11.

A Bayesian formal test of the null hypothesis

$$H: \lambda_1 = .1 \text{ versus the alternative } A: \lambda_1 \neq .1. \tag{1.12}$$

is conducted with the result that the posterior probability of the null hypothesis is very close to 1. The data used for the test are the 14 waiting times for the first Poisson process with $\lambda_1 = .1$, the value used to generate the data! It should be noted that in order to conduct a formal Bayesian test, the prior probabilities of the null hypothesis have to be specified as well as the prior distribution for λ_1 under the alternative. This section of Chapter 6 ends with a derivation of future waiting times for bus number 1.

The Poisson process is generalized to the so-called thinning Poisson processes. A thinning Poisson event (arrival) can be one of several types, each occurring with some nonzero probability. The initial process has a given rate λ, but the subsequent thinned processes have rates smaller than λ induced by the thinning probabilities of the component processes.

A good example of this is the birth of humans, where the overall birth rate is, say, λ, and the birth rates for males and females are, say, $p\lambda$ and $(1 - p)\lambda$, respectively, where p is the probability of a male birth.

If the overall birth process follows a Poisson process, it can be shown that the male births follow a Poisson process, as do female births; that is to say, both component processes have stationary and independent increments. Several examples are used to illustrate the idea of thinning processes. For example, R is used to generate birth rates for males with Poisson processes and, then based on those data, the Bayesian estimation of p, the proportion of males, and the overall birth rate λ. Also presented is another interesting example of a thinned Poisson process concerning earthquakes in Italy, where p is the proportion of major quakes and λ is the overall rate of earthquakes. The Bayesian analysis is based on actual earthquake information. Additional Bayesian inferences are conducted including the test of hypotheses concerning the equality of the interarrival rates of earthquakes in three areas of Italy. Lastly, the Bayesian predictive mass function of future earthquakes is derived.

An important generalization is that the spatial Poisson process is a generalization of the one-dimensional process studied in Section 6.4 to two or higher dimensions, and there are many examples: models of location of trees in a forest, the distribution of galaxies in the universe, and clusters of disease epidemics. Let dimension $d > 1$ and subset $A \subset R^d$. Suppose that the random variable $N(A)$ counts the number of points in subset A and $|A|$ denotes the size of A (in one dimension, $|A|$ would be length, and in two, it would be area, etc.), then the spatial Poisson process $\{N(A), A \subset R^d\}$ is defined as follows:

1. For each set A, $\{N(A), A \subset R^d\}$ is a spatial process with parameter $\lambda|A|$.
2. Whenever A and B are disjoint sets, $N(A)$ and $N(B)$ are independent random $\lambda|A|$.

Note how properties 1 and 2 generalize the Poisson process to higher dimensions, where property 1 is the generalization of stationary increments and property 2 describes independent increments. Consider the following problem, in two dimensions with parameter $\lambda = 1/3$, then what is the probability that a circle of radius 2 centered at (3,4) contains three points? The spatial Poisson process is a generalization of the one-dimensional process studied in Section 6.4 to two or higher dimensions.

Consider the following statistical problem, in two dimensions with parameter $\lambda = 1/3$, then what is the probability that a circle of radius 2 centered at (3,4) contains five points? To answer this question, spatial Poisson process data are generated with R, then parameter λ is estimated by Bayesian techniques including estimation, and tests of hypotheses about λ are executed with WinBUGS.

Three versions of concomitant Poisson processes are considered: (1) independence, (2) complete similarity, and (3) partial similarity.

Suppose k Poisson processes $N_i(t)$ with parameter λ_i, $i = 1, 2, \ldots, k$, where n_i events are observed over the interval $(0, t_i]$. The processes could be related.

For example, consider the case where the different processes correspond to different intersections in a large city where for the ith intersection, $N_i(t)$ counts the number of accidents at intersection i and the average number of accidents per day (over a 24-hour period) is denoted by λ_i. One should know enough information about the network of intersections, such as their proximity to each other and their location to busy businesses etc. In large cities, traffic communication centers continuously monitor the accidents at each intersection in the networks. Thus, there are k homogeneous Poisson processes, and we present Bayesian inferences about parameters λ_i for the cases, namely, of independence of and complete and partial similarity between the k processes. As before, R is used to generate accident data at the various intersections; then Bayesian inferences for the three cases of concomitant processes (independence, partial similarity, and complete similarity) are provided. Concomitant processes lead us into the ideas of considering Poisson processes with covariates.

Covariates can be incorporated into the Poisson model in a variety of ways, but only two are considered here, namely, (1) the direct approach and (2) as part of the prior distribution of the rate λ.

The goal is to determine relationships between several Poisson processes via their covariates. For example, consider the previous example concerning traffic accidents at several intersections. We may want to see what factors (covariates) affect the accident rates between intersections. If the two intersections share the same covariates, one would compare the two accident rates by comparing the effects of the covariates on the accident rates.

A simple case is presented: consider two intersections 1 and 2 with a common covariate, say, the population density of the neighborhoods, which encompasses those neighborhoods (an area surrounding the intersections). Let $\{N_1(t), t > 0\}$ and $\{N_2(t), t > 0\}$ be the Poisson processes that count the number of accidents for intersections 1 and 2 with accident rates λ and $\lambda\mu$, respectively. Thus, the parameter μ modifies the accident rate λ, and one would think that λ and $\lambda\mu$ reflect the population density of intersections 1 and 2, respectively. R generates the accident data, and then Bayesian inferences about λ and μ are carried out with noninformative priors for the parameters. Another way to incorporate covariates into a Poisson process is introduced by assigning the Poisson rate λ a gamma prior distribution, where the parameters of the gamma directly depend on the vector of covariates.

Chapter 6 concludes with a presentation of the nonhomogeneous Poisson process and how to choose its intensity function.

1.7 Examples of Continuous-Time Markov Chains

In Chapter 6, details of the Poisson process, an important case of continuous-time Markov chains (CTMCs), are presented. In Chapter 7, Bayesian inferences for general CTMC are presented. First to be considered are the important concepts involved in the study of such processes. For example, the ideas of transition rates, holding times, embedded chain, and transition probabilities are defined and explained. Such concepts are illustrated by using R to compute the transition function and to generate observation from the CTMC. This is followed by a presentation of Bayesian inferences of estimation and testing of hypotheses about the unknown parameters of the process.

Also developed is the Bayesian predictive distribution for future observations of the CTMC.

As with discrete time chains, the understanding of stationary distributions, absorbing states, mean time to absorption, and time reversibility is an important concept. It is important to know the Markov property for a continuous time chain and the idea of time homogeneity.

Let $\{X(t), t > 0\}$ be a continuous-time stochastic process, and then it is called a CTMC if

$$\begin{aligned} &P[X(t+s) = j | X(s) = i, X(u) = x(u), 0 \le u < s] \\ &= P[X(t+s) = j | X(s) = i] \end{aligned} \tag{1.13}$$

for all states i, j, and $x(u)$, where $0 \le u < s$. Of course, it is understood that the state space is countable.

Also, a CTMC is time homogenous, that is to say

$$\begin{aligned} &P[X(t+s) = j | X(s) = i] = P[X(t) = j | X(0) = i] \\ &= P_{ij}(t). \end{aligned} \tag{1.14}$$

Thus, the probabilistic properties of a CTMC over the interval $[s,t+s]$ are the same as that over the interval $[0,t]$. Or to express it another way, when the chain visits state i, its forward behavior from that time toward the future is the same as if the process started in i at time $t = 0$. Note that the function $P_{ij}(t)$ is called the transition function of the process. Recall that for a Poisson process, the interarrival times are identically exponentially distributed; however, for the CTMC, there is a difference as follows. Let T_i be the holding time of the process, that is, the time the process occupies state i before switching to another state, then it can be shown that T_i has an exponential distribution. It is important that one knows the fundamental ideas for continuous chains, including transition rates, holding times, and transition probabilities. An alternative to describing a CTMC is by transition rates between pairs of states. When the process is in state, there is a chance that it will change to one of the other possible states, that is, state i is paired with state j for all states $j \ne i$. If j can be reached from i, one can associate an alarm that is activated after a time that has an exponential distribution with parameter q_{ij}. When state i is first occupied, the alarms are all started at the same time, and the first alarm that is activated determines the next state to be occupied. If alarm (i,j) is first activated and the process moves to state j, a new set of alarms are activated with exponential transition rates $q_{j1}, q_{j2}, \ldots$. Thus, to repeat, the first alarm that is activated determines the next state to be occupied etc. The q_{ij} are called transition rates, and from them, the transition probabilities and holding time parameters can be determined.

Suppose the process starts at i, then the alarms are initiated and the first one that is activated determines the next transition; therefore, the time of the first alarm is the minimum of independent exponential random variable with parameters $q_{i1}, q_{i2}, \ldots$, which are exponential random variables with parameter $\sum_k q_{ik}$. Thus, the process remains in state i for a holding time which has an exponential distribution with parameter $\sum_k q_{ik} = q_i$. From i, the chain moves to state j if alarm (i,j) is first activated which occurs with probability

$$p_{ij} = q_{ij}/q_i, \tag{1.15}$$

which is the transition probability of moving from state i to state j of the embedded chain. One sees from Equation 1.15 that the transition probabilities of the chain are completely determined by the transition rates q_{ij}.

Consider the matrix Q with off-diagonal elements and transition rates q_{ij}, that is, $q_{ij} = Q_{ij}$ and $i \neq j$; and the diagonal entries are $-q_i$; thus, each row of Q has a sum of 0. The infinitesimal generator matrix Q plays an important role for the Bayesian analysis of the system.

We now see the role that equation $p_{ij} = q_{ij}/q_i$ plays in determining the transition probability matrix $P(t)$, which is a solution to the forward Kolmogorov equation:

$$P'(t) = P(t)Q,$$

where Q is the infinitesimal matrix and $P'(t)$ is the derivative matrix of the transition probability matrix.

This is also expressed as

$$P'_{ij}(t) = \sum_k p_{ik}(t)q_{kj} = -p_{ij}(t)q_j + \sum_{k \neq j} p_{ik}(t)q_{kj}. \tag{1.16}$$

It is obvious that solution $P(t)$ to Equation 1.16 is given by the matrix equation

$$P(t) = \exp(tQ), t \geq 0, \tag{1.17}$$

where $P(0) = I$.

Also, the solution can be written as

$$P'(t) = (d/dt)e^{tQ} = \sum_{n=0}^{n=\infty} (1/n!)(tQ)^n = I + tQ + t^2Q^2/2 + t^3Q^3/3! + \ldots.$$

It is important to know that R can be used to find the transition probability of the embedded chain if one knows the infinitesimal generator matrix Q. Consider a four-state chain with generator matrix

$$Q = \begin{pmatrix} -3, 1, 1, 1 \\ 2, -6, 2, 2 \\ 3, 3, -9, 3 \\ 4, 4, 4, -12 \end{pmatrix}. \tag{1.18}$$

It is known that $\sum_i \pi_i Q_{ij} = 0, \ \forall j$, which implies that the probability transition matrix of the embedded chain is

$$P = \begin{pmatrix} 0, 1/3, 1/3, 1/3 \\ 1/3, 0, 1/3, 1/3 \\ 1/3, 1/3, 0, 1/3 \\ 1/3, 1/3, 1/3, 0 \end{pmatrix}. \tag{1.19}$$

Chapter 7 continues by demonstrating that R can be used to compute Equation 1.19 based on the Q matrix (Equation 1.18).

As with discrete time chains, the ideas of stationary and limiting distributions and time reversibility are defined for the continuous case. See Section 7.3. The first example to be examined is the deoxyribonucleic acid (DNA) evolution model, which was studied for the discrete time case in Section 5.7, where the chain had four states, namely, the four base nucleotides: (1) adenine, (2) guanine, (3) cytosine, and (4) thymine. In the Jukes–Cantor model, the transition rates are all the same with infinitesimal generator matrix:

$$Q = \begin{pmatrix} -3r, r, r, r \\ r, -3r, r, r \\ r, r, -3r, r \\ r, r, r, -3r \end{pmatrix}, \tag{1.20}$$

where the first row and column correspond to adenine; the second row and column, to guanine; the third, to cytosine; and the last row and column, to thymine.

Thus, the corresponding transition matrix is

$$P(t) = \exp(tQ)$$

$$= (1/4)\begin{pmatrix} 1+3e^{-4rt}, 1-e^{-4rt}, 1-e^{-4rt}, 1-e^{-4rt} \\ 1-e^{-4rt}, 1+3e^{-4rt}, 1-e^{-4rt}, 1-e^{-4rt} \\ 1-e^{-4rt}, 1-e^{-4rt}, 1+3e^{-4rt}, 1-e^{-4rt} \\ 1-e^{-4rt}, 1-e^{-4rt}, 1-e^{-4rt}, 1+3e^{-4rt} \end{pmatrix}. \tag{1.21}$$

Thus, at time t, the probability that the DNA base adenine is replaced by quinine (or by cytosine or by thymine) is $(1/4)(1 - e^{-4rt})$, etc. On the other hand, the probability that adenine at time t is not replaced by another base is $(1/4)(1 + 3e^{-4rt})$.

The statistical problem is to make inferences about the rate r based on observing the evolution at various times.

Based on Equation 1.21, the infinitesimal rates are

$$q_{ij} = r, i \neq j, \; i, j = 1, 2, 3, 4;$$

thus, the holding time exponential parameters are

$$q_i = 3r, i = 1,\ 2,\ 3,\ 4.$$

Recall that the transition probabilities are given by

$$p_{ij} = q_{ij}/q_i = r/3r = 1/3, i \neq j, i, j = 1,\ 2,\ 3,\ 4.$$

If one assigns a value to r, one can make inferences about the holding time parameters and the transition probabilities; however, one knows that the transition probabilities are all 1/3; thus, only the holding time exponential parameters will be of interest.

In practice, one would observe the holding times of the various states and then from those observations, estimate r. Using WinBUGS, we will assume a value of r then generate the exponential holding times. Consider the holding time for occupying the first-state adenine and assume that its mean time is 2 time units. A WinBUGS analysis is based on 50 observations of the holding time for adenine with an average holding time of 2 time units. The main objective is to estimate the average holding time for adenine. WinBUGS is used to generate the exponentially distributed observations and is used to determine the posterior distribution of the average holding time of staying on the first state. Also explained is a formal Bayesian test of hypothesis about the mean holding time:

$$H_0: \lambda = 1/6 \quad \text{versus} \quad H_1: \lambda \neq 1/6.$$

A similar Bayesian analysis (estimation, hypothesis testing, and prediction) is performed for a generalization from the Jukes–Cantor model to the Kimura model, where R is employed to generate the relevant holding times and WinBUGS to execute the posterior distributions of the mean holding times of adenine.

The last of the DNA evolution models is the Felsenstein–Churchill model that allows for specialized DNA substitutions called transversions and transcriptions. The Bayesian analysis is much the same as that carried out for the Juke–Cantor and the Kimura models.

Chapter 7 revisits many of the DTMC examples examined earlier but now for their continuous-time cousins.

The discrete-time version of birth and death processes are presented in Section 5.4, and the continuous-time analogue will be described in Chapter 7. Such processes are time-reversible Markov chains that are quite valuable in a variety of scientific disciplines.

For example, consider the birth and death process in continuous time. Let the present state of the chain be i, and then the process can gain one unit corresponding to a birth or decrease one unit corresponding to a death, where $X(t)$, $t > 0$ is the size of the population at time t. Since this is a continuous-time process, the chain is defined in terms of the infinitesimal rate matrix:

$$Q = \begin{pmatrix} -\lambda_0, \lambda_0, 0, 0, \ldots\ldots\ldots\ldots \\ \mu_1, -(\lambda_1 + \mu_1), \lambda_1, \ldots\ldots.. \\ 0, \mu_2, -(\lambda_2 + \mu_2), \lambda_2, \ldots. \\ 0, 0, \mu_3, -(\lambda_3 + \mu_3), \lambda_3, . \\ \ldots \\ .. \\ .. \end{pmatrix}, \tag{1.22}$$

corresponding to the state space $S = \{0, 1, 2, \ldots\}$. State 0 is an absorbing barrier; that is, once the population size reaches 0, it dies out.

Thus, if the population size is 1, the holding time distribution of state 1 until the next death has an exponential parameter μ_1, while on the other hand, the holding time distribution of state 1 until a birth is exponential with parameter λ_1. Note the birth rates λ_i and death rates μ_i depend on the present size of the population.

Since the process is time reversible, one can derive the stationary distribution via the local balance equations

$$\pi_i \lambda_i = \pi_{i+1} \mu_{i+1}, \quad i = 0, 1, 2, \ldots$$

In the infinite case, notice that all states are transient except the absorbing state 0.

Now consider a finite state space with infinitesimal rate matrix and state space $S = \{0,1,2,3,4\}$:

$$Q = \begin{pmatrix} -\lambda_0, \lambda_0, 0.0, 0.0, 0.00 \\ \mu_1, -(\mu_1 + \lambda_1), \lambda_1, 0, 0 \\ 0, \mu_2, -(\mu_2 + \lambda_2), \lambda_2, 0 \\ 0, 0, \mu_3, -(\mu_3 + \lambda_3), \lambda_3 \\ 0.0, 0.0, 0.0, \mu_4, -\mu_4 \end{pmatrix} \tag{1.23}$$

For the purpose of illustration, suppose $\mu_1 = 2$, $\lambda_1 = 1$, $\mu_2 = 3$, $\lambda_2 = 2$, $\mu_3 = 4$, and $\lambda_3 = 3$.

Using these values, I generated 20 holding times for 6 exponentially distributed holding times: from states 1 to 0 with parameter 2, from states 1 to 2 with parameter 1, from states 2 to 1 with parameter 3, from states 2 to 2 with parameter 2, from states 3 to 2 with parameter 4, and from states 3 to 4 with parameter 3. These 6 holding times are used as data for the Bayesian analysis. Note that the birth rate for a population of size i is $\lambda_i/(\mu_i + \lambda_i)$ and the corresponding death rate is $\mu_i/(\mu_i + \lambda_i)$, i =1, 2, 3. A Bayesian analysis is executed with 45,000 observations for the simulation and 5,000 for the burn-in. What is the time to extinction for this process? This section of Chapter 7 explains the Bayesian way to estimate this important parameter. Other versions of the birth and death process such as the random walk and Yule process are investigated with the Bayesian approach to inference.

From the least complex, such as the Yule process, to the more complicated, we now study the so-called birth and death process with immigration. Suppose the immigration has a rate ν included in the simple birth and death process, then the infinitesimal rate matrix is

$$Q = \begin{pmatrix} -\nu, \nu, 0, 0, 0, 0, 0, 0, 0, \ldots\ldots\ldots \\ \mu, -(\nu+\lambda+\mu), \nu+\lambda, 0, 0, 0, 0, \ldots\ldots \\ 0, 2\mu, -(\nu+2(\lambda+\mu)), \nu+2\lambda, 0, 0, 0, 0, \ldots \\ 0, 0, 3\mu, -(\nu+3(\lambda+\mu)), \nu+3\lambda, 0, 0, 0\ldots \\ . \\ . \\ . \end{pmatrix}, \tag{1.24}$$

where the rates μ, ν, and λ are positive and are the parameters of the relevant holding times, which have an exponential distribution.

Referring to Equation 1.24, it is apparent that if the population is 0, it can increase by one person if one person immigrates to the population. Our goal is to estimate the parameters and to test hypotheses about those parameters. In order to perform a Bayesian analysis, let $\mu = 1/2$, $\nu = 1/3$, and $\lambda = 1$; that is, if the population is size i ($i = 1,2, \ldots$), the average number of immigrants per day is 3, the average number of birth is 1, and the average number of deaths is 2 per day. Note the birth and death rates depend in the present population size, but the immigration rate does not. The Bayesian analysis is relatively straightforward and is executed with noninformative prior gamma (.01, .01) distributions and using 35,000 observations for the simulation and a burn-in of 5,000.

Chapter 7 ends with the introduction of the continuous-time version of epidemic models and is a generalization of the discrete version of stochastic epidemic model of Sections 5.6.1 and 5.6.2. Recall the dynamics of SI and SIS models, where the number of susceptible people at time t is denoted by $S(t)$ and the number of infected individuals is given by $I(t)$. Infected individuals are also infectious; that is, there is no latent period, and the total population size $N = I(t) + S(t)$ remains constant over the period of observation.

This SI model has been used to explain such diseases as the common cold and influenza where the epidemic is best described by the system of differential equations:

$$dS(t)/dt = -(\beta/N)S(t)I(t)$$

and (1.25)

$$dI(t)/dt = (\beta/N)S(t)I(t),$$

where $S(0) + I(0) = N$ and β is the transmission rate, the number of contacts per unit time that result in an infection of a susceptible individual. The main parameter of interest is the transmission rate β, which is estimated by Bayesian methods where the analysis is executed with WinBUGS. Chapter 7 reveals the details necessary to understand Bayesian inferences for stochastic epidemics.

1.8 Bayesian Methods for Gaussian Processes

Previous chapters have employed Bayesian inferences for stochastic processes with discrete time and discrete state space, and in Chapter 8, inferential techniques will be applied to the most general case where time and state space are both continuous.

Bayesian inferential methods of estimation, testing of hypotheses, and forecasting will reveal interesting aspects of stochastic processes that are unique. Remember that the standard course in stochastic processes does not emphasize inference but instead solely focuses on the probabilistic properties of the model.

Subjects to be presented are the properties of the Wiener process (or the Brownian motion) and Bayesian estimation of its variance and covariance function. The Wiener process and the random walk will be generalized to a continuous state space, and formal Bayesian testing methods demonstrated with the parameters of the random walk.

The Brownian motion is a special case of normal stochastic processes where the joint distribution of the variables of model has a mean vector and variance covariance as parameters. The Bayesian approach uses the inverse normal Wishart distribution as a prior for the mean vector and precision matrix, with the result that the marginal posterior distribution of the mean vector has a multivariate t distribution. This in turn provides easily implemented Bayesian techniques of inference.

Certain mapping or transformations of the Wiener process are of interest and have many applications. Translations, reflections, rescaling, and inversions of the Brownian motion will be presented and lead to such concepts as stopping times, first hitting times, and determining the zeros of the Wiener process. Such mapping applied to the Brownian motion produces other types of models that are amenable to Bayesian inferences, which will be implemented with WinBUGS and R.

Certain variations of the Brownian motion will be studied including the Brownian motion with drift, the Brownian bridge, and the Ornstein–Uhlenbeck process. Bayesian inferences are especially interesting when applied to a financial example of stock options.

In 1927, Robert Brown, a botanist, described a biological scene where pollen particles suspended in water exhibited erratic behavior with continuous movement of tiny particles ejected from the grain in the water. In 1905, Albert Einstein developed, on the basis of the laws of physics, a mathematical description of the phenomena explained by Brown. Einstein showed that the position of the particle denoted by y at time t is described by the partial differential heat equation:

$$(\partial / \partial t)f(y,t) = (1/2)(\partial^2 / \partial y^2)f(y,t), \tag{1.26}$$

where $f(y, t)$ represents the number of particles per unit volume at position y at time t. It can be shown that the solution to Equation 8.1 is

$$f(y,t) = \left(1/\sqrt{2\pi t}\right)e^{-y^2/2t}; \tag{1.27}$$

that is, the solution is the density of a normal distribution with mean 0 and variance t. Thus, the process called the Brownian motion is a continuous-time continuous-state stochastic process. Albert[1] investigated the properties of this process, and the model is sometimes called a Wiener process, and he showed that the sample functions of the process are continuous almost everywhere, but the process is not differentiable at any time point. The Wiener process $\{B(t), t > 0\}$ or standard Brownian motion is defined as follows:

1. For all $t > 0$, $B(t)$ has a normal distribution with mean zero and variance t.
2. The process has stationary increments, namely, for $s,t > 0$; $B(t + s) - B(s)$ has the same distribution as $B(t)$.
3. The process has independent increments; that is, for $0 \le q < r \le s < t$, $B(t) - B(s)$ and $B(r) - B(q)$ are independent.
4. The function $B(t)$ is continuous with probability 1.

Note that we use the notation that $B(t) \sim N(0,t)$ and recall that the Wiener process can be interpreted as the movement of a particle that diffuses randomly along a line, where at time, the location of the particles is normally distributed about the line with standard deviation $\sqrt{t}$. Wiener's fundamental contribution to our knowledge was to prove the existence of such a process as the Brownian motion. Bayesian inferences will be based on independent increments; because each increment $B(t) - B(s)$ has a normal distribution with mean 0 and variance $t - s$ for $s < t$, it is easy to write down the likelihood function. Independent increments make it easier to evaluate complicated probabilities involving the Brownian motion.

Also, it is easy to use R to generate Brownian motion values.

The following R Code shows how to generate standard Brownian motion variables over the interval [0,50], with adjacent observations one unit apart:

```
t<-1:50
sig2<-1
x<-rnorm(n=length(t)-1,sd=sqrt(sig2))
x<-c(0,cusum(x, t))
plot(t,x,type= "1", ylim=c(-10,10))
```

In the preceding code, sig2 is the variance of process which is one for standard Brownian motion, and the corresponding standard deviation is sd. Brownian motion-generated values are designated by the vector x, and the plot command has an abscissa range from 1 to 50 with time units of length one and with the ordinate range from −10 to 10.

On the other hand, the following R Code generates 30 Wiener variables with $\sigma^2 = 4$.

```
t<-0:50
sig2<-4
x<-rnorm(n=length(t)-1,sd=sqrt(sig2))
x<-c(0,cusum(x, t))
plot(t,x,type= "1", ylim=c(-8,8))
```

The plot command produces a plot of the 30 Wiener values:

```
0.0000000, -4.4847302, 0.4643178, -0.3437283, -0.4905968,
0.4821004, 1.8126740, -1.2513397, -1.1408787, -1.5291112,
-3.0735822, -2.9174191, -4.2614190, -6.6813461, -4.7708805,
-2.1957203, -1.3201780, 1.4937973, 1.4242956, 1.6672101,
2.7032072, 3.8358068, 4.2725358, 5.5445094, 6.5348815,
6.7983973, 5.5906146, 6.1619978, 8.8013248, 7.8247243.
```

Our goal is to use the Bayesian approach to estimate σ^2.

Let

$$B(2i) - B(2i-1) = X(i), \quad i = 1, 2, .., 15, \tag{1.28}$$

where the $B(i)$ are the 30 Wiener values generated with the R Code shown earlier, then the $X(i)$ have independent increments and are distributed as

$$X(i) \sim B(2i) - B(2i-1) \sim B(1) \sim N(0, \sigma^2) \tag{1.29}$$

The 15 observed increments provide the sample information for the Bayesian analysis which is executed with 45,000 observations and a burn-in of 5,000. Noninformative prior distributions are assigned to μ and τ, and the posterior distributions show that the posterior mean of σ^2 is quite close to the value of 4 used to generate the data. The mean μ should be zero; thus, it is of interest to test in a formal Bayesian way the null hypothesis H: $\mu = 0$ versus the alternative A: $\mu \neq 0$ and is conducted with the result that the posterior probability of the null hypothesis is very close to 1, implying that the simulated independent increments do indeed have a mean of 0.

Chapter 8 continues by showing the connection between various Brownian motions, and the Bayesian analysis focuses on the estimation of the parameters (the variance) of the Brownian motion whose values are generated with R. For example, the following R Code computes the maximum of a symmetric random walk using 50 replications, where the generating process is the sequence of binary random variables with values plus one and minus one with equal probability:

```
n<-50
sim<-replicate(50,
max(cumsum(sample(c(-1,1),n, replace =T))))
max(sim)
mean(sim)
sd(sim)
s<-seq(1,50,1)
plot(s,sim)
```

The 50 generated random walk values are as follows:

```
0, 1, 5, 1, 12, 4, 7, 19, 14, -1, 5, 4, -1, 4, 9, 4, 2, 2, 9, 3, 1, 5, 3,
6, 7, 5, 5, 10, 7, 5, 3, 5, 2, 2, 9, 15, 2, 0, 11, 7, 2, 7, 9, 4, 1, 5,
14, 7, 9.
```

Based on this sample information, Bayesian inferences for the parameters of the Brownian motion are easily determined, but the details are found in Chapter 8.

One interesting problem for the Wiener process is to determine the time until the process reaches a particular value, say, the real number v. The first hitting time is described and defined as follows:

Let T_v be the random variable starting from 0 until reaching v, then

$$T_v = \min[t : B(t) = v] \tag{1.30}$$

If T_v is thought of as the stopping time, then the process begins with the translated process at time v.

For a standard Brownian motion at time t, the process is equally likely to be above zero as to be below zero. Assume $v > 0$, then for the process beginning at time v, at any time $t > v$, the process is equally likely to be above the horizontal line v units above the line $y = 0$. In symbols,

$$P[B(t) > v|T_v < t] = P[B(t) > 0] = 1/2, \tag{1.31}$$

and from this, it follows that the distribution function of T_v is

$$\begin{aligned} P[T_v < t] &= 2P[B(t) > v] = 2\int_v^{\infty}\left(1/\sqrt{2\pi t}\right)\left[\exp -x^2/2t\right]dt \\ &= 2\int_{v/\sqrt{t}}^{\infty}\left(1/\sqrt{2\pi}\right)\left[\exp -x^2/2\right]dx. \end{aligned} \tag{1.32}$$

Upon differentiating with respect to t, the corresponding density function is

$$f(t) = \left(v/\sqrt{2\pi t^3}\right)\exp -v^2/2t, t > 0$$

Based on a realization of the general Brownian motion with unknown mean and variance, the goal of the Bayesian approach will be to estimate these two parameters as well as the preceding moments and to test hypotheses about μ and σ^2. R easily generates observations from the Brownian motion process with known parameters, and then based on the corresponding sample increments, a Bayesian posterior analysis is easily conducted. Remember that the main focus is to estimate the average hitting time of the target value of 10. The Bayesian analysis finishes with a test of the hypothesis that the variance of the Brownian motion is indeed 0.016, the value used in generating the sample information.

The Wiener process reaches a particular level, regardless of how small or large with certainty, and reaches the level 0 infinitely often. The times the process assumes the value 0 are called the zeros of the process. Of course, the process has infinitely many zeros occurring in the interval $(0, \varepsilon)$, regardless of how small $\varepsilon > 0$.

Our goal is to develop Bayesian inferences for the zeros of the Brownian motion and the last time that the origin is visited by the process. This as will be seen is related to a coin tossing experiment. Our analysis will focus on the probability $z_{r,t}$ that standard Brownian motion has at least one zero in the interval (r, t), namely,

$$z_{r,t} = (2/\pi)\arccos\left(\sqrt{r/t}\right), 0 \le r < t \tag{1.33}$$

Thus, at least one zero in $(0, \varepsilon)$ is

$$z_{0,\varepsilon} = (2/\pi)\arccos(0) = 1 \tag{1.34}$$

This is related to the random variable L_t; the time to the last 0 is shown to be related to an interesting experiment in coin tossing with two players 1 and 2. If the coin lands heads, player 1 pays player 2 $1, and if tails occurs, player 2 pays player 1 $1. If the coin is flipped a large number of times, when would you expect the players to be even? The data for the Bayesian analysis depend on R as follows: The R Code generates 10,000 tosses of a fair coin, and the histogram is the last time the two players are even. Of course, this is related to the number of zeros occurring over a given time interval, and the details of the Bayesian analysis are described in Chapter 8. A useful generalization of the Brownian motion is when drift is included.

Consider the Brownian motion with drift defined as

$$X(t) = \mu t + \sigma B(t), \tag{1.35}$$

where $\sigma > 0$, μ is any real number, and $\{B(t), t > 0\}$ is standard Brownian motion. It can be demonstrated that this is a normal process with mean μt and variance σ^2 and that the process has independent and stationary increments $[X(t+s) - X(t)]$ with mean μs and variance $\sigma^2 s$, where $s,t > 0$.

Consider the following application of the Brownian motion with a drift, where the objective is to estimate the home field advantage by determining the probability that the home team leads by y points after a fraction t $(0 < t < 1)$ of the game is played. To evaluate this probability, let $X(t)$ denote the difference in scores between the home team and the visiting team after $100t\%$ of the game is played.

This approach is based on a Brownian motion process $\{X(t),\ 0 < t < 1\}$, where μ denotes the magnitude of the home field advantage.

The probability that the home team wins, conditional on the fact they have a y point lead at time t, is

$$\begin{aligned}
p(y,t) &= P[X(1) > 0|X(t) = y] \\
&= P[X(1) - X(t) = -y] \\
&= P[X(1-t) = -y] \\
&= P[\mu(1-t) + \sigma B(1-t) > y] \\
&= P\left[B(t) < \left(\sqrt{t}(y + \mu(1-t))\right)/\sigma\sqrt{1-t}\right]
\end{aligned} \tag{1.36}$$

In order to evaluate this probability, data from the 1992 National Basketball Association season with 493 games were used, which gave a home field advantage estimated as $\mu = 4.87$ and an estimated standard deviation of $\sigma = 15.82$. Instead of substituting these two estimates into Equation 1.36, the Bayesian approach will be based on data generated (via R) with the Brownian motion model and will use the estimates $\mu = 4.87$ and $\sigma = 15.82$, and then find the posterior distribution of μ and σ^2 and finally estimate the desired probability (Equation 1.36) with WinBUGS. The Bayesian analysis computed a posterior mean of .5025 for the probability that the home team wins when the teams are tied at half time.

Next to be considered in Chapter 8 are two generalizations of the Brownian motion, namely, geometric Brownian motion and the Brownian bridge.

Geometric processes are used to model exponential growth and decay and are often used in finance to model stock prices and represent the return from stock options.

The process is defined as

$$G(t) = G(0)\exp[X((t))], \quad t \geq 0, \quad G(0) > 0, \tag{1.37}$$

where $\{X(t), t > 0\}$ is the Brownian motion with drift μ and variance σ^2.

It can be shown that the first two moments for $G(t)$ are

$$E(G(t)) = G(0)\exp\left(t(\mu + \sigma^2/2)\right)$$

and

$$Var(G(t)) = \left(e^{t\sigma^2} - 1\right)G^2(0)\exp 2t\left(\mu + \sigma^2/2\right). \tag{1.38}$$

It is obvious that the mean of the process exhibits exponential growth with growth rate $(\mu + \sigma^2/2)$.

The goal is to develop Bayesian inferences for the parameters of the geometric Brownian motion using an example about stock prices.

Realizations for the geometric version are easily generated with R by transforming standard Brownian motion. This is done by transforming the Brownian motion with trend $\mu = 1$ and variance $\sigma^2 = 0.5$, then based on those values, conduct a Bayesian analysis for μ and σ^2 based on noninformative priors. The last example for Chapter 8 is based on a model for the selection of stock option on geometric Brownian motion.

An option is a contract that gives its owner the right to buy shares of a stock sometime in the future at a fixed price.

The following approach is not based on the Black–Scholes model of Section 8.6. Assume the stock is selling for \$90 per share, and under the terms of contract, in 80 days, you may buy a share of stock for \$110. Once having bought the option, several alternatives need to be considered. Assume that in 80 days, the price of the stock exceeds \$110, so if you exercise the option and buy the stock for \$110 and sell it at the current prize, your payoff would be $G(80/365) - 110$, where $G(80/365)$ is the price of the stock in 80 days.

However, the other alternative has to be considered, namely, that in 80 days, the stock price is less than \$110, and the option would not be exercised; consequently, you receive nothing. In general, the two alternatives imply that the payoff is $\max\{G(80/365)-110,0\}$. What is the profit for such a situation? It is the payoff minus the cost of the option of \$15. Assuming that the stock price follows the geometric Brownian motion, the following explanation of how to find the future average payoff is repeated here.

Let $G(0)$ denote the current stock price, and t, the future time the option is exercised. Suppose k denotes the strike price, the price you can buy the stock if you exercise the option. Note for this illustration that $G(0) = 90$, $t = 80/365$, and $k = \$110$. Our goal is to determine the average profit of the option, namely,

$$\begin{aligned} E[\max\{G(t) - k, 0\}] &= E[\max\{G(0)\exp(\mu t + \sigma B(t)) - k, 0\}] = \\ G(0)\exp t(\mu + \sigma^2/2)\Pr&[Z > (\beta - \sigma t)/\sqrt{t}] - k\Pr[Z > \beta/\sqrt{t}], \end{aligned} \tag{1.39}$$

where

$$\beta = [\ln(k/G(0) - t\mu)]/\sigma.$$

Notice the similarity in the expected payoff of an option and the price of stock given by the Black–Scholes approach. The two parameters μ and σ^2 are the drift and variance of the geometric Brownian motion process, respectively. R easily generates 365 daily values for the stock process with parameters $\sigma^2 = 0.25$ and $\mu = 0.1$ for the underlying Brownian motion process, and the Bayesian analysis is executed using these generated values as sample information. The posterior mean of the expected payoff (Equation 1.39) for the stock option is $52. The exercises at the end of Chapter 8 increase the knowledge of using Bayesian methods for Markov chains in continuous times.

1.9 Bayesian Inference for Queues and Time Series

Chapter 9 presents Bayesian inferences for two types of stochastic processes not considered in the previous chapters. For example, the first process to be studied is a large useful family of models referred to as queuing models, and the second class, elementary types of time series.

As in previous chapters, R will be implemented in order to simulate the various stochastic processes where the parameters of the processes are known, then using those observations generated by R as the sample information and assuming that the parameters are now not known, a Bayesian analysis will be executed with WinBUGS that will allow one to make Bayesian inferences about those unknown parameters. Bayesian inferences consist of three phases: estimation, testing hypotheses, and prediction of future observations.

It is important to remember that Bayesian inferences depend on prior information about the unknown parameters, the sample information expressed by the likelihood function and the posterior distribution about those parameters.

Of course, all Bayesian inferences are based on the posterior distribution.

The fundamental properties of queues are described, and then Section 9.1 will define the various models for the queuing process, including the $M/M/1$ model, then progressing to $G/M/1$, and, finally, the $M/G/1$ model. For each model, using the WinBUGS package and sometimes R, data will be generated for the arrival times and the service times, where the parameters are known, then based on those observations, Bayesian inferences are described and implemented using WinBUGS.

There are many situations where the model for a queue can reveal interesting behavior. For example, customers at a shopping mall or people who need a heart transplant or other organs all have the same concern: the time to wait while in line to be served and, once being served, the time it takes to complete being served. Section 9.1 will employ R to generate observations for the queuing model, such as the arrival time of the customers entering the queue and the service time of those doing their transactions.

Based on those observations generated by R with known parameters for the queuing model, Bayesian inferences for those parameters (now assumed unknown) will be implemented with WinBUGS.

In what is to follow, the fundamental properties of a general queuing model is described, then focus is centered on the special case of the $M/M/1$ system, followed by explanations of non-Markov processes.

Generally speaking, a queuing system is a family of several stochastic processes describing the waiting and service times of the people in the queue. They arrive according to some process (which can be represented by deterministic or stochastic mechanisms), and then they have to wait if required before being attended to by one or more servers.

Analytical results for the queuing model are often difficult to determine; thus, interest will be focused on a case where it is possible, namely, with the $M/M/1$ process, where the arrival process is Poisson with parameter λ; hence, with exponentially distributed independent interarrival times with mean $1/\lambda$ and independent exponentially distributed service times with mean $1/\mu$. The system is denoted by $M(\lambda)/M(\mu)/1$. Notice the similarity of the queuing process to a birth and death process, where the arrival of a customer is interpreted as a birth, and the completion of service, as a death. An important parameter is the traffic intensity that is

$$\rho = \lambda/\mu; \tag{1.40}$$

hence, the system is stable if the arrival rate is less than that of the service rate. From previous considerations, the equilibrium distributions exist, and the limiting distribution for the number of people in the system is geometric:

$$\rho \sim Ge(1-p), \tag{1.41}$$

with mean $E[N] = \rho/(1-\rho)$ and that the number of clients in the queue waiting for service has the following mass function:

$$\begin{aligned} P\left[N_q = n\right] &= P[N=0] + P[N=1], \quad n = 0 \\ &= P[N = n+1], \quad n \geq 1. \end{aligned} \tag{1.42}$$

In addition, the limiting distribution for the W spent by an arriving customer in the system is

$$W \sim \exp(\mu - \lambda), \tag{1.43}$$

and mean $E(W) = 1/(\mu - \lambda) = 1/(\mu(1-\rho))$.

Lastly, the limiting distribution of the time W_q has cumulative distribution of

$$P\left[W_q \leq t\right] = 1 - \rho e^{-(\mu-\lambda)t}, \quad t \geq 0, \tag{1.44}$$

and that the idle period time J of a server has a density of

$$f_J(t) = \lambda e^{-\lambda t}, \quad t > 0. \tag{1.45}$$

The purpose of a Bayesian analysis will be to provide inferences for the unknown parameters μ and λ. Remember that in practice, one would have data organized as follows: For each customer, the time of arrival to the queue is recorded, the waiting time also recorded, and the service time for that client would be noted.

For the statistician, a distribution needs to be assigned to the waiting times and service times. These are determined empirically with goodness-of-fit tests, etc., then classical inferences such as maximum likelihood made for parameters μ and λ.

For the M/M/1 process, one is assuming that the interarrival times and service times are exponential, but one must remember that this assumption needs to be justified. For this case, Bayesian inferences are quite simple. Suppose one has the following information: the total time t_a taken for the first n_a arrivals and the total time t_s taken to service the first n_s. It is obvious that the likelihood function is

$$L((\mu, \lambda)|data) \propto \lambda^{n_a} e^{-\lambda t_a} \mu^{n_s} e^{-\mu t_s}, \quad \lambda, \mu > 0. \tag{1.46}$$

Prior distributions must be assigned to λ and μ; thus, consider the improper prior

$$\xi(\mu, \lambda) = 1/\lambda\mu, \quad \lambda, \mu > 0, \tag{1.47}$$

then the posterior distribution of λ is gamma (n_a,t_a), and that of μ is gamma (n_s,t_s), and λ and μ are independent. Thus, Bayesian inferences are somewhat straightforward if one knows the sufficient statistics n_a, t_a, n_s, and t_s; however, it should be remembered that these depend on the individual waiting and service times.

Suppose it is assumed that the arrival rate to the queue is Poisson with $\lambda = 2$ and the service rate is Poisson with $\mu = 4$, then WinBUGS generates 32 interarrival times with parameter $\lambda = 2$ and 24 service times with $\mu = 4$, where the data generated so far is the sample information for Bayesian inferences. Vector y is composed of the 32 interarrival times, while vector x denotes the 24 service times.

The main focus of the Bayesian analysis is on the traffic intensity ρ and the probability the queue is stable; that is, $\Pr[\rho < 1|\text{data}]$, and the Bayesian analysis computes a posterior mean of .9838. Chapter 9 continues with Bayesian analyses for the G/M/1 and G/G/1 queues, where the general interarrival and service time distributions are gamma.

The bulk of Chapter 9 deals with Bayesian inferences for the time series, beginning with the fundamental properties of a time series, including the three basic components: the trend, the seasonal effects, and the noise or errors of the process. An R function called decompose will estimate these three components by graphically delineating a time series into plots of the trend over time, the seasonal effects over times, and errors versus time. Next to be considered is the R function acf, which plots the autocorrelations of lag 1, 2, 3, ... and computes estimates of the lagged autocorrelations. The first process to be studied is the autoregressive process AR(p), with p autoregressive coefficients that determine the autocorrelation function.

For the AR(1) model,

$$Y(t) = \theta Y(t-1) + W(t), \tag{1.48}$$

the first two moments are

$$\mu(t) = 0$$

and covariance function with lag $k = 1, 2, \ldots$

$$\gamma_k = \theta^k \sigma^2 / (1 - \theta^2), \tag{1.49}$$

and autocorrelation function

$$\rho_k = \theta^k, \quad |\theta| < 1. \tag{1.50}$$

It is obvious from Equation 1.50 that the autocorrelations are nonzero and decay exponentially with k. Another important second-order property of the AR(p) process is that the partial correlation function at lag k, which is the correlation that results after removing the effects of correlations of terms with lags less than k. Note the condition $|\theta| < 1$ guarantees that the process is stationary.

The following R Code generates 100 values from the AR(1) process with autoregression coefficient $\theta = 0.6$. The autocorrelation function is acf while that for the partial autocorrelation is pacf. Note that the variance of white noise is $\sigma^2 = 1$.

```
set.seed(1)
y<-w<-rnorm(100)
for( t in 2:100) y[t]<-.6*y[t-1]+w[t]
time <- 1:100
plot(time,y)
acf(y)
pacf(y)
```

Based on the first 50 observations generated earlier, a Bayesian analysis is performed assuming noninformative prior information for the autocorrelation θ and Gaussian noise variance σ^2. The Bayesian analysis showed that the posterior mean of these two parameters is very close to those values used to generate the data.

Chapter 9 continues where various regression models are introduced. First to be considered is the estimation of the trend in linear models with autocorrelated errors, then later to provide inferences for the seasonal effects using harmonic and latent variables. Time series regression models are different from the usual regression models in that the errors are autocorrelated. One of the first linear models to be studied is simple linear regression with errors following an AR(1) process. What is the effect of autocorrelation on the usual estimates of the regression coefficients? If the correlation is positive, the estimated standard errors of the estimates tend to be less than the estimated standard errors of the estimates assuming no correlation. Of course, a corresponding scenario holds for the Bayesian estimates of the regression coefficients.

As a first example, consider the linear regression model

$$Y(t) = 50 + 3t + Z(t), \quad t = 1, 2, \ldots, 100, \tag{1.51}$$

where the $Z(t)$ follows an AR(1) process with autocorrelation θ.

R generates 100 observations from the linear regression model in Equation 1.51. The model is assumed to have a standard deviation of $\sigma = 10$ for the Gaussian white noise of the

AR(1) process, and the autocorrelation coefficient is assigned with the value of $\theta = .6$, while the regression coefficients are assumed to be 3 for the slope and 50 for the intercept.

The first 20 observations are used as the sample information and are considered vectors with a multivariate normal distribution with mean vector consisting of 20 values of $50 + 3t$, $t = 1, 2, \ldots, 20$, and a 20×20 precision matrix, which is the inverse of the variance–covariance matrix with components specified by Equation 1.49, the covariance matrix of an AR(1) process. Thus, in WinBUGS, one must use the multivariate normal distribution with mean vector and variance covariance matrix as described earlier. The Bayesian analysis reported that the posterior means were very close to the corresponding values used to generate the sample information. Another regression model is analyzed where the trend is quadratic, and seasonal effects are included resulting in six autoregressive parameters, as well the autocorrelation coefficient θ and variance σ^2 of the white noise errors. The Bayesian analysis is based on noninformative prior distributions for these parameters, resulting in posterior medians very close to the corresponding values used to generate the data.

Next to be considered is the Bayesian analysis of nonlinear regression models with AR(1) errors, then the moving average class of time series is studied.

A moving average process MA(q) is defined as

$$Y(t) = W(t) + \beta_1 W(t-1) + \ldots\ldots. + \beta_q W(t-q), \tag{1.52}$$

where $W(t), W(t-1), \ldots, W(t-q)$ is a sequence of independent white noise random variables with variance σ^2 and β_i, $i = 1, 2, \ldots, q$, are unknown real parameters. It is obvious that the mean value function of the process is zero; the variance is

$$\mathrm{Var}[Y(t)] = \sigma^2 \left(1 + \sum_{i=1}^{i=q} \beta_i^2\right), \tag{1.53}$$

and autocorrelation of lag k is

$$\rho(k) = \left(\sum_{i=0}^{i=q-k} \beta_i \beta_{i+k} \Big/ \sum_{i=0}^{i=q} \beta_i^2\right). \tag{1.54}$$

As an example, consider the MA(1) process:

$$Y(t) = W(t) + \beta_1 W(t-1),\ t = 2, 3, 4, \ldots\ldots, 100, \tag{1.55}$$

where $\beta_1 = 0.8$ and $\sigma^2 = 1$.

The following R Code generates 1000 observations from the MA(1) process with $\beta_1 = 0.8$ and $\sigma^2 = 1$.

```
set.seed(1)
b<-c(.8)
x<-w<-rnorm(1000)
for ( t in 2:1000){
for ( j in 1:1) x[t]<-w[t]+b[j]*w[t-j]}
```

Based on these first 20 values generated by R, the Bayesian analysis will estimate the parameters of the MA(1) process (Equation 1.55), where the posterior analysis is executed with 35,000 observations for the simulation and a burn-in of 5,000. Noninformative prior distributions are used for the parameters: β_1 is normal (.8, .01), while σ^2 is gamma (.001, .001). Note that it is assumed that the data vector of 20 observations has a multivariate normal distribution with mean vector zero and a variance–covariance matrix appropriate for an MA(1) process. Refer to Chapter 9 for the details of the Bayesian analysis, which reveals that the posterior means are quite close to the corresponding values used to generate the sample information. In a similar way, Bayesian inferences are developed for the ARMA(1,1) process and regression models with ARMA(1,1) errors. In each case, observations are generated for the appropriate model with known parameter values, then based on those observations, Bayesian inferences are developed for those same parameters.

1.10 R Package

1.10.1 Introduction to R

The following information about R can be downloaded at https://www.r-project.org/about.html.

"R is a language and environment for statistical computing and graphics. It is a GNU project which is similar to the S language and environment which was developed at Bell Laboratories (formerly AT&T, now Lucent Technologies) by John Chambers and colleagues. R can be considered as a different implementation of S. There are some important differences, but much code written for S runs unaltered under R.

R provides a wide variety of statistical (linear and nonlinear modelling, classical statistical tests, time-series analysis, classification, clustering, …) and graphical techniques, and is highly extensible. The S language is often the vehicle of choice for research in statistical methodology, and R provides an Open Source route to participation in that activity.

One of R's strengths is the ease with which well-designed publication-quality plots can be produced, including mathematical symbols and formulae where needed. Great care has been taken over the defaults for the minor design choices in graphics, but the user retains full control.

R is available as Free Software under the terms of the Free Software Foundation's GNU General Public License in source code form. It compiles and runs on a wide variety of UNIX platforms and similar systems (including FreeBSD and Linux), Windows and MacOS."

1.10.2 The R Environment

R is an integrated suite of software facilities for data manipulation, calculation and graphical display. It includes

- An effective data handling and storage facility,
- A suite of operators for calculations on arrays, in particular matrices,
- A large, coherent, integrated collection of intermediate tools for data analysis,
- Graphical facilities for data analysis and display either on-screen or on hardcopy, and
- A well-developed, simple and effective programming language which includes conditionals, loops, user-defined recursive functions and input and output facilities.

The term "environment" is intended to characterize it as a fully planned and coherent system, rather than an incremental accretion of very specific and inflexible tools, as is frequently the case with other data analysis software.

R, like S, is designed around a true computer language, and it allows users to add additional functionality by defining new functions. Much of the system is itself written in the R dialect of S, which makes it easy for users to follow the algorithmic choices made. For computationally-intensive tasks, C, C++ and Fortran code can be linked and called at run time. Advanced users can write C code to manipulate R objects directly.

Many users think of R as a statistics system. We prefer to think of it as an environment within which statistical techniques are implemented. R can be extended (easily) via *packages*. There are about eight packages supplied with the R distribution and many more are available through the CRAN family of Internet sites covering a very wide range of modern statistics.

1.10.3 Use of R for Stochastic Processes

For this book, R has primarily been used to generate observations from stochastic processes with known parameters. For example, suppose one wants to generate a realization from an ARMA(1,1) time series. The following code is from the excellent book about time series with R by Cowpertwait and Metcalfe (p. 29).[2]

Consider the ARMA(1,1) process

$$Y(t) = -0.6Y(t-1) + 0.5W(t-1) + W(t), \quad t = 2,3,..., \tag{1.56}$$

with autoregressive parameter $\theta = 0.5$, moving average parameter $\beta = 0.5$, and where $W(t) \sim \text{normal}(0, \sigma^2)$ with variance $\sigma^2 = 1$. The following two R statements will generate 1000 observations from Equation 1.56:

```
>set.seed(1)
>x<-arma.sim(n=1000,list(ar=.5, ma=0.5))
```

Vector x contains the 1000 values generated from Equation 1.56.

For additional information about R, refer to Jones, Maillardet, and Robinson[3] and to Verzani.[4] The first reference is a good introduction to using R for simulation, and the book contains many examples applicable to stochastic processes, while the second reference is an introductory statistics book that heavily relies on R. For a Bayesian approach to statistical computation using R, Albert[1] presents an excellent account. Copies of the code will be hosted at lbroemeling.com. Of course, if the reader is not familiar with R, I recommend reading the book by Albert[1] because it is introductory and is presented from a Bayesian point of view.

1.11 WinBUGS Package

1.11.1 Main Body

The main body of the software is a WinBUGS document, which contains the program statements, the major part of the document, and a list statement or statements, which include the data values and some initial values for the MCMC simulation of the posterior distribution. The document is given a title and saved as a WinBUGS file, which can be accessed as needed.

In order to illustrate the essential features of the WinBUGS package, refer to the program WinBUGS Code 9.15 found in Chapter 9 about estimating the parameters of an ARMA (1,1) model, with $\theta = 0.5$, $\beta = 0.5$, and $\sigma^2 = 1$, where the data were generated in Section 1.10.3. The first 24 observations are as follows:

```
Y=(2.00439112, 0.57587660, -2.23738188, -1.10110996, -
0.03302313, -0.05516863, 0.90815676, 1.74721768, 1.87812076,
2.15498841, 2.31911919, 1.62519273, -1.13947284, -0.94458652, -
0.21850913, -0.29311444, -1.69520736, -2.06112993, -0.85169843,
1.14180112, 1.14745261, 0.91000405, 0.59503279, -1.10644568, -
1.65674718)
```

WinBUGS Code 9.15

```
model;
{
theta~dbeta(5,5)
beta~dbeta(5,5)
v~dgamma(.01,.01)

for ( t in 1:25){ mu[t]<-0}

Y[1,1:25]~dmnorm(mu[],tau[,])

for( i in 1:25){Sigma[i,i]<-v+v*pow(theta+beta,2)/(1-theta*theta)}

for( i in 1:24){for(j in i+1:25){Sigma[i,j]<-(theta+beta)*pow(theta,j-1)*v+
pow(theta+beta,2)*v*pow(theta,j)/(1-theta*theta)}}

for( i in 2:25){ for ( j in 1:i-1){Sigma[i,j]<-(theta+beta)*pow(theta,j-1)*v+
pow(theta+beta,2)*v*pow(theta,j)/(1-theta*theta)}}
tau[1:25,1:25]<-inverse(Sigma[,])
}
 list(Y=structure(.Data=c(2.00439112, 0.57587660, -2.23738188,
-1.10110996, -0.03302313,
-0.05516863, 0.90815676, 1.74721768, 1.87812076, 2.15498841,
2.31911919,1.62519273, -1.13947284, -0.94458652, -0.21850913,
-0.29311444, -1.69520736, -2.06112993, -0.85169843, 1.14180112,
1.14745261,0.91000405, 0.59503279, -1.10644568, -1.65674718
),.Dim=c(1,25)))

list(theta=.5,v=1,beta=.5)
```

The main body of this program consists of three components, the main statements below the model and lasting until the first list statement.

1.11.2 List Statements

List statements allow the program to incorporate certain necessary information that is required for successful implementation. For example, experimental or study information is usually input with a list statement. The sample information is contained in the first list statement in program WinBUGS Code 9.15. Remember that the 24 values of the first list statement were generated with R in Section 1.10.3.

The last list statement contains the initial values or the MCMC simulation, assigning $\theta = 0.5$, $\beta = 0.5$, and $\sigma^2 = 1$.

1.11.3 Executing the Analysis

MCMC can analyze complex statistical models, and the following describes the use of drop-down menus from the tool bar for executing the posterior analysis.

1.11.4 Specification Tool

The tool bar of WinBUGS is labeled as follows, from left to right: File, Edit, Attributes, Tools, Info, Model, Inference, Doodle, Maps, Text, Windows, Examples, Manuals, and Help, and I have highlighted the model and inference labels. When the user clicks on one of the labels, a pop-up menu appears. In order to execute the program, the user clicks on Model, then clicks on Specification, and the specification tool appears. Refer to the specification tool shown in Figure 1.1.

The specification tool is used together with the WinBUGS document as follows: (1) click on the word *model* of the document, (2) click on the check model box of the specification tool, (3) click on the compile box of the specification tool, (4) click on the word *list* of the list statement of the document, and lastly, (5) click on load units box of the tool. Now close the specification tool and go to the next step as follows.

1.11.5 Sample Monitor Tool

The sample tool is activated by first clicking on the inference menu of the tool bar and then clicking on sample, and the sample monitor tool appears. Type *beta* then click on Set, then

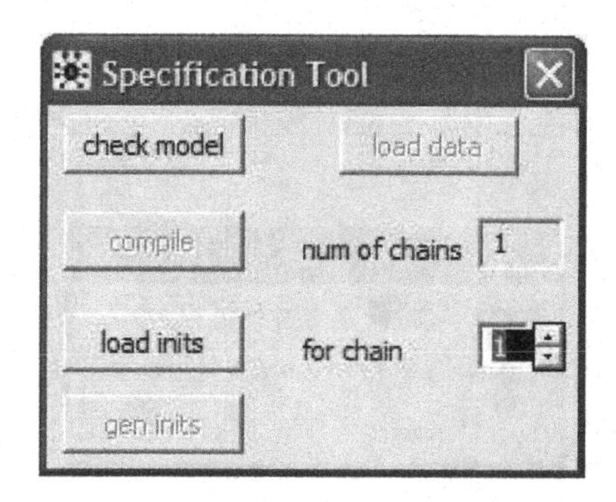

FIGURE 1.1
Specification tool.

type *theta* in the node box and click on Set; lastly, type v then click on Set, and finally, type an * in the node box. Type *5000* in the beg box, which means the first 5001 observations generated for the posterior distribution of the beta coefficients. The 5000 observations typed in beg are referred to as the *burn-in*. The menu for the monitor tool is depicted in Figure 1.2.

1.11.6 Update Tool

In order to activate the update tool, click on the Model menu of the tool bar, and then click on Updates. Note Figure 1.3 for the update tool.

Suppose you want to generate 45,000 observations from the posterior distributions of the three parameters, using the statements which are listed in the preceding document, then type *45000* in the Updates box, and 100 for refresh. In order to execute the simulation using the program statements in the document, click on Update of the Update tool.

1.11.7 Output

Table 1.1 reports the results of the Bayesian analysis for the ARMA(1,1) process that was executed by WinBUGS.

Figure 1.4 shows the posterior density of the autoregressive coefficient.

A good reference for learning WinBUGS for Bayesian modeling is Ntzoufras,[5] while the books by Congdon[6–8] use WinBUGS to some extent.

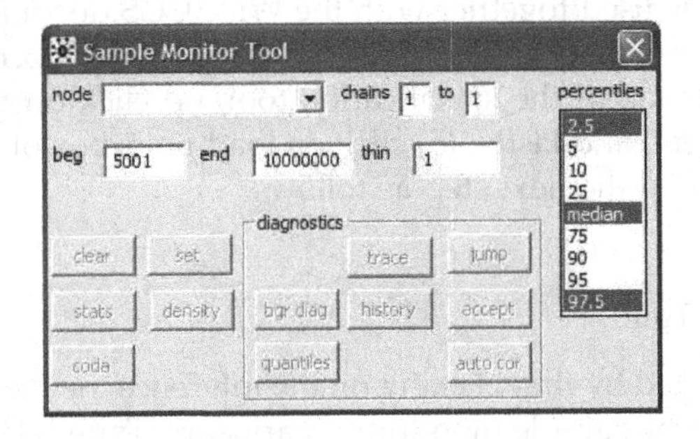

FIGURE 1.2
Sample monitor tool.

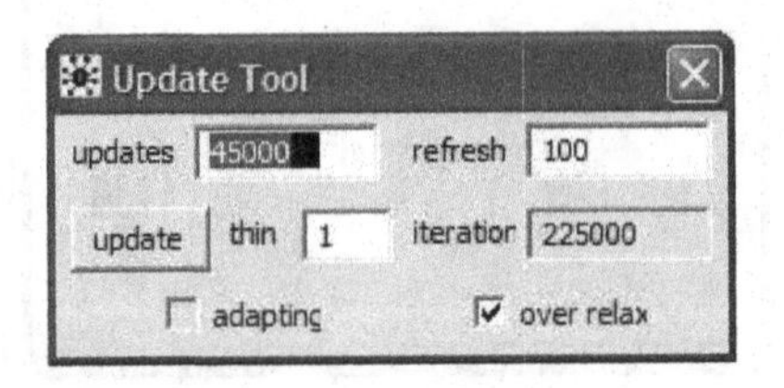

FIGURE 1.3
Update tool.

TABLE 1.1

Bayesian Analysis for the ARMA Process

Parameter	Value	Mean	SD	Error	2 1/2	Median	97 1/2
β	.5	.4979	.1512	.00071	.2094	.4978	.7857
θ	.5	.4604	.1407	.00081	.1956	.459	.7339
σ^2	1	1.01	0.4248	0.00386	0.4092	0.9369	2.05

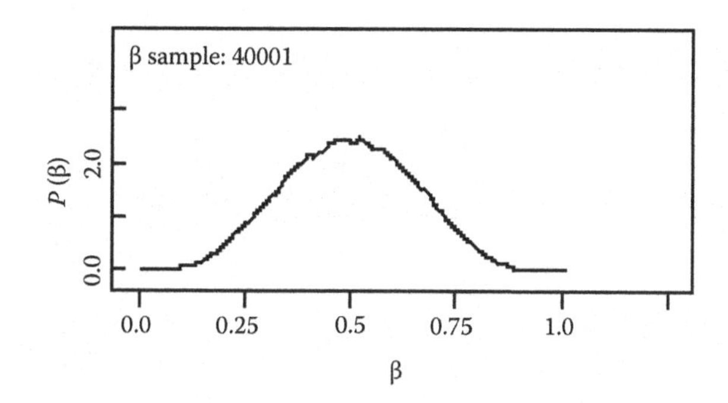

FIGURE 1.4
Posterior analysis autoregressive coefficient.

References

1. Albert, J. 2007. *Bayesian Computation*, New York: Springer-Verlag.
2. Cowpertwait, R. S. P., and Metcalfe, A. V. 2009. *Introductory Times Series with R*. New York: Springer-Verlag.
3. Jones, O., Maillardet, R., and Robinson, A. 2014. *Introduction to Scientific Programming and Simulation Using R*, Second Edition. Boca Raton, FL: CRC Press/Taylor & Francis.
4. Verzani, J. 2005. *Using R for Introductory Statistics*. Boca Raton, FL: Chapman and Hall/CRC Press.
5. Ntzoufras, I. 2009. *Bayesian Modeling with WinBUGS*. New York: John Wiley & Sons.
6. Congdon, P. 2001. *Bayesian Statistical Modeling*. New York: John Wiley & Sons.
7. Congdon, P. 2003. *Applied Bayesian Modeling*. New York: John Wiley & Sons.
8. Congdon, P. 2005, *Bayesian Modeling for Categorical Data*. New York: John Wiley & Sons.

2

Bayesian Analysis

2.1 Introduction

Bayesian methods will be employed to make inferences for stochastic processes, and this chapter will introduce the theory that is necessary in order to describe those procedures. The Bayes theorem, the foundation of the subject, is first introduced and followed by an explanation of the various components of the Bayes theorem: prior information; information from the sample given by the likelihood function; the posterior distribution, which is the basis of all inferential techniques; and lastly, the Bayesian predictive distribution. A description of the main three elements of inference, namely, estimation, tests of hypotheses, and forecasting future observations follows.

The remaining sections refer to the important standard distributions for Bayesian inference, namely, the Bernoulli, beta, multinomial, Dirichlet, normal, gamma, normal–gamma, multivariate normal, Wishart, normal–Wishart, and multivariate t-distributions. As will be seen, the relevance of these standard distributions to inferential techniques is essential for understanding the statistical analysis of stochastic processes

As will be seen, the multinomial and Dirichlet are the foundation for the Bayesian analysis of Markov chains and Markov jump processes. For normal stochastic processes such as the Wiener process and the Brownian motion, the multivariate normal and the normal-Wishart play a key role in determining Bayesian inferences.

Of course, inferential procedures can only be applied if there is adequate computing available. If the posterior distribution is known, often, analytical methods are quite sufficient to implement Bayesian inferences and will be demonstrated for the binomial, multinomial, and Poisson populations and several cases of normal populations. For example, when using a beta prior distribution for the parameter of a binomial population, the resulting beta posterior density has well-known characteristics, including its moments. In a similar fashion, when sampling from a normal population with unknown mean and precision and with a vague improper prior, the resulting posterior t-distribution for the mean has known moments and percentiles, which can be used for inferences.

Posterior inferences by direct sampling methods are easily done if the relevant random number generators are available. On the other hand, if the posterior distribution is quite complicated and not recognized as a standard distribution, other techniques are needed. To solve this problem, Monte Carlo Markov chain (MCMC) techniques have been developed and have been a major success in providing Bayesian inferences for quite complicated problems. This has been a great achievement in the field and will be described in later sections.

Minitab, S-Plus, WinBUGS, and R are packages that provide random number generators for direct sampling from the posterior distribution for many standard distributions, such as binomial, gamma, beta, and t-distributions. On occasion, these will be used; however, my preferences are WinBUGS and R, because they have been adopted by other Bayesians. This is also true for indirect sampling, where WinBUGS and R are excellent packages and are preferred for this book. Many institutions provide special-purpose software for specific Bayesian routines. For example, at MD Anderson Cancer Center, where Bayesian applications are routine, several special-purpose programs are available for designing (including sample size justification) and analyzing clinical trials.

Inferences for stochastic processes consist of testing hypotheses about unknown population parameters, estimation of those parameters, and forecasting future observations. When a sharp null hypothesis is involved, special care is taken in specifying the prior distribution for the parameters. A formula for the posterior probability of the null hypothesis is derived, via the Bayes theorem, and illustrated for Bernoulli, Poisson, and normal populations. If the main focus is the estimation of parameters, the posterior distribution is determined, and the mean, median, standard deviation, and credible intervals are found, either analytically or by computation with WinBUGS. For example, when sampling from a normal population with unknown parameters and using a conjugate prior density, the posterior distribution of the mean is a t and will be derived algebraically. On the other hand, for making Bayesian inferences for Markov chains with finite state space, the posterior distributions are beta for the individual transition probabilities and are Dirichlet for the joint distribution of the row transition probabilities. These posterior inferences are provided both analytically and numerically with WinBUGS. Of course, all analyses should be preceded by checking to determine if the model is appropriate, and this is where the predictive distribution comes into play. By comparing the observed results of the experiment with those predicted, the model assumptions are questioned. The most frequent use of the Bayesian predictive distribution is for forecasting future observations of stochastic processes such as for certain Markov chains, Markov jump processes, and time series processes.

2.2 Bayes Theorem

The Bayes theorem is based on the conditional probability law:

$$P[A|B] = P[B|A]P[A]/P[B], \tag{2.1}$$

where $P[A]$ is the probability of A before one knows the outcome of event B, $P[B \mid A]$ is the probability of B assuming what one knows about event A, and $P[A \mid B]$ is the probability of A knowing that event B has occurred. $P[A]$ is called the prior probability of A, while $P[A \mid B]$ is called the posterior probability of A.

Another version of the Bayes theorem is to suppose X is a continuous observable random vector, and $\theta \in \Omega \subset R^m$ is an unknown parameter vector, and suppose the conditional density of X given θ is denoted by $f(x|\theta)$. If $x = (x_1, x_2, \ldots x_n)$ represents a random sample of size n from a population with density $f(x|\theta)$, and $\xi(\theta)$ is the prior density of θ, then the Bayes theorem expresses the posterior density as

$$\xi(\theta|x) = c\prod_{i=1}^{i=} f(x_i|\theta)\xi(\theta), \quad x_i \in R \text{ and } \theta \in \Omega, \tag{2.2}$$

where the proportionality constant is c, and the term $\prod_{i=1}^{i=n} f(x_i|\theta)$ is called the likelihood function. Density $\xi(\theta)$ is the prior density of θ and represents the knowledge one possesses about the parameter before one observes X. Such prior information is most likely available to the experimenter from other previous related experiments. Note that θ is considered a random variable and that the Bayes theorem transforms one's prior knowledge of θ, represented by its prior density, to the posterior density and that the transformation is the combining of the prior information about θ with the sample information represented by the likelihood function.

"An essay toward solving a problem in the doctrine of chances" by the Reverend Thomas Bayes[1] is the beginning of our subject. He considered a binomial experiment with n trials, assumed that the probability θ of success was uniformly distributed (by constructing a billiard table), and presented a way to calculate $\Pr(a \le \theta \le b|x = p)$, where x is the number of successes in n independent trials. This was a first in the sense that Bayes was making inferences via $\xi(\theta|x)$, the conditional density of θ given x. Also, by assuming the parameter as uniformly distributed, he was assuming vague prior information for θ. This type of prior information, where very little is known about the parameter, is called noninformative or vague information.

It can well be argued that Laplace[2] is the greatest Bayesian because he made many significant contributions to inverse probability (he did not know of Bayes), beginning in 1774 with "Memorie sur la probabilite des causes par la evenemens," with his own version of the Bayes theorem, and, over a period of some 40 years, culminating in "Theorie analytique des probabilites." See Stigler[3] and Chapters 9–20 of Hald[4] for the history of Laplace's contributions to inverse probability.

It was in modern times that Bayesian statistics began its resurgence with Lhoste,[5] Jeffreys,[6] Savage,[7] and Lindley.[8] According to Broemeling and Broemeling,[9] Lhoste was the first to justify noninformative priors by invariance principals, a tradition carried on by Jeffreys. Savage's book was a major contribution in that Bayesian inference and decision theory was put on a sound theoretical footing as a consequence of certain axioms of probability and utility, while Lindley's two volumes showed the relevance of Bayesian inference to everyday statistical problems and was quite influential and set the tone and style for later books such as those by Box and Tiao,[10] Zellner,[11] and Broemeling.[12] Books by Box and Tiao and Broemeling were essentially works that presented Bayesian methods for the usual statistical problems of the analysis of variance and regression, while Zellner focused Bayesian methods primarily on certain regression problems in econometrics. During this period, inferential problems were solved analytically or by numerical integration. Models with many parameters (such as hierarchical models with many levels) were difficult to use because at that time, numerical integration methods had limited capability in higher dimensions. For a good history of inverse probability, see Chapter 3 of Stigler[3] and Hald,[13] who present a comprehensive history and are invaluable as references. Dale[14] gives a complete and very interesting account of Bayes' life.

The last 20 years is characterized by the discovery and development of resampling techniques, where samples are generated from the posterior distribution via MCMC methods, such as Gibbs sampling. Large samples generated from the posterior make it possible to make statistical inferences and to employ multilevel hierarchical models to solve complex, but practical problems. See Leonard and Hsu,[15] Gelman et al.,[16] Congdon,[17–19] Carlin and Louis,[20] and Gilks, Richardson, and Spiegelhalter,[21] who demonstrate the utility of MCMC techniques in Bayesian statistics.

2.3 Prior Information

2.3.1 Binomial Distribution

Where do we begin with prior information, a crucial component of the Bayes theorem rule? Bayes assumed that the prior distribution of the parameter is uniform, namely,

$$\xi(\theta) = 1, \quad 0 \le \theta \le 1,$$

where θ is the common probability of success in n independent trials and

$$f(x|\theta) = \binom{n}{x} \theta^x (1-\theta)^{n-x}, \tag{2.3}$$

where x is the number of successes (= 0, 1, 2, …, n). The distribution of X, the number of successes, is binomial and is denoted by $X \sim \text{Binomial}(\theta, n)$. The uniform prior was used for many years; however, Lhoste[5] proposed a different prior, namely,

$$\xi(\theta) = \theta^{-1}(1-\theta)^{-1}, \quad 0 \le \theta \le 1, \tag{2.4}$$

to represent information which is noninformative and is an improper density function. Lhoste based the prior on certain invariance principles, quite similar to Jeffreys.[6] Lhoste also derived a noninformative prior for the standard deviation σ of a normal population with density

$$f(x|\mu,\sigma) = \left(1/\sqrt{2\pi}\sigma\right) \exp -(1/2\sigma)(x-\mu)^2, \quad \mu \in R \text{ and } \sigma > 0 \tag{2.5}$$

He used invariance as follows: he reasoned that the prior density of σ and the prior density of $1/\sigma$ should be the same, which leads to

$$\xi(\sigma) = 1/\sigma. \tag{2.6}$$

Jeffreys' approach is similar to that in developing noninformative priors for binomial and normal populations, but he also developed noninformative priors for multiparameter models, including the mean and standard deviation for the normal density as

$$\xi(\mu,\sigma) = 1/\sigma, \quad \mu \in R \text{ and } \sigma > 0. \tag{2.7}$$

Noninformative priors where ubiquitous from the 1920s to the 1980s and were included in all the textbooks of that period. For example, see Box and Tiao,[10] Zellner,[11] and Broemeling.[12] Looking back, it is somewhat ironic that noninformative priors were almost always used, even though informative prior information was almost always available. This limited the utility of the Bayesian approach, and people saw very little advantage over the conventional way of doing business. The major strength of the Bayesian way is that it is a convenient, practical, and logical method of utilizing informative prior information. Surely, the investigator knows informative prior information from previous related studies.

How does one express informative information with a prior density? For example, suppose one has informative prior information for the binomial population. Consider

$$\xi(\theta) = [\Gamma(\alpha+\beta)/\Gamma(\alpha)\Gamma(\beta)]\theta^{\alpha-1}(1-\theta)^{\beta-1}, \quad 0 \le \theta \le 1, \tag{2.8}$$

as the prior density for θ. The Beta density with parameters α and β has a mean of $[\alpha/(\alpha+\beta)]$ and a variance of $[\alpha\beta/(\alpha+\beta)^2(\alpha+\beta+1)]$ and can express informative prior information in many ways.

As for prior information for the binomial, consider the analysis of Markov processes, namely, in estimating the transition probability matrix of a stationary finite-state Markov chain. Consider the 5 × 5 transition matrix P with components

$$P = \left(p_{ij}\right), \text{ where}$$

$$P = \begin{pmatrix} .2,.2,.2,.2,.2 \\ .2,.2,.2,.2,.2 \\ .2,.2,.2,.2,.2 \\ .2,.2,.2,.2,.2 \\ .2,.2,.2,.2,.2 \end{pmatrix}. \tag{2.9}$$

Note that

$$p_{ij} = \Pr[X_{n+1} = j | X_n = i], \tag{2.10}$$

where $n = 0, 1, 2, \ldots$.

That is to say X_n is a discrete-time Markov with state space $S = \{1,2,3,4,5\}$. Note that p_{ij} are the one-step transition probabilities of the Markov chain X_n, where the first row is the conditional distribution (given $X_0 = 1$) of a discrete random variable with mass points 1, 2, 3, 4, and 5, with probabilities .2, .2, .2, .2, and .2. The second row is the conditional distribution (given $X_0 = 2$) of a discrete random variable with mass points 1, 2, 3, 4, and 5 with probabilities .2, .2, .2, .2, .2, etc. This is an example of a Markov chain where each state is recurrent; that is, it is possible to reach any state from any other state, using a multinomial distribution with probability mass function

$$f(n|P) \propto \prod_{i,j=1}^{i,j=5} p_{ij}^{n_{ij}} \tag{2.11}$$

where $\sum_{i,j=1}^{i,j=5} p_{ij} = 1$ and $\sum_{i,j=1}^{i,j=5} n_{ij} = n$.

Ninety-eight n_{ij} values are generated from the chain with the following result:

```
1 2 5 1 4 1 5 3 3 2 4 4 2 5 5 5 3 2 2 2 3 2 1 4 5 3 2 1 2 3 4 3 3 4 2 4 5 1 5 1 1 4 5 1
4 1 3 4 4 2 5 1 5 1 3 1 2 4 1 3 1 3 5 5 4 1 4 1 2 1 2 5 4 4 4 2 4 2 5 5 4 5 2 4 3 2 2 3
2 1 2 1 2 4 2 3 1 4 4 2
```

R Code 2.1 is used to simulate the 98 observations from the Markov chain with transition matrix P:

R Code 2.1

```
   MC.sim<-function(n,P,x1){
sim<-as.numeric(n)
   m<-ncol(P)
   if (missing(x1)){
sim[1]<-sample(1:m,1)# random start
} else {sim[1]<-x1}
for ( i in 2:n){
newstate<-sample(1:m,1,prob=P[sim[i-1],])
sim[i]<-newstate
}
sim
}
P<-matrix(c(.2, .2, .2, .2, .2, .2, .2, .2, .2, .2, .2, .2, .2, .2, .2, .2, .2, .2,
.2, .2, .2, .2, .2, .2, .2),nrow=5,ncol=5,byrow=TRUE)
MC.sim(100,P,1)
```

These cell frequencies can be displayed with the 5 × 5 matrix:

$$N = \begin{pmatrix} 1, 7, 4, 6, 2 \\ 5, 3, 4, 6, 5 \\ 3, 6, 2, 3, 1 \\ 5, 7, 2, 5, 4 \\ 6, 1, 3, 3, 4 \end{pmatrix}. \tag{2.12}$$

Thus, there is one one-step transition from 1 to 1, seven transitions from 1 to 2, and, lastly, two one-step transitions from 1 to 5. Since the simulation was based on the multinomial distribution, it is known that the marginal distribution of the cell frequency n_{ij} is binomial with parameters p_{ij} and $n = 98$. In order to perform a Bayesian analysis, a prior distribution is assigned to the unknown cell frequencies: The conjugate distribution to the multinomial is the Dirichlet, which induces a beta prior to the individual cell frequencies. This results in a Dirichlet for the posterior distribution of the transition probabilities p_{ij} and, consequently, a beta for the individual transition probabilities. For the Dirichlet, the density is

$$f(p_{11}, p_{12}, ..p_{55}) \propto \prod_{i,j=1}^{i,j=5} p_{ij}^{\alpha_{ij}-1}, \tag{2.13}$$

where $\sum_{i,j=1}^{i,j=5} p_{ij} = 1$ and α_{ij} are positive.

Later in this chapter, a posterior analysis for estimating the transition probabilities will be presented.

2.3.2 Normal Distribution

Of course, the normal density plays an important role as a model in stochastic processes. For example, as will be seen in future chapters, the normal distribution will model certain normal stochastic processes such as the Brownian motion. How is informative prior information expressed for the parameters μ and σ (the mean and standard deviation, respectively)? Suppose a previous study has m observations $X = (x_1 x_2, ..., x_m)$, then the density of X given μ and σ is

$$f(x|\mu,\sigma) \propto \left[\sqrt{m}/\sqrt{2\pi\sigma^2}\right] \exp -(m/2\sigma^2)(\bar{x}-\mu)^2$$

$$\left[(2\pi)^{-(n-1)/2}\sigma^{-(n-1)}\right] \exp -(1/2\sigma^2) \sum_{i-1}^{i=m}(x_i-\bar{x})^2. \tag{2.14}$$

This is a conjugate density for the two-parameter normal family and is called the normal-gamma density. Note that it is the product of two functions, where the first, as a function of μ and σ, is the conditional density of μ given σ, with mean $\bar{x}$ and variance σ^2/m, while the second is a function of σ only and is an inverse gamma density. Or equivalently, if the normal is parameterized with μ and precision $\tau = 1/\sigma^2$, the conjugate distribution is as follows: (1) the conditional distribution of μ given τ is normal with mean $\bar{x}$ and precision $m\tau$ and (2) the marginal distribution of τ is gamma with parameters $(m + 1)/2$ and $\sum_{i=1}^{i=m}(x_i - \bar{x})^2/2 = (m-1)S^2/2$, where S^2 is the sample variance. Thus, if one knows the results of a previous experiment, the likelihood function for μ and τ provides informative prior information for the normal population.

The normal distribution is very important in describing a normal stochastic process $\{X(t), t \geq 0\}$, where for each t, $X(t)$ has a normal distribution. The Weiner process (Brownian motion) is defined as follows:

Consider the time points $0 \leq t_1 < t_2 < ... < t_n$, where n is a positive integer and suppose that for each t,

1. $X(t)$ is normal with mean of 0 and variance of $\sigma^2 t$.
2. The process $\{X(t), t \geq 0\}$ has independent increment; that is, for all n and choice of time points, the $n - 1$ differences $X(t_i) - X(t_{i-1})$, $i = 2, ..., n$, are independent normal random variables with mean of 0 and variance of $\sigma^2(t_i - t_{i-1})$. We let $X(t_0) = X(0) = 0$.

If the process is observed at these n time points, the joint density of the increments $d_i = X(t_i) - X(t_{i-1})$ is

$$f(d_1, d_2, \ldots, d_n|\sigma^2) \propto \left[1/\sigma^n \prod_{i=1}^{i=n} (t_i - t_{i-1})^{1/2}\right] \exp -(1/2\sigma^2) \sum_{i=1}^{i=n} (d_i^2/(t_i - t_{i-1})) \tag{2.15}$$

where $\sigma^2 > 0$.

Often, it is convenient to parameterize the likelihood in terms of the precision $\tau = 1/\sigma^2$; thus, the joint density (the likelihood function) is given as

$$f(d_1, d_2, \ldots, d_n|\tau) \propto \left(\tau^{n/2} / \prod_{i=1}^{i=n} (t_i - t_{i-1})^{1/2}\right) \exp -(\tau/2) \sum_{i=1}^{i=n} (d_i^2/(t_i - t_{i-1})), \tag{2.16}$$

where $\tau > 0$.

Of course, the goal of the Bayesian analysis is to estimate the process variance σ^2 or precision $\tau > 0$. For the Bayesian analysis, a prior distribution must be assigned to τ, in which case, the conjugate distribution, which is a gamma, can be used. The posterior analysis for the Wiener process will be demonstrated in Section 2.4. Please note that the Wiener process and the Brownian motion are the same mathematical object.

2.4 Posterior Information

2.4.1 Binomial Distribution

The preceding section explains how prior information is expressed in an informative or in a noninformative way. Several examples are given and will be revisited as illustrations for the determination of the posterior distribution of the parameters. Suppose a uniform prior distribution for the transition probability (of the five-state Markov chain) p_{ij} is used. What is the posterior distribution of p_{ij}?

By the Bayes theorem,

$$f\left(p_{ij}|N\right) \propto \binom{n}{n_{ij}} p_{ij}^{n_{ij}} \left(1 - p_{ij}\right)^{n-n_{ij}}, \tag{2.17}$$

where n_{ij} is the observed transitions from state i to state j and n is the total cell counts for the 5×5 cell frequency matrix N. Of course, this is recognized as a Beta $(n_{ij} + 1, n - n_{ij} + 1)$ distribution, and the posterior mean is $(n_{ij} + 1/n + 2)$. On the other hand, if the Lhoste[5] prior density (Equation 2.4) is used, the posterior distribution of p_{ij} is Beta $(n_{ij}, n - n_{ij})$ with mean of n_{ij}/n, which is the usual estimator of p_{ij}.

2.4.2 Normal Distribution

Consider a random sample $X = (x_1, x_2, ..., x_n)$ of size n from a normal $(\mu, 1/\tau)$ population, where $\tau = 1/\sigma^2$ is the inverse of the variance, and suppose the prior information is vague and the Jeffreys–Lhoste prior $\xi(\mu, \tau) \propto 1/\tau$ is appropriate, then the posterior density of the parameters is

$$\xi(\mu,\tau|\text{data}) \propto \tau^{n/2-1} \exp -(\tau/2)\left[n(\mu - \bar{x})^2 + \sum_{i=1}^{i=n}(x_i - \bar{x})^2\right]. \tag{2.18}$$

Using the properties of the gamma density, τ is eliminated by integrating the joint density with respect to τ to give

$$\begin{aligned}&\xi(\mu,\tau|\text{data})\\ &\propto \left\{\Gamma(n/2)n^{1/2}/(n-1)^{1/2}S\pi^{1/2}\Gamma(n-10/2)\right\}/\left[1 + n(\mu-\bar{x})^2/(n-1)S^2\right]^{(n-1+1)/2},\end{aligned} \tag{2.19}$$

which is recognized as a t-distribution with $n - 1$ degrees of freedom, location $\bar{x}$, and precision n/S^2. Transforming to $(\mu - x)\sqrt{n}/S$, the resulting variable has a Student's t-distribution with $n - 1$ degrees of freedom. Note that the mean of μ is the sample mean, while the variance is $[(n - 1)/n(n - 3)]$, $n > 3$.

Eliminating μ from Equation 2.12 results in the marginal distribution of τ as

$$\xi(\tau|S^2) \propto \tau^{[(n-1)/2]-1} \exp -\tau(n-1)S^2/2, \quad \tau > 0, \tag{2.20}$$

which is a gamma density with parameters $(n - 1)/2$ and $(n - 1)S^2/2$. This implies that the posterior mean is $1/S^2$ and the posterior variance is $2/(n - 1)S^4$.

Now consider a Brownian motion example where $\sigma^2 = 0.01$, then R can be used to generate the following 100 observations:

```
X=(0.00000000, -0.12066594, 0.12261439, 0.30593504,
0.25192900, 0.15859211,0.15995869, 0.19497827, 0.23087514,
0.10417925,0.01373721, -0.12641825, -0.10337710, -0.14188594,
-0.05242343, -0.13633928, -0.18213174, -0.17070495,-
0.18482716, -0.09181292, 0.10854388, 0.10409353,
0.25193405, 0.39038363, 0.37192310, 0.24942431,
0.34064631, 0.54417604, 0.64985999, 0.66225207, 0.63390233,
0.57778938, 0.64035220, 0.74328382, .88498606, 0.98498418,
1.01933431,0.99682087, 0.86299816, 0.74439032, 0.80420728,
0.84194706, 0.86085161, 0.75355936, 0.70584299,
0.79916880, 0.89982912, 0.89759060, 0.81825987,
0.81468989, 0.60323491, 0.69081813, 0.55690599, 0.60584469,
0.62862193, 0.52824828, 0.66174973, 0.93504614, 0.89310704,
0.80477246, 0.76667270, 0.99644731, 1.04091656, 1.06369006,
1.23021060, 1.15540209, 1.10621274, 1.37052053, 1.43990730,
1.46114072, 1.47904886, 1.61770116, 1.46674563, 1.48799118,
1.47145937, 1.64238708, 1.57441723, 1.67824584, 1.75466133,
1.76300545, 1.75221320, 1.82687502, 1.83908812, 1.98250229,
2.07598916, 1.90912144, 1.92619380, 1.62350822, 1.34434067,
```

1.27522009, 1.16668612, 1.10504900, 1.10992517, 1.18635928, 1.26900223, 1.346168, 1.391436, 1.418384, 1.482995, 1.62940).

The R Code used for the simulation is given by R Code 2.2.

R Code 2.2

```
t <- 0:100
sig2 <- 0.01
x <- rnorm(n=length(t)-1,sd=sqrt(sig2))
x<-c(0,cumsum(x))
plot(t,x,type="l",ylim=c(-2,2))
```

The vector x contains the 100 simulated values for Brownian motion.

Based on the 100 values generated from the Brownian motion, our goal is to estimate the variance of Brownian motion using the posterior distribution of τ given by Equation 2.16. Note that

$$\sum_{i=1}^{i=100} d_i^2 = 1.648375 \tag{2.21}$$

Thus, the posterior distribution τ is gamma with parameters

$$\alpha = (n+1)/2 = 101/2 = 50.5 \tag{2.22}$$

and

$$\beta = \sum_{i=1}^{i=100} d_i^2/2 = 0.824187 \tag{2.23}$$

Consequently, the posterior mean of τ and of σ^2 are

$$\begin{aligned} &E(\tau|x) = \alpha/\beta = 61.2 \\ &\text{and} \\ &E(\sigma^2|x) = \beta/(\alpha-1) = 0.0166, \end{aligned} \tag{2.24}$$

respectively. Recall that $\sigma^2 = 0.01$ is the "true" variance of the Brownian motion, which is to be compared to the estimated value of 0.0166. Is this a reasonable estimate? It should be noted that

$$t_i - t_{i-1} = 1, \quad i = 1, 2, \ldots, 100; \tag{2.25}$$

that is, the Brownian motion process was sampled at equal time units of length 1.

2.4.3 Poisson Distribution

The Poisson distribution often occurs as a population for a discrete random variable with mass function

$$f(x|\theta) = e^{-\theta}\theta^{x}/x!, \tag{2.26}$$

where the gamma density

$$\xi(\theta) = [\beta^{\alpha}/\Gamma(\alpha)]\theta^{\alpha-1}e^{-\theta\beta} \tag{2.27}$$

is a conjugate distribution that expresses informative prior information. For example, in a previous experiment with m observations, the prior density would be gamma with the appropriate values of alpha and beta. Based on a random sample of size n, the posterior density is

$$E(\theta|data) \propto \theta^{\sum_{i=1}^{i=n} x_i + \alpha - 1} e^{-\theta(n+\beta)} \tag{2.28}$$

which is identified as a gamma density with parameters $\alpha' = \sum_{i=1}^{i=n} x_i + \alpha$ and $\beta' = n + \beta$. Remember that the posterior mean is α'/β'; median, $(\alpha' - 1)/\beta'$; and variance, $\alpha'/(\beta')^2$.

One of the most important Markov jump (continuous time and countable state space) is the Poisson process.

Recall from Chapter 1 that the introduction to the Poisson process $N(t)$ with parameter $\lambda > 0$ is defined as follows:

1. $N(t)$ is the number of events occurring over time 0 to t with $N(0) = 0$, and the process has independent increments.
2. For all $t > 0$, $0 < P[N(t) > 0] < 1$; that is to say, for all intervals, no matter how small, there is a positive probability that an event will occur, but it is not certain that an event will occur.
3. For all $t \geq 0$,

   $$\lim\{P[N(t+h) - N(t) \geq 2]/P[N(t+h) - N(t) = 1]\},$$

   where the limit is as h approaches 0. This implies that events cannot occur simultaneously.
4. The process has stationary independent increments; thus, for all points $t > s \geq 0$ and $h > 0$, the two random variables $N(t+h) - N(s+h)$ and $N(t) - N(s)$ are identically distributed and are independent.

Based on these four axioms, one may show that for all $t > 0$, there exists a $\lambda > 0$ such that $N(t)$ has a Poisson distribution with mean λt. Thus, the average number of events occurring over $[0,t)$ is λt, and the average number of events occurring per unit time is λ. The Poisson process is a counting process (it counts the number of events occurring over time) and has many generalizations that will be introduced in Chapter 7. An interesting feature of the Poisson process is that the time between the occurrences of two adjacent events has an exponential distribution. In particular, if $N(t)$, $t \geq 0$ is a Poisson process with parameter λ, then the successive interarrival times are independent and have an exponential distribution with mean $1/\lambda$; thus, the Poisson process can be simulated via the exponential distribution. For example, consider a Poisson process with parameter $\lambda = 5$, and suppose a realization of 50 using the exponential distribution with mean 1/5 = .2 is to be generated using WinBUGS Code 2.1

WinBUGS Code 2.1

```
model {
for (i in 1 : 1000) {
y[i] ~ dexp(.2)
}
}
```

The 50 successive interarrival times are given by the vector I:

```
I=(.403, 11.00, 23.11, 1.92, .25, 4.34, 3.53, .10, 1.59,
.05, 3.11, 2.21, 3.03, 5.96, 7.22, .96, 8.75, 2.23, 29.84,
2.96, 2.41, 2.86, .5411.48, 2.10, 1.43, 8.99, 6.87, 1.73,
6.76, 14.91, 11.90, 1.21, 12.08, 4.49, 4.14, 1.94, 1.30,
1.86, 4.86, .21, 13.27, .42, 1.60, 3.38, 3.39, 2.97, 9.97,
7.03, 2.54).
```

The 50 corresponding waiting times are the components of the following vector W:

```
W=(.40, 11.40, 34.51, 36.43, 36.68, 41.02, 44.55, 44.65,
46.24, 46.29, 49.40, 51.61, 54.64, 60.60, 67.82, 68.78,
77.53, 79.76, 109.60, 112.56, 114.97, 117.83, 118.37,
129.85, 131.95, 133.38, 142.37, 149.24, 150.97, 157.73,
172.64, 184.54, 185.75, 197.83, 202.32, 206.46, 208.40,
209.70, 211.56, 216.42, 216.63, 229.90, 230.32, 231.92,
235.30, 238.69, 241.66, 251.63, 258.66, 261.20).
```

Thus, the first event occurred at time 0.403 time units, and the second, at 11.40 time units, and the last, at 261.2 units.

Let T_n be the nth interarrival time and W_n be the corresponding waiting time, then

$$W_n = T_1 + T_2 + \ldots + T_n, \tag{2.29}$$

where $n = 0, 1, 2, \ldots$; thus, we know that

$$T_n \sim \exp(\lambda) \tag{2.30}$$

and

$$W_n \sim \text{gamma}(n, \lambda). \tag{2.31}$$

That is, the interarrival times have a common exponential distribution with parameter λ, and nth waiting time has a gamma distribution with parameters n and λ. In the next section on inference, based on the preceding interarrival and waiting times, Bayesian inferences for intensity λ will be performed.

2.5 Inference

2.5.1 Introduction

In a statistical context, by inference, one usually means the estimation of parameters, the testing of hypotheses, and the prediction of future observations. With the Bayesian approach, all inferences are based on the posterior distribution of the parameters, which in turn is based on the sample, via the likelihood function and the prior distribution. We have seen the role of the prior density and likelihood function in determining the posterior distribution and, presently, will focus on the determination of point and interval estimation of the model parameters and will later emphasize what is the effect of the posterior distribution on a test of hypothesis. Lastly, the role of the predictive distribution in testing hypotheses and in goodness of fit will be explained.

When the model has only one parameter, one would estimate that parameter by listing its characteristics, such as the posterior mean, media, and standard deviation, and plotting the posterior density. On the other hand, if there are several parameters, one would determine the marginal posterior distribution of the relevant parameters and, as mentioned earlier, calculate its characteristics (e.g., mean, median, mode, standard, and deviation) and plot the densities. Interval estimates of the parameters are also usually reported and are called credible intervals.

Suppose we want to estimate

$$P_{ij} = P[X_{n+1} = j|X_n = i],$$

the one-step transition probability of the binomial example of Section 2.4, where the matrix of cell frequencies

$$N = \begin{pmatrix} 1, 7, 4, 6, 2 \\ 5, 3, 4, 6, 5 \\ 3, 6, 2, 3, 1 \\ 5, 7, 2, 5, 4 \\ 6, 1, 3, 3, 4 \end{pmatrix} \tag{2.32}$$

was generated using the transition probability matrix

$$P = \begin{pmatrix} .2, .2, .2, .2, .2 \\ .2, .2, .2, .2, .2 \\ .2, .2, .2, .2, .2 \\ .2, .2, .2, .2, .2 \\ .2, .2, .2, .2, .2 \end{pmatrix}. \tag{2.33}$$

Recall that each row is the conditional probability distribution. For example, for the first row, the conditional probability of a transition from state 1 to states 1, 2, 3, 4, and 5 is .2, .2,

.2, .2, and .2. Thus, the number of transitions from state 1 to state 2 is 7. Our objective is to estimate the transition probabilities ϕ_{ij}, based on the cell counts of transitions given by the matrix N. Note that ϕ_{ij} is the one-step conditional (given i) probability of j.

Consider the ith row of the matrix n; then the ith row cell frequencies n_{ij}, $j = 1, 2, 3, 4, 5$, have a multinomial distribution with parameters ϕ_{ij}, where $j = 1, 2, 3, 4, 5$, and $n_{i.}$ = row total for row i, where $i = 1, 2, 3, 4, 5$. Now assuming a uniform prior for the ϕ_{ij}, $j = 1, 2, 3, 4, 5$, the posterior distribution of the ϕ_{ij} is Dirichlet with parameters $n_{ij} + 1$, $j = 1, 2, 3, 4, 5$. Thus, the posterior density of $\phi_{i1}, \phi_{i2}, \phi_{i3}, \phi_{i4}, \phi_{i5}$ is

$$f(\phi_{i1}, \phi_{i2}, \phi_{i3}, \phi_{i4}, \phi_{i5} | n_{i1}, n_{i2}, n_{i3}, n_{i4}, n_{i5}) \propto \prod_{j=1}^{j=5} \phi_{ij}^{n_{ij}}, \tag{2.34}$$

where $\sum_{j=1}^{j=5} \phi_{ij} = 1$ and $\sum_{j=1}^{j=5} n_{ij} = n_{i.}$, $j = 1, 2, 3, 4, 5$.

In order to estimate the ϕ_{ij}, consider the posterior mean

$$E(\phi_{ij}|\text{data}) = (n_{ij} + 1)/(n_{i.} + 5), \tag{2.35}$$

$j = 1, 2, 3, 4, 5$.

Thus, in particular,

$$E(\phi_{12}|\text{data}) = (n_{12} + 1)/(n_{1.} + 5) = 8/25 = .32 \tag{2.36}$$

On the other hand, assuming the noninformative improper prior

$$f(\phi_{i1}, \phi_{i2}, \phi_{i3}, \phi_{i4}, \phi_{i5}) \propto \prod_{j=1}^{j=5} \phi_{ij}^{-1}, \tag{2.37}$$

where $\sum_{j=1}^{j=5} \phi_{ij} = 1$ and $\phi_{ij} > 0$, for $j = 1, 2, 3, 4, 5$, the posterior distribution of the ϕ_{ij}, where $j = 1, 2, 3, 4, 5$, is Dirichlet with parameters n_{ij}, $j = 1, 2, 3, 4, 5$; thus, the posterior mean of ϕ_{ij} is

$$E(\phi_{ij}|\text{data}) = n_{ij}/n_{i.}, \tag{2.38}$$

where $j = 1, 2, 3, 4, 5$, which is the usual estimator of the transition probabilities. Therefore,

$$E(\phi_{12}|\text{data}) = n_{12}/n_{1.} = 7/20 = .28 \tag{2.39}$$

is the estimate of $\phi_{12} = .22$. Note that, assuming the improper prior density (Equation 2.37), the posterior variance of ϕ_{ij} is

$$\text{Var}(\phi_{ij}|\text{data}) = n_{ij}(n_{i.} - n_{ij})/n_{i.}^2(n_{i.} + 1). \tag{2.40}$$

Thus, in particular,

$$\text{Var}(\phi_{12}|\text{data}) = n_{12}(n_{1.} - n_{12})/n_{1.}^2(n_{1.} + 1) \tag{2.41}$$

or

$$\text{Var}(\phi_{12}|\text{data}) = 7(20 - 7)/20^2(20 + 1) = .010833. \tag{2.42}$$

2.5.2 Estimation

What are the 95% credible intervals for the transition probabilities?

Inferences for the normal (μ, τ) population are somewhat more demanding, because both parameters are unknown. Assuming the vague prior density $\xi(\mu, \tau) \propto 1/\tau$, the marginal posterior distribution of the population mean μ is a t-distribution with $n - 1$ degrees of freedom, mean $\bar{x}$, and precision n/S^2; thus, the mean and the median are the same and provide a natural estimator of μ, and because of the symmetry of the t-density, a $(1 - \alpha)$ credible interval for μ is $x \pm t_{\alpha/2,n-1} S/\sqrt{n}$, where $t_{\alpha/2,n-1}$ is the upper $100\alpha/2$ percent point of the t-distribution with $n - 1$ degrees of freedom. To generate values from the $t(n-1, \bar{x}, n/S^2)$ distribution, generate values from Student's t-distribution with $n - 1$ degrees of freedom, multiply each by $S/\sqrt{n}$, and then add $\bar{x}$ to each. Suppose $n = 30$,

X = (7.8902, 4.8343, 11.0677, 8.7969, 4.0391, 4.0024, 6.6494, 8.4788, 0.7939, 5.0689, 6.9175, 6.1092, 8.2463, 10.3179, 1.8429, 3.0789, 2.8470, 5.1471, 6.3730, 5.2907, 1.5024, 3.8193, 9.9831, 6.2756, 5.3620, 5.3297, 9.3105, 6.5555, 0.8189, 0.4713), then $\bar{x}$ = 5.57 and S = 2.92.

Using the same dataset, WinBUGS Code 2.2 is used to analyze the problem:

WinBUGS Code 2.2

```
Model;
{ for( i in 1:30) { x[i]~dnorm(mu,tau) }
mu~dnorm (0.0,.0001)
tau ~dgamma( .0001,.0001) (4.17)
sigma <- 1/tau }
list( x =
c(7.8902, 4.8343, 11.0677, 8.7969, 4.0391, 4.0024, 6.6494, 8.4788,
0.7939, 5.0689, 6.9175, 6.1092, 8.2463, 10.3179, 1.8429, 3.0789, 2.8470,
5.1471, 6.3730, 5.2907, 1.5024, 3.8193, 9.9831, 6.2756, 5.3620, 5.3297,
9.3105, 6.5555, 0.8189, 0.4713))
list( mu = 0, tau = 1)
```

Note that a somewhat different prior was employed here, compared to the previous one, in that μ and τ are independent and assigned as proper, but noninformative distributions. The corresponding analysis gives the data found in Table 2.1.

Upper and *Lower* refer to the upper and lower 2.5 percent points of the posterior distribution. Note that a 95% credible interval for μ is (4.47, 6.65), and the estimation error is 0.003566. See Chapter 1 for the details on executing the WinBUGS statements mentioned earlier.

TABLE 2.1

Posterior Distribution of μ and $\sigma = 1/\sqrt{\tau}$

Parameter	Mean	Std Dev	Markov Chain Error	Median	Lower	Upper
μ	5.572	0.5547	0.003566	5.571	4.4790	6.656
σ	9.15	2.570	0.01589	8.733	5.359	15.37

The program generated 30,000 samples from the joint posterior distribution of μ and σ using a Gibbs sampling algorithm and used 29,000 for the posterior moments and graphs, with a refresh of 100.

2.5.3 Testing Hypotheses

An important feature of inference is testing hypotheses. Often in stochastic processes, the scientific hypothesis can be expressed in statistical terms, and a formal test, implemented. Suppose $\Omega = \Omega_0 \cup \Omega_1$ is a partition of the parameter space, then the null hypothesis is designated as H_0: $\theta \in \Omega_0$, and the alternative, as H_1: $\theta \in \Omega_1$, and a test of H_0 versus H_1 consists of rejecting H_0 in favor of H_1 if the observations $x = (x_1, x_2, \ldots, x_n)$ belong to a critical region C. In the usual approach, the critical region is based on the probabilities of type I errors, namely, $\Pr(C|\theta)$, where $\theta \in \Omega_0$, and of type II errors, $1 - \Pr(C|\theta)$, where $\theta \in \Omega_1$. This approach to testing hypothesis was developed by Neyman and Pearson and can be found in many of the standard references, such as Lehmann.[22] Lee[23] presents a good elementary introduction to testing and estimation in a Bayesian context.

In the Bayesian approach, the posterior probabilities

$$p_0 = \Pr(\theta \in \Omega_0|\text{data}) \tag{2.43}$$

and

$$p_1 = \Pr(\theta \in \Omega_1|\text{data}) \tag{2.44}$$

are required, and on the basis of the two, a decision is made whether or not to reject H in favor of A or to reject A in favor of H. Of course, also required are the two corresponding prior probabilities:

$$\pi_0 = \Pr(\theta \in \Omega_0) \tag{2.45}$$

and

$$\pi_1 = \Pr(\theta \in \Omega_1). \tag{2.46}$$

Now consider the prior odds π_0/π_1 and posterior odds p_0/p_1. In turn, consider the Bayes factor B in favor of H_0 relative to H_1, namely,

$$B = (p_0/p_1)/(\pi_0/\pi_1). \tag{2.47}$$

Then, the posterior probabilities p_0 and p_1 can be expressed in terms of the Bayes factor; thus,

$$p_0 = 1/\left[1 + (\pi_1/\pi_1)B^{-1}\right], \tag{2.48}$$

and the Bayes factor is interpreted as the odds in favor of H_0 relative to H_1 as implied by the information from the data.

When the hypotheses are simple, that is, $\Omega_0 = \{\theta_0\}$ and $\Omega_1 = \{\theta_1\}$, the odds ratio can be expressed as the likelihood ratio.

$$B = p(x|\theta_0)/p(x|\theta_1). \tag{2.49}$$

This interpretation is not valid when Ω_0 and Ω_1 are composite. Consider the restriction of the prior density $p(\theta)$ to Ω_0, namely,

$$p_0(\theta) = p(\theta)/\pi_0, \quad \theta \in \Omega_0, \tag{2.50}$$

and its restriction to Ω_1, namely,

$$p_1(\theta) = p(\theta)/\pi_1, \quad \theta \in \Omega_1. \tag{2.51}$$

Note that the integral of $p_0(\theta)$ with respect to θ over Ω_0 is 1.

Now it can be shown that the posterior probability of the null hypothesis is

$$p_0 = \pi_0 \int p(x|\theta)p_0(\theta)d\theta, \tag{2.52}$$

where the integral is taken over Ω_0.

In a similar way, the posterior probability of H_1 is

$$p_1 = \pi_1 \int p(x|\theta)p_1(\theta)d\theta, \tag{2.53}$$

where the integral is taken over Ω_1.

Now the Bayes factor can be expressed as

$$\begin{aligned} B &= (p_0/p_1)/(\pi_0/\pi_1) \\ &= \int_{\theta\in\Omega_0} p(x|\theta)p_0(\theta)d\theta \Big/ \int_{\theta\in\Omega_1} p(x|\theta)p_1(\theta)d\theta, \end{aligned} \tag{2.54}$$

which is the ratio of weighted likelihood functions, weighted by the prior probability densities restricted to Ω_0 and Ω_1.

An important aspect of testing hypotheses is when the null hypothesis is a point null hypothesis and the alternative is composite; thus, consider

$$H_0: \theta = \theta_0 \tag{2.55}$$

versus

$$H_1: \theta \neq \theta_0, \tag{2.56}$$

where θ_0 is known. How does one assign prior information to this case? A reasonable approach is to assign a positive probability π_0 for the null hypothesis, and for the alternative, assign a prior density $\pi_1 p_1(\theta)$, where

$$\int_{\theta \neq \theta_0} p_1(\theta) d\theta = 1. \tag{2.57}$$

Thus, $\pi_0 + \pi_1 = 1$, and it is seen that the prior probability of the alternative is π_1, and for values $\theta \neq \theta_0$, p_1 is the density of a continuous random variable that expresses the prior knowledge one has for the alternative hypothesis.

Let

$$p(x) = \pi_0 p(x|\theta_0) + \pi_1 \int p_1(\theta) p(x|\theta) d\theta, \tag{2.58}$$

where x is the vector of observations with conditional density $p(x|\theta)$ and where $p(x)$ is the marginal density of the observations.

By letting

$$p_1(x) = \int_{\theta \neq \theta_0} p_1(\theta) p(x|\theta) d\theta, \tag{2.59}$$

the marginal density (Equation 2.58) can be expressed as

$$p(x) = \pi_0 p(x|\theta_0) + \pi_1 p_1(x), \tag{2.60}$$

and then the posterior probabilities of the null and alternative hypotheses can be expressed as

$$\begin{aligned} p_0 &= \pi_0 p(x|\theta_0)/[\pi_0 p(x|\theta_0) + \pi_1 p_1(x)] \\ &= \pi_0 p(x|\theta_0)/p(x). \end{aligned} \tag{2.61}$$

In a similar manner,

$$p_1 = \pi_1 p_1(x)/p(x) \tag{2.62}$$

for the posterior probability of the alternative hypothesis. If one desires to use the Bayes factor, then one may show

$$B = p(x|\theta_0)/p_1(x). \tag{2.63}$$

The preceding derivation of the posterior probabilities in the context of hypothesis testing closely follows Lee.[23]

In summary, for testing hypotheses via the Bayesian approach, the following is required:

1. The prior probabilities of the null and alternative hypotheses, namely, π_0 and π_1.
2. The prior density $p_1(\theta)$ for values of $\{\theta : \theta \neq \theta_0\}$.
3. The likelihood function, that is, the joint conditional density of the observations $x = (x_1, x_2, \ldots, x_n)$ given θ, for all values of θ in the parameter space.

For the first example in testing hypotheses when the null is simple but the alternative is composite, consider Section 2.5.2, the example involving a Markov chain with five states, 1, 2, 3, 4, and 5, and an observed transition count matrix

$$N = \begin{pmatrix} 1, 7, 4, 6, 2 \\ 5, 3, 4, 6, 5 \\ 3, 6, 2, 3, 1 \\ 5, 7, 6, 5, 4 \\ 6, 1, 3, 3, 4 \end{pmatrix}. \tag{2.32}$$

Recall that this corresponds to the one-step transition matrix

$$\Phi = \begin{pmatrix} \phi_{11}, \phi_{12}, \phi_{13}, \phi_{14}, \phi_{15} \\ \phi_{21}, \phi_{22}, \phi_{23}, \phi_{24}, \phi_{25} \\ \phi_{31}, \phi_{32}, \phi_{33}, \phi_{34}, \phi_{35} \\ \phi_{41}, \phi_{42}, \phi_{43}, \phi_{44}, \phi_{45} \\ \phi_{51}, \phi_{52}, \phi_{53}, \phi_{54}, \phi_{55} \end{pmatrix}, \tag{2.64}$$

where ϕ_{ij} is the one-step transition probability from state i to state j. It is important to remember that the first row of Φ is the conditional probability distribution of the five states given $i = 1$. That is, given $i = 1$, ϕ_{1j} is the probability (in one step) of going from state $i = 1$ to state j, where $j = 1, 2, 3, 4, 5$. It is important to note that if $\phi_{11} > 0$, it is possible to remain in the same state as the initial state $i = 1$.

The goal is to test the hypothesis that

$$H_0: \phi_{11} = .2 \tag{2.65}$$

versus the alternative

$$H_1: \phi_{11} \neq .20. \tag{2.66}$$

Of course, any ϕ_{ij} could have been used to illustrate the Bayesian testing procedure.

The posterior probabilities p_0 and p_1 given by Equations 2.61 and 2.62, respectively, are required; thus, consider first

$$p_0 = \pi_0 p(n_{11}|\phi_{11} = .20)/p(n_{11}), \tag{2.67}$$

where π_0 is the prior probability of the null hypothesis and the probability mass function of n_{11} (given the null hypothesis) is

$$p(n_{11}|\phi_{11} = .20) = \binom{n_{1.}}{n_{11}}(.7)^{n_{11}}(.3)^{n1.-n_{11}}, \tag{2.68}$$

where $n_{11} = 1,2,\ldots,n_{1.}$.

In addition,

$$p_1(n_{11}) = \pi_1 \int p_1(\phi_{11})p(n_{11}|\phi_{11})d\phi_{11} \tag{2.69}$$

and

$$\xi(n_{11}) = \pi_0\xi(n_{11}|\phi_{11} = .2) + \pi_1\xi_1(n_{11}), \tag{2.70}$$

where π_1 is the prior probability of the alternative hypothesis and $\pi_0 + \pi_1 = 1$.

Note that the marginal mass function of n_{11} is given by Equation 2.69 and that $p_1(\phi_{11})$ is the prior density of ϕ_{11} over interval [0, 1], that is, over the values specified by the alternative hypothesis. It is convenient to choose

$$\phi_{11} \sim \text{beta}(\alpha, \beta), \tag{2.71}$$

where $0 \le \phi \le 1$ and α and β are positive parameters; thus, the prior distribution of ϕ_{11} is a beta with parameters α and β. Now it can be shown that the marginal distribution of n_{11} (Equation 2.69) is

$$p_1(n_{11}) = \pi_1[\Gamma(\alpha+\beta)/\Gamma(\alpha)\Gamma(\beta)]\binom{n_{1.}}{n_{11}}[\Gamma(n_{11}+\alpha)\Gamma(n_{1.}-n_{11}+\beta)/\Gamma(n_{1.}+\alpha+\beta)], \tag{2.72}$$

where α and β must be chosen to reflect the prior information about the null hypothesis. Combining Equations 2.68, 2.70, and 2.72 allows one to evaluate the posterior probability p_0 of the null hypothesis. For the problem at hand, let

$$\begin{aligned} \pi_1 &= \pi_0 = 1/2, \\ \alpha &= 2, \\ \beta &= 8, \\ n_{11} &= 1, \text{ and} \\ n_1 &= 20, \end{aligned} \tag{2.73}$$

and one can show that the posterior probability of the null hypothesis is $p_0 = .552484379$, and for the alternative, $p_1 = .44751562$.

Note that the values used to evaluate the posterior probabilities depend on the values of Equation 2.73, and these values are somewhat arbitrary. I chose $\alpha = 2$ and $\beta = 8$ for the parameters of prior beta distribution for values ϕ_{11} of the alternative hypothesis, because the prior mean

$$E(\phi_{11}) = \alpha/(\alpha + \beta) = .2 \tag{2.74}$$

is centered at the value of .2 of the null hypothesis.

I also chose equal values for the prior probabilities π_0 and π_1, and one would expect the posterior probabilities to be somewhat sensitive to α, β, π_0, and π_1. The other values specified by Equation 2.73 depend on the data n_{11} and the first row total $n_{1.}$. It is left as an exercise for the student to verify the value for p_0 = .552484379 and to investigate the sensitivity of p_0 to the prior information.

An earlier and more informal approach (see Lindley[8]) to testing hypotheses is to reject the null hypothesis if the 95% credible region for θ does not contain the set of all θ such that $\theta \in \Omega_0$. In the special case that H: $\theta = \theta_0$ versus the alternative A: $\theta \neq \theta_0$, where θ is a scalar, H is rejected when the 95% confidence interval for θ does not include θ_0. However, there are some logical problems with this approach. If a continuous prior density is used for the entire parameter space, the prior probability of the null hypothesis is zero, which implies a posterior probability of zero for the null hypothesis, thus implying an illogical approach for this way of testing hypotheses!

2.6 Predictive Inference

2.6.1 Introduction

Our primary interest in the predictive distribution is to check for model assumptions. Is the adopted model for an analysis the most appropriate?

What is the predictive distribution of a future set of observations Z? It is the conditional distribution of Z given $X = x$, where x represents the past observations, which when expressed as a density is

$$g(z|x) = \int_{\Omega} f(z|\theta)\xi(\theta|x)d\theta, \quad z \in R^m, \tag{2.75}$$

where the integral is with respect to θ, and $f(x|\theta)$ is the density of $X = (x_1, x_2 ..., x_n)$, given θ. This assumes that given θ, Z and X are independent. Thus, the predictive density is the posterior average of $f(z|\theta)$ with respect to the posterior distribution of θ.

The posterior predictive density will be derived for the binomial and normal populations.

2.6.2 Binomial Population

Suppose the binomial case is again considered, where the posterior density of the binomial parameter θ is

$$\xi(\theta|x) = [\Gamma(\alpha + \beta)\Gamma(n + 1)/\Gamma(\alpha)\Gamma(\beta)\Gamma(x + 1)\Gamma(n - x + 1)]\theta^{\alpha+x-1}(1 - \theta)^{\beta+n-x-1}. \tag{2.76}$$

A beta with parameters $\alpha + x$ and $n - x + \beta$ and x is the sum of the set of n observations. The population mass function of a future observation Z is $f(z|\theta) = \theta^z(1-\theta)^{1-z}$; thus, the predictive mass function of Z, called the beta-binomial, is

$$g(z|x) = \Gamma(\alpha+\beta)\Gamma(n+1)\Gamma\left(\alpha + \sum_{i=1}^{i=n} x_i + z\right)\Gamma(1+n+\beta-x-z) \div \Gamma(\alpha)\Gamma(\beta)\Gamma(n-x+1)\Gamma(x+1)\Gamma(n+1+\alpha+\beta), \tag{2.77}$$

where $z = 0, 1$. Note that this function does not depend on the unknown parameters because they were averaged with respect to the posterior distribution and that the n past observations are known, and that if $\alpha = \beta = 1$, one is assuming a uniform prior density for θ.

As an example, consider the predictive distribution of the binomial distribution; recall Section 2.4.2 with

$$N = \begin{pmatrix} 1, 7, 4, 6, 2 \\ 5, 3, 4, 6, 5 \\ 3, 6, 2, 3, 1 \\ 5, 7, 2, 5, 4 \\ 6, 1, 3, 3, 4 \end{pmatrix} \tag{2.32}$$

for the transition counts for a five-state Markov chain and

$$\Phi = \begin{pmatrix} \phi_{11}, \phi_{12}, \phi_{13}, \phi_{14}, \phi_{15} \\ \phi_{21}, \phi_{22}, \phi_{23}, \phi_{24}, \phi_{25} \\ \phi_{31}, \phi_{32}, \phi_{33}, \phi_{34}, \phi_{35} \\ \phi_{41}, \phi_{42}, \phi_{43}, \phi_{44}, \phi_{45} \\ \phi_{51}, \phi_{52}, \phi_{53}, \phi_{54}, \phi_{55} \end{pmatrix} \tag{2.64}$$

as the one-step transition matrix.

Our focus is on forecasting the number of transitions Z_{11} from 1 to 1, that is, the number of times the chain remains in state 1, assuming a total of m replications for the first row of the chain, that is, $Z_{11} = 0, 1, 2, \ldots, m$. Using Equation 2.77, one may show that the predictive mass function of Z_{11} is

$$g(z|n_{11} = 1) = \binom{m}{z}\Gamma(\alpha+\beta)\Gamma(n+1)\Gamma(\alpha+z+n_{11})\Gamma(\beta+n-z-n_{11}) / \Gamma(\alpha)\Gamma(\beta)\Gamma(n_{11}+1)\Gamma(n-n_{11}+1)\Gamma(\alpha+\beta+n). \tag{2.78}$$

The relevant quantities of Equation 2.78 are $n = 20$ and $n_{11} = 1$. Also remember that α and β are the parameters of the prior distribution of ϕ_{11}, the probability of remaining in state 1, and that predictive inferences are conditional on $n = 20$, the total transition counts for the first row of the one-step transition matrix of the chain.

2.6.3 Forecasting for a Normal Population

Moving on to the normal density with both parameters unknown, what is the predictive density of Z, with a noninformative prior density of

$$\xi(\mu, \tau) = 1/\tau, \quad \mu \in R \text{ and } \tau > 0? \tag{2.79}$$

The posterior density is

$$\xi(\mu, \tau|\text{data}) = \left[\tau^{n/2-1}/(2\pi)^{n/2}\right] \exp -(\tau/2)\left[n(\mu - x)^2 + (n-1)S_x^2\right], \tag{2.80}$$

where $\bar{x}$ and S_x^2 are the sample mean and variance, respectively, based on a random sample of size n, $x = (x_1, x_2, \ldots, x_n)$. Suppose z is a future sample $z = (z_1, z_2, \ldots, z_m)$ of size m, then the predictive density of Z is

$$\begin{aligned} g(z|x) &= \int\int \left[\tau^{(n+m)/2-1}/(2\pi)^{(n+m)/2}\right] \exp -(\tau/2) \\ &\left[n(\mu - \bar{x})^2 + (n-1)S_x^2 + m\left(\bar{z} - \mu\right)^2 + (m-1)S_z^2\right], \end{aligned} \tag{2.81}$$

where the integration is with respect to $\mu \in R$ and $\tau > 0$.

m is the number of future observations, $\bar{z}$ is the sample mean of the m future observations, and S_z^2 is the corresponding sample variance.

It can be shown that the predictive density of z is

$$g(z|x) \propto \Gamma((n+m-1)/2)/\left[1 + \zeta(z-x)^2/(n+m-3)\right]^{(n+m-3+1)}, \tag{2.82}$$

where

$$\zeta = nm/(n+m)k(n+m-3), \tag{2.83}$$

$$k = (n-1)S_x^2 + (m-1)S_z^2 + n^2\left(\bar{x}\right)^2/(n+m). \tag{2.84}$$

This density is recognized as a noncentral t-distribution with $n + m - 3$ degrees of freedom, location $\bar{x}$, and precision ζ.

The predictive distribution can be used as an inferential tool to test hypotheses about future observations, to estimate the mean of future observations, and to find confidence bands for future observations. In the context of stochastic processes, the predictive distribution for future normal observations will be employed to generate future values from various stochastic processes.

Of interest in the context of the Brownian motion is the predictive distribution of z when $\mu = 0$, that is, when the posterior density is

$$g(z|x) \propto \left[\tau^{(n+m)/2-1}/(2\pi)^{(n+m)/2}\right] \exp-(\tau/2)\left[\sum_{i-1}^{i=n} x_i^2 + \sum_{i=1}^{i=m} z_i^2\right]. \tag{2.85}$$

This assumes the improper prior for τ as

$$\zeta(\tau) = 1/\tau, \quad \tau > 0 \tag{2.86}$$

For the Wiener process of Section 2.4, the process was sampled at times t_1, t_2, t_n with $t_1 < t_2 <, < t_n$ and with independent increments

$$d_i = X_i - X_{i-1} \tag{2.87}$$

and where $t_i - t_{i-1} = 1$ and $i = 1, 2, ..., n$.

Also $t_0 = 0$.

Consider m future increments

$$z_i = x_i - x_{i-1} \tag{2.88}$$

for i = $n + 1, \dots, n + m$ and time points satisfying $t_i - t_{i-1} = 1$.

Now assume the improper prior density (Equation 2.84) for τ, then the joint density of d_i and z_i is

$$g(d, z|\tau) \propto \tau^{(n+m)/2-1} \exp(-\tau/2)\left[\sum_{i=1}^{i=n} d_i^2 + \sum_{i=n+1}^{i=n+m} z_i^2\right], \quad \tau > 0 \tag{2.89}$$

Thus, the predictive distribution of the m future independent increments is

$$g(z|d) \propto \Gamma((n+m)/2)/\left[1 + (\zeta/(n+m-1)) \sum_{i=n+1}^{i=n+m} z_i^2\right]^{(n+m-1+1)/2}. \tag{2.90}$$

This is recognized as a noncentral t-density with $n + m - 1$ degrees of freedom, location at the zero vector, and precision $\zeta = (n + m - 1)/\sum_{i=1}^{i=n} d_i^2$.

Recall the Brownian motion example of Section 2.4.2 with a variance of 0.01 and where 100 observations from the process is designated by x. Using Equation 2.89 with $m = 100$, $n = 100$, $\sum_{i=1}^{i=100} d_i^2 = 1.6458374, \zeta = 120.64514$, then based on WinBUGS Code 2.3, the 100 predicted z values appear as the vector z in Equation 2.91:

WinBUGS Code 2.3

```
model {
for (i in 1 : 100) {
y[i] ~ dt(0, 120.645, 199)
}
```

```
z = c(
-0.05473,-0.119,0.1056,-0.129,0.05168,
-0.06759,0.07575,0.04806,0.003992,-0.1113,
-0.154,-0.009263,0.09279,-0.06837,-0.07757,
-0.1289,0.03088,0.09818,0.01693,-0.04028,
-0.1602,0.09864,-0.05848,0.002767,-0.1908,
-0.1578,-0.004863,0.04017,-0.05318,0.08215,                (2.91)
-0.0231,0.1652,0.01179,0.151,-0.2395,
0.00945,-0.05023,-0.09512,-0.04164,0.09382,
-0.01882,-0.1193,0.03329,0.02761,-0.07163,
-0.05162,0.04595,0.108,0.01209,0.09053,
-0.08401,0.08781,-0.05834,-0.09858,0.1072,
0.1007,0.04107,0.222,0.1023,-0.003405,
-0.002853,0.1584,0.05611,0.05067,0.04823,
0.02001,0.1747,-0.1451,-0.0137,-0.1187,
0.04217,-0.01667,-0.04725,0.00841,0.09915,
-0.05576,0.02669,0.04407,0.03509,0.06624,
0.05622,-0.05857,-0.1255,-0.03296,-0.128,
-0.0193,-0.05927,-0.1122,0.06573,0.06395,
0.044,0.04435,0.04717,-0.1504,0.06941,
-0.03644,-0.04695,-0.1194,-0.003718,-0.08247))
```

To see the accuracy of the preceding predicted values, the sample mean and variance should be computed. How close to zero is the sample mean and how close to 0.01 is the sample variance of the Brownian motion example?

2.7 Checking Model Assumptions

2.7.1 Introduction

It is imperative to check model adequacy in order to choose an appropriate model and to conduct a valid study. The approach taken here is based on many sources, including Chapter 6 of Gelman et al.,[16] Chapter 5 of Carlin and Louis,[20] and Chapter 10 of Congdon.[17] Our main focus will be on the likelihood function of the posterior distribution, and not on the prior distribution, and to this end, graphical representations such as histograms, box plots, and various probability plots of the original observations will be compared to those of the observations generated from the predictive distribution. In addition to graphical methods, Bayesian versions of the overall goodness of fit type operations are taken to check model validity. Methods presented at this juncture are just a small subset of those presented in more advanced works, including Gelman et al., Carlin and Louis, and Congdon.

Of course, the prior distribution is an important component of the analysis, and if one is not sure of the "true" prior, one should perform a sensitivity analysis to determine the robustness of posterior inferences to various alternative choices of prior information. See Gelman et al. or Carlin and Louis for details of performing a sensitivity study for prior information. Our approach is to either use informative or vague prior distributions, where the former is done when prior relevant experimental evidence determines the prior, or the latter is taken if there are none or very few germane experimental studies. In scientific studies, the most likely scenario is that there are relevant experimental studies providing informative prior information.

2.7.2 Sampling from an Exponential, but Assuming a Normal Population

Consider a random sample of size 30 from an exponential distribution with mean 3. An exponential distribution is often used to model the survival times of a screening test.

```
X =
(1.9075, 0.7683, 5.8364, 3.0821, 0.0276, 15.0444, 2.3591, 14.9290, 6.3841,
7.6572, 5.9606, 1.5316, 3.1619, 1.5236, 2.5458, 1.6693, 4.2076, 6.7704, 7.0414,
1.0895, 3.7661, 0.0673, 1.3952, 2.8778, 5.8272, 1.5335, 7.2606, 3.1171, 4.2783,
0.2930).
```

The sample mean and standard deviation are 4.13 and 3.739, respectively. Assume the sample is from a normal population with unknown mean and variance, with an improper prior density $\xi(\mu, \tau) = 1/\tau$, $\mu \in R$ and $\tau > 0$; the posterior predictive density is a univariate t with $n - 1 = 29$ degrees of freedom, mean $\bar{x} = 3.744$, standard deviation of 3.872, and precision $p = 0.645$. This is verified from the original observations x and the formula for the precision. From the predictive distribution, 30 observations are generated:

```
Z =
(2.76213, 3.46370, 2.88747, 3.13581, 4.50398, 5.09963, 4.39670, 3.24032,
3.58791, 5.60893, 3.76411, 3.15034, 4.15961, 2.83306, 3.64620, 3.48478,
2.24699, 2.44810, 3.39590, 3.56703, 4.04226, 4.00720, 4.33006, 3.44320,
5.03451, 2.07679, 2.30578, 5.99297, 3.88463, 2.52737),
```

which gives a mean of $\bar{z} = 3.634$ and standard deviation $S = 975$. The histograms for the original and predicted observations should be computed.

One will see that the histograms obviously are different, where for the original observations, a right skewness is depicted; however, this is lacking for the histogram of the predicted observations, which is for a t-distribution. Although the example seems trivial, it would not be for the first time that exponential observations were analyzed as if they were generated from a normal population! Of course, we have seen the relevance of the exponential distribution to Markov jump processes such as the Poisson process of Section 2.4.3, where the interarrival times are independent and identically distributed with a mean that is the reciprocal of the mean rate of event happenings.

It would be interesting to generate more replicate samples from the predictive distribution in order to see if these conclusions hold firm.

2.7.3 Poisson Population

It is assumed that the sample is from a Poisson population; however, it is actually generated from a uniform discrete population over the integers from 0 to 10. The sample of size 25 is

$X = (8, 3, 8, 2, 6, 1, 0, 2, 4, 10, 7, 9, 5, 4, 8, 4, 0, 9, 0, 3, 7, 10, 7, 5, 1)$, with a sample mean of 4.92 and standard deviation of 3.278. When the population is Poisson $P(\theta)$ and an uninformative prior $\xi(\theta) = 1/\theta,\ \theta > 0$ is appropriate, the posterior density is gamma with parameters alpha = $\sum_{i=1}^{i=25} x_i = 123$ and beta = $n = 25$. Observations z from the predictive distribution are generated by taking a sample θ from the gamma posterior density, then selecting a z from the Poisson distribution $P(\theta)$. This was repeated 25 times to give $z = (2, 5, 6, 2, 4, 3, 5, 3, 2, 3, 3, 6, 7, 5, 5, 3, 1, 5, 7, 3, 5, 3, 6, 4, 5)$, with a sample mean of 4.48 and standard deviation of 1.896.

The most obvious difference show a symmetric sample from the discrete uniform population, but on the other hand, box plots of the predicted observations reveal a slight skewness to the right. The largest difference is in the interquartile ranges being (2, 8) for the original observations and (3, 5.5) for the predictive sample. Although there are some differences, to declare that the Poisson assumption is not valid might be premature. Of course, to reiterate, the Poisson process is one of the most important Markov jump processes, see Section 2.4.3 for a review of the topic.

2.7.4 Wiener Process

Refer to Section 2.4.2, where an example of a normal stochastic process is introduced with $n = 100$ observations simulated with R from a Wiener process with $\sigma = 0.01$ as the parameter. Vector D is the corresponding 100 increment values with mean 0 and variance of .01:

```
D=(-.1206600, 2432790, .1833210, -.0540060, -.0933370, .0013660,
.0350202, .0358970, -.1266960, -.0904420, -.1401550, .0230410,
-.0385080,.0894620, -.0839162, -.0457920, .3528350, -.3555310,
.0930150, .2003558, .0044500, .1478410, .1384490, .0184600,
.1224990, .0912220, .2035300, .1056839, .0123930, -.0283500,
-.0561130, .0625630, .1029310, .1417030, .0999980, .0343503,
-.0225143, -.1338220, -.1186080, .0298170, .0377400,
.0186340, -.1070220, -.0477170, .0933260, .1006602, -.0022484,
-.0793380, -.0035692, -.2114558, .0875840, -.1339130,
.0489390, .0227763, -.1003730, .1335010, .2732970, -.0419390,
-.0883350, -.0381000, .2297750, .4127186, .0145250, .1665200,
-.0748080, -.0491901, .2643080, .0693873, .0212337, .0179080,
.1386532, -.1509560, .0212460, -.0165320, .1709280, -.0679698,
.1038280, .0764160, .0083440, -.07920, .0746620, .0122130,
.1434140, .0934870, -.1668680, .0170720, -.3026848, -.2791680,
-.0691200, -.1085340, -.0616370, .0048761, .0871080, .0826430,
.0771660, . 0452680, .0269480, .0646110, .1461450, .0801460).
```

These 100 observations should be from a normal population with mean 0 and variance 0.01. What does the histogram of these values indicate about the normal assumption?

The sample mean is $\bar{x} = 0.0204995$ and the sample standard deviation is $s = 0.1237805$ or a sample variance of $s^2 = 0.01622$. These values are fairly close to the values 0 and 0.01, respectively, a good indication that the normality distribution of the increments is a reasonable assumption.

2.7.5 Testing the Multinomial Assumption

Consider a multinomial distribution where n_i is the number of times state i occurs, where $i = 0, 1, 2, 3, 4$, and let p_i be the corresponding probability that state i occurs, then the probability mass function for the multinomial is

$$f(n_1, n_2, n_3, n_4, n_5 | p_1, p_2, p_3, p_4, p_5) \propto \prod_{i=0}^{i=4} p_i^{n_i}, \tag{2.92}$$

where $\sum_{i=0}^{i=4} p_i = 1$ and $\sum_{i=0}^{i=4} n_i = n$ with n known and fixed. The vector

$$n = \begin{pmatrix} n_1 \\ n_2 \\ n_3 \\ n_4 \\ n_5 \end{pmatrix} \tag{2.93}$$

is said to have a multinomial distribution with parameters n and

$$p = \begin{pmatrix} p_1 \\ p_2 \\ p_3 \\ p_4 \\ p_5 \end{pmatrix}. \tag{2.94}$$

Note that the constraint $\sum_{i=0}^{i=4} n_i = n$ implies that the components of n are correlated, and it can be shown that

$$\text{cov}\left(n_i, n_j\right) = -np_i p_j. \tag{2.95}$$

We have seen that the multinomial distribution can be used to generate values from a Markov chain. Thus, consider the following R Code that allows one to generate observations from a multinomial:

```
N<-1000
prob<-c(.2, .2, .2, .2, .2)
rmultinom(1,n,prob)
```
(2.96)

$n = 1000$ is the sample size and prob is the vector of multinomial probabilities .2, .2, .2, .2, .2 referring to the probabilities of the five categories.

The simulation gives $n_0 = 215$, $n_1 = 174$, $n_2 = 205$, $n_3 = 204$, and $n_4 = 202$, that is, the number of zeros is 214; the number of ones, 174; the number of twos, 205; the number of threes, 204; and the number of fours, 202. Based on the multinomial probability vector prob = (.2, .2, .2, .2, .2), one would expect to have 200 outcomes for each of the five categories.

Our concern is what is the accuracy of the R Code in generating the five possible values; that is, does the preceding realization (Equation 2.96) represent a sample from the multinomial distribution?

Note that one could use the chi-square goodness of fit test to test the hypothesis

$$H: p_i = .2 \tag{2.97}$$

versus the alternative

$$A: p_i \neq .2$$

for $i = 0, 1, 2, 3, 4$.

The test statistic is

$$\chi^2 = \sum_{i=0}^{i=4} (n_i - 200)^2/200 = 1.125 + 3.38 + 0.125 + 0.08 + 0.02 = 4.91,$$

which when compared to the fifth percentile of the chi-square distribution with 4 degrees of freedom, $\chi^2_{4,.05} = 0.71$, implies the simulated values (Equation 2.96) were indeed generated from a multinomial distribution with probabilities given by Equation 2.97. There is not enough evidence to reject the null hypothesis.

Thus, in the case of stochastic processes, the best way to determine the validity of the multinomial model is to know the details of how the study was designed and conducted. It is often the case that the details of the study are not available. The other important aspect of a multinomial population is that the probability of a particular outcome is constant over all cases. One other statistical way to check is to look for runs in the sequence, etc.

2.8 Computing

2.8.1 Introduction

This section introduces the computing algorithms and software that will be used for the Bayesian analysis of problems encountered in agreement investigations. In the previous sections of the chapter, direct methods (noniterative) of computing the characteristics of the posterior distribution were demonstrated with some standard one sample and two sample problems. An example of this is the posterior analysis of a normal population, where the posterior distribution of the mean and variance is generated from its posterior distribution by the *t*-distribution random number generator in Minitab. In addition to some direct methods, iterative algorithms are briefly explained.

MCMC methods (an iterative procedure) of generating samples from the posterior distribution are introduced, where the Metropolis–Hasting algorithm and Gibbs sampling are explained and illustrated with many examples. WinBUGS uses MCMC methods such as the Metropolis–Hasting and Gibbs sampling techniques, and many examples of a Bayesian analysis are given. An analysis consists of graphical displays of various plots of the posterior density of the parameters, by portraying the posterior analysis with tables that list the posterior mean, standard deviation, median, and lower and upper 2 ½ percentiles, and of other graphics that monitor the convergence of the generated observations.

2.8.2 Monte Carlo Markov Chain

2.8.2.1 Introduction

MCMC techniques are especially useful when analyzing data with complex statistical models. For example, when considering a hierarchical model with many levels of parameters, it is more efficient to use an MCMC technique such as Metropolis–Hasting or Gibbs sampling iterative procedure in order to sample from the many posterior distributions. It is very difficult, if not impossible, to use noniterative direct methods for complex models.

A way to draw samples from a target posterior density $\xi(\theta|x)$ is to use Markov chain techniques, where each sample only depends on the last sample drawn. Starting with an approximate target density, the approximations are improved with each step of the sequential procedure. Or in other words, the sequence of samples is converging to samples drawn at random from the target distribution. A random walk from a Markov chain is simulated, where the stationary distribution of the chain is the target density, and the simulated values converge to the stationary distribution or the target density. The main concept in a Markov chain simulation is to devise a Markov process whose stationary distribution is the target density. The simulation must be long enough so that the present samples are close enough to the target. It has been shown that this is possible and that convergence can be accomplished. The general scheme for a Markov chain simulation is to create a sequence θ_t, $t = 1, 2, \ldots$ by beginning at some value θ_0, and at the tth stage, select the present value from a transition function $Q_t(\theta_t|\theta_{t-1})$, where the present value θ_t only depends on the previous one, via the transition function. The value of the starting value θ_0 is usually based on a good approximation to the target density. In order to converge to the target distribution, the transition function must be selected with care. The account given here is a summary of Chapter 11 of Gelman et al.,[16] who presents a very complete account of MCMC. Metropolis–Hasting is the general name given to methods of choosing appropriate transition functions, and two special cases of this are the Metropolis algorithm and the other is referred to as Gibbs sampling.

2.8.2.2 Metropolis Algorithm

Suppose the target density $\xi(\theta|x)$ can be computed, then the Metropolis technique generates a sequence θ_t, $t = 1, 2, \ldots$ with a distribution that converges to a stationary distribution of the chain. Briefly, the steps taken to construct the sequence are as follows:

1. Draw the initial value θ_0, where $\xi(\theta_0|x) > 0$, from some approximation to the target density, say, $p_0(\theta|x)$.
2. For $t = 1, 2, \ldots,$
    a. Sample a candidate point θ_* from a jumping distribution at time t, $J_t(\theta_*|\theta_t)$. The jumping distribution is symmetric, that is, $J_t(\theta_a|\theta_b) = J_t(\theta_b|\theta_a)$, $\forall\ \theta_a, \theta_b$, and t.

b. Calculate the ratio

$$r = p(\theta_*|x)/p(\theta_{t-1}|x)$$

c. Let $\theta_t = \theta_*$ with probability min(r,1); otherwise, let $\theta_t = \theta_{t-1}$.

Given the current value θ_{t-1}, the Markov transition chain function is a mixture of the jumping distribution $J_t(\theta_t|\theta_{t-1})$ and a point mass at $\theta_t = \theta_{t-1}$. One must show that the sequence generated is a Markov chain with a unique stationary density that converges to the target distribution. This discussion of the Hasting algorithm is quite brief; thus, the reader should refer to pages 323–325 of Gelman et al.[16] for additional information.

2.8.2.3 Gibbs Sampling

Another MCMC algorithm is Gibbs sampling that is quite useful for multidimensional problems and is an alternating conditional sampling way to generate samples from the joint posterior distribution. Gibbs sampling can be thought of as a practical way to implement the fact that the joint distribution of two random variables is determined by the two conditional distributions.

The two-variable case is first considered by starting with a pair (θ_1, θ_2) of random variables. The Gibbs sampler generates a random sample from the joint distribution of θ_1 and θ_2 by sampling from the conditional distributions of θ_1 given θ_2 and from θ_2 given θ_1. The Gibbs sequence of size k

$$\theta_2^0, \theta_1^0; \theta_2^1, \theta_1^1; \theta_2^2, \theta_1^2; \ldots; \theta_2^k, \theta_1^k \tag{2.98}$$

is generated by first choosing the initial values θ_2^0, θ_1^0 while the remaining are obtained iteratively by alternating values from the two conditional distributions. Under quite general conditions, for large enough k, the final two values θ_2^k, θ_1^k are samples from their respective marginal distributions. To generate a random sample of size n from the joint posterior distribution, generate the preceding Gibbs sequence n times. Having generated values from the marginal distributions with large k and n, the sample mean and variance will converge to the corresponding mean and variance of the posterior distribution of (θ_1, θ_2).

Gibbs sampling is an example of a MCMC because the generated samples are drawn from the limiting distribution of a 2 × 2 Markov chain. See Casella and George[24] for a proof that the generated values are indeed values from the appropriate marginal distributions. Of course, Gibbs sequences can be generated from the joint distribution of three, four, and more random variables.

The Gibbs sampling scheme is illustrated with the case of three random variables for the common mean of two normal populations.

2.8.2.4 Common Mean of Normal Populations

Gregurich and Broemeling[25] describe the various steps in Gibbs sampling to determine the posterior distribution of the parameters in independent normal populations with a common mean.

The Gibbs sampling approach can best be explained by illustrating the procedure using two normal populations with a common mean θ. Thus, let y_{ij}, $j = 1,2, \ldots, n_i$ be a random sample of size n_i from a normal population for $i = 1, 2$.

The likelihood function for θ, τ_1, and τ_2 is

$$L(\theta, \tau_1, \tau_2|\text{data}) \propto$$
$$\tau_1^{\frac{n_1}{2}} \exp -\frac{\tau_1}{2}\left[(n_1-1)s_1^2 + n_1(\theta-\bar{y}_1)^2\right] \times \tau_2^{\frac{n_2}{2}} \exp -\frac{\tau_2}{2}\left[(n_2-1)s_2^2 + n_2(\theta-\bar{y}_2)^2\right],$$

where, $\theta \in \mathfrak{R}$, $\tau_1 > 0$, $\tau_2 > 0$, $s_1^2 = \sum_{j=1}^{n_1}(y_{1j}-\bar{y}_1)^2/(n_1-1)$, and $s_2^2 = \sum_{j=1}^{n_2}(y_{2j}-\bar{y}_2)^2/(n_2-1)$.

The prior distribution for parameters θ, τ_1, and τ_2 is assumed to be a vague prior defined as

$$g(\theta, \tau_1, \tau_2) \propto \frac{1}{\tau_1}\frac{1}{\tau_2}, \quad \tau_i > 0.$$

Then, combining the preceding equations gives the posterior density of the parameters as

$$P(\theta, \tau_1, \tau_2|\text{data}) \propto \prod_{i=1}^{2} \tau_i^{\frac{n_i-1}{2}} \exp -\frac{\tau_i}{2}\left[(n_i-1)s_i^2 + n_i(\theta-\bar{y}_i)^2\right].$$

Therefore, the conditional posterior distribution of τ_1 and τ_2 given θ is such that

$$\tau_i|\theta \sim \text{Gamma}\left[\frac{n_i}{2}, \frac{(n_i-1)s_i^2 + n_i(\theta-\bar{y}_i)^2}{2}\right], \tag{2.99}$$

for $i = 1, 2$ and given θ, τ_1 and τ_2 are independent.

The conditional posterior distribution of θ given τ_1 and τ_2 is normal. It can be shown that

$$\theta|\tau_1, \tau_2 \sim N\left[\frac{n_1\tau_1\bar{y}_1 + n_2\tau_2\bar{y}_2}{n_1\tau_1 + n_2\tau_2}, (n_1\tau_1 + n_2\tau_2)^{-1}\right]. \tag{2.100}$$

Given the starting values $\tau_1^{(0)}$, $\tau_2^{(0)}$, and $\theta^{(0)}$, where $\tau_1^{(0)} = 1/s_1^2$, $\tau_2^{(0)} = 1/s_2^2$, and $\theta^{(0)} = \frac{n_1\bar{y}_1 + n_2\bar{y}_2}{n_1 + n_2}$, draw $\theta^{(1)}$ from the normal conditional distribution of θ, given $\tau_1 = \tau_1^{(0)}$ and $\tau_2 = \tau_2^{(0)}$. Then draw $\tau_1^{(1)}$ from the conditional gamma distribution (43), given $\theta = \theta^{(1)}$. And lastly, draw $\tau_2^{(1)}$ from the conditional gamma distribution of τ_2 given $\theta = \theta^{(1)}$. Then generate

$$\theta^{(2)} \sim \theta/\tau_1 = \tau_1^{(1)}, \tau_2 = \tau_2^{(1)}$$
$$\tau_1^{(2)} \sim \tau_1/\theta = \theta^{(2)}$$
$$\tau_2^{(2)} \sim \tau_2/\theta = \theta^{(2)}$$

Continue this process until there are t iterations $(\theta^{(t)}, \tau_1^{(t)}, \tau_2^{(t)})$. For large t, $\theta^{(t)}$ would be one sample from the marginal distribution of θ, $\tau_1^{(t)}$ from the marginal distribution of τ_1, and $\tau_2^{(t)}$ from the marginal distribution of τ_2.

Independently repeating the preceding Gibbs process m times produces m 3-tuple parameter values $(\theta_j^{(t)}, \tau_{1j}^{(t)}, \tau_{2j}^{(t)})$, $j = 1, 2, \ldots, m$ which represents a random sample of size m from the joint posterior distribution of (θ, τ_1, τ_2). The statistical inferences are drawn from the m sample values generated by the Gibbs sampler.

The statistical inferences can be drawn from the m sample values generated by the Gibbs sampler. The Gibbs sampler will produce three columns, where each row is a sample drawn from the posterior distribution of (θ, τ_1, τ_2). The first column is the sequence of sample m, the second column is a random sample of size m from the poly-t-distribution of θ; the third and fourth columns are also random samples of size m but from the marginal posterior distributions of τ_1 and τ_2, respectively.

Note that the posterior mean is

$$E(\theta|\text{data}) = \sum_{j=1}^{j=\infty} \theta_j^t / m$$

and is the mean of the posterior distribution of θ. The sample variance

$$(m-1)^{-1} \sum_{j=1}^{m} \left[\theta_j^t - \bar{\theta}\right]^2$$

is the variance of the posterior distribution of θ.

Additional characteristics such as the median, mode, and the 95% credible region of the posterior distribution of parameter can be calculated from the samples generated by the Gibbs technique. Hypothesis testing can also be performed. Similar characteristics of parameters τ_1, τ_2, τ_k can be calculated from the samples resulting from the Gibbs method.

2.8.2.5 Example

The example is from page 481 of Box and Tiao.[10] It is referred to as *the weighted mean problem*. It has two sets of normally distributed independent samples with a common mean and different variances. Samples from the posterior distributions were generated from Gibbs sequences. The final value of each sequence was used to approximate the marginal posterior distribution of the parameters θ, τ_1, τ_k. All Gibbs sequences were generated holding the value of t equal to 50. Each example has the results of the parameters using four different Gibbs sampler sizes, where the sample size m is equal to 250, 500, 750, and 1500.

The weighted mean problem has two sets of normally distributed independent observations with a common mean and different variances. The estimated values of m determined by the Gibbs sampling method are reported in Table 2.2. The mean value of the posterior distribution of θ generated from the 250 Gibbs sequences is 108.42 with 0.07 as the standard error of the mean. The mean value of θ generated from 500 and 750 Gibbs sequences have the same value of 108.31, and the standard errors of the mean equal 0.04 and 0.03, respectively. The mean value of θ generated from 1500 Gibbs sequences is 108.36 and a standard error of the mean of 0.02. Box and Tiao determined the posterior distribution of θ using the t-distribution as an approximation to the target density. They estimated the value of θ to be 108.43. This is close to the value generated using the Gibbs sampler method. The exact posterior distribution of θ is the poly-t-distribution. The effect of m appears to be minimal, indicating that 500–750 iterations of the Gibbs sequence are sufficient.

TABLE 2.2
Results from Gibbs Sampler for θ Box and Tiao's "The Weighted Mean Problem"

m	Mean	STD	SEM	95% Credible Region Lower	Upper
250	108.42	1.04	0.07	106.03	110.65
500	108.31	0.94	0.04	106.35	110.21
750	108.31	0.90	0.03	106.64	110.15
1500	108.36	0.94	0.02	106.51	110.26

2.9 Comments and Conclusions

Beginning with the Bayes theorem, introductory material for the understanding of Bayesian inference is presented in this chapter. Many examples from stochastic processes illustrate the Bayesian methodology. For example, Bayesian methods for the analysis of Markov chains are analyzed using the theory and methods unique with the Bayesian approach. Inference for the standard populations is introduced. The most useful population for Markov chains is the binomial population, which models the number of one-step transitions among the states of a finite chain. It is shown how to estimate the one-step transition probabilities of the Markov chain, where the posterior distribution of the transition probabilities is a beta. For the analysis of normal processes, the focus is on estimating the variance parameter of the Brownian motion process, where Bayesian estimation and hypothesis testing methods are demonstrated with the gamma distribution. The R package and WinBUGS software allows one to implement the Bayesian methods that are developed in this chapter.

There are many books that introduce Bayesian inference and the computational techniques that will execute a Bayesian analysis, and the reader is encouraged to read Ntzoufras.[26] The material found in Ntzoufras is an excellent introduction to Bayesian inference and to WinBUGS, the computing software that is employed in this book. WinBUGS is also introduced in Chapter 1; thus, together with Ntzoufras, the reader should be able to execute the Bayesian analyses for various inferential investigations of stochastic processes.

The current chapter presents two particular types of stochastic processes: (1) the Markov chain with a finite number of state and (2) a Markov jump process with a normal distribution for the state variable. For a good introduction to Markov chains, the student is referred to Parzen[27] and to Karlin and Taylor,[28] while for normal processers such as the Brownian motion, see Allen.[29] Lastly, when using Bayesian inference for stochastic processes, I referred to the recent book by Insua, Ruggeri, and Wiper.[30]

2.10 Exercises

1. For the Beta density (Equation 2.8) with parameters α and β, show that the mean is $[\alpha/(\alpha+\beta)]$ and the variance is $[\alpha\beta/(\alpha+\beta)^2(\alpha+\beta+1)]$.

2. From Equation 2.14, show the following. If the normal distribution is parameterized with μ and the precision $\tau = 1/\sigma^2$, the conjugate distribution is as follows: (a) the conditional distribution of μ given τ is normal with mean $\bar{x}$ and precision $n\tau$ and (b) the marginal distribution of τ is gamma with parameters $(n-1)/2$ and $\sum_{i=1}^{i=n}(x_i - x)^2/2 = (n-1)S^2/2$, where S^2 is the sample variance.
3. Verify Table 2.1, which reports the Bayesian analysis for the parameters of a normal population.
4. Verify the following statement:

   To generate values from the $t(n-1, \bar{x}, n/S^2)$ distribution, generate values from Student's t-distribution with $n-1$ degrees of freedom and multiply each by $S/\sqrt{n}$ and then add $\bar{x}$ to each.
5. Verify Equation 2.82, the predictive density of a future observations Z from a normal population with both parameters unknown.
6. Suppose $x_1, x_2, ..., x_n$ are independent and that $x_i \sim Gamma(\alpha_i, \beta)$ and show that $y_i = x_i/(x_1 + x_2 + ... + x_n)$ jointly have a Dirichlet distribution with parameter $(\alpha_1, \alpha_2, ..., \alpha_n)$. Describe how this can be used to generate samples from the Dirichlet distribution.
7. Suppose $(X_1, X_2, ..., X_k)$ is multinomial with parameters n and $(\theta_1, \theta_2, ..., \theta_k)$, where $\sum_{i=1}^{i=k} X_i = n$, $0 < \theta_i < 1$, and $\sum_{i=1}^{i=k} \theta_i = 1$. Show that $E(X_i) = n\theta_i$, $\text{Var}(X_i) = n\theta_i(1-\theta_i)$, and $\text{cov}(X_i, X_j) = -n\theta_i\theta_j$. What is the marginal distribution of θ_i?
8. Suppose $(\theta_1, \theta_2, ..., \theta_k)$ is Dirichlet with parameters $(\alpha_1, \alpha_2, ..., \alpha_k)$, where $\alpha_i > 0$, $\theta_i > 0$, and $\sum_{i=1}^{i=k} \theta_i = 1$. Find the mean and variance of θ_i and covariance between θ_i and θ_j, $i \neq j$.
9. Show that the Dirichlet family is conjugate to the multinomial family.
10. Suppose $(\theta_1, \theta_2, ..., \theta_k)$ is Dirichlet with parameters $(\alpha_1, \alpha_2, ..., \alpha_k)$. Show that the marginal distribution of θ_i is beta and give the parameters of the beta. What is the conditional distribution of θ_i given θ_j?
11. For the exponential density

$$f(x|\theta) = \theta \exp - \theta x, \quad x > 0,$$

   where x is positive and θ is a positive unknown parameter, suppose the prior density of θ is

$$g(\theta) \propto \theta^{\alpha-1} \exp - \beta\theta, \quad \theta > 0,$$

   what is the posterior density of θ? In Markov jump processes, the exponential distribution is the distribution of the interarrival times between events.

12. Refer to the R Code given by R Code 2.1. Using the function MC.sim, generate 100 observations from a two-state Markov chain with probability transition matrix

$$P = \begin{pmatrix} .5, .5 \\ .5, .5 \end{pmatrix}.$$

What are the number of transitions from 1 to1? From 1 to 2? From 2 to 1? And from 2 to 2? Is this Markov chain irreducible?

13. Refer to Section 2.3.2 and define a Wiener process with parameter σ^2.
14. Refer to Section 2.4.2. Using R Code 2.2, generate 50 observations from a Brownian motion process with parameter $\sigma^2 = 1$.
15. Refer to Section 2.4.3. Based on WinBUGS Code 2.1, generate 100 observations from an exponential distribution with mean 2.
16. Refer to Section 2.5.2 on testing hypotheses. Let θ be the parameter of a Bernoulli distribution and consider a test of the null hypothesis $H_0: \theta = .5$ versus the alternative $H_1: \theta \neq .5$. Assume that the null hypothesis has prior probability $\pi_0 = .5$ and assume that under the alternative, the prior density is Beta (3,3) over the interval [0,1]. If the sample consists of

$$x = (1,1,1,0,0,1,0,0,1,1,1,0,0,0,0,0,0,0,),$$

compute the posterior probability p_0 of the null hypothesis. Refer to Equation 2.48.

17. Verify the predictive probability mass function (Equation 2.77) for the binomial.
18. Verify the predictive distribution (Equation 2.89) of m future observations from a Brownian motion process.
19. Based on Equation 2.89 and WinBUGS Code 2.2, generate 100 observations from the predictive distribution (Equation 2.89) and compare to the 100 predictive values labeled by Equation 2.90.
20. Based on the multinomial mass function (Equation 2.91) and the R Code

```
N<-1000
Prob <- c(.2,.2,.2,.2,.2)
rmultinom(1,n,prob),
```

generate 1000 observations from the multinomial similar to Equation 2.96.

21. Were the 1000 observations of Equation 2.96 generated from a multinomial distribution with parameters $n = 1000$ and $p_i = .2$, for $i = 0, 1, 2, 3$?

References

1. Bayes, T. 1764. An essay towards solving a problem in the doctrine of chances, *Philosophical Transactions of the Royal Society London* 53:370.
2. Laplace, P. S. 1778. Memorie des les probabilities. *Memories de l'Academie des sciences de Paris*, p. 227.
3. Stigler, M. 1986. *The History of Statistics: The Measurement of Uncertainty before 1900*. Cambridge, MA: Belknap Press of Harvard University Press.
4. Hald, A. 1990. *A History of Mathematical Statistics from 1750–1930*. London: Wiley Interscience.
5. Lhoste, E. 1923. Le calcul des probabilities appliqué a l'artillerie, lois de probabilite a prior. *Revu d'artillerie*, Mai, p. 405.
6. Jeffreys, H. 1939. *An Introduction to Probability*. Oxford: Clarendon Press.
7. Savage, L. J. 1954. *The Foundation of Statistics*. New York: John Wiley & Sons.
8. Lindley, D. V. 1965. *Introduction to Probability and Statistics from a Bayesian Viewpoint, Volumes I and II*. Cambridge, UK: Cambridge University Press.
9. Broemeling, L. D., and Broemeling, A. L. 2003. Studies in the history of probability and statistics XLVIII: The Bayesian contributions of Ernest Lhoste, *Biometrika* 90(3):728–731.
10. Box, G. E. P., and Tiao, G. C. 1973. *Bayesian Inference in Statistical Analysis*. Reading, MA: Addison–Wesley.
11. Zellner, A. 1971. *An Introduction to Bayesian Inference in Econometrics*. New York: JohnWiley & Sons.
12. Broemeling, L. D. 1984. *The Bayesian Analysis of Linear Models*. New York: Marcel Dekker.
13. Hald, A. A. 1998. *History of Mathematical Statistics before 1750*. London: Wiley Interscience.
14. Dale, A. I. 1991. *A History of Inverse Probability from Thomas Bayes to Karl Pearson*. Berlin: Springer-Verlag.
15. Leonard, T., and Hsu, J. S. J. 1999. Bayesian methods. *An Analysis for Statisticians and Interdisciplinary Researchers*. Cambridge, UK: Cambridge University Press.
16. Gelman, A., Carlin, J. B., Stern, H. S., and Rubin, D. B. 1997. *Bayesian Data Analysis*, New York: Chapman and Hall/CRC Press.
17. Congdon, P. 2001. *Bayesian Statistical Modeling*. London: John Wiley & Sons.
18. Congdon, P. 2003. *Applied Bayesian Modeling*. New York: John Wiley & Sons.
19. Congdon, P. 2005. *Bayesian Models for Categorical Data*. New York: John Wiley & Sons.
20. Carlin, B. P., and Louis, T. A. 1996. *Bayes and Empirical Bayes for Data Analysis*. New York: Chapman and Hall/CRC Press.
21. Gilks, W. R., Richardson, S., and Spiegelhalter, D. J. 1996. *Markov Chain Monte Carlo in Practice*. Boca Raton, FL: Chapman and Hall/CRC Press.
22. Lehmann, E. L. 1959. *Testing Statistical Hypotheses*. New York: John Wiley & Sons.
23. Lee, P. M. 1997. *Bayesian Statistics: An Introduction, Second Edition*: London: Edward Arnolds.
24. Casella, G., and George, E. I. 2004. Explaining the Gibbs sampler, *The American Statistician* 46:167–174.
25. Gregurich, M. A., and Broemeling, L. D. 1997. A Bayesian analysis for estimating the common mean of independent normal populations using the Gibbs sampler, *Communications in Statistics* 26 (1):35–51.
26. Ntzoufras, I. 2009. *Bayesian Modeling Using WinBUGS*. Hoboken, NJ: John Wiley & Sons.
27. Parzen, E. 1962. *Stochastic Processes*. San Francisco: Holden Day Wilson.
28. Karlin, S., and Taylor, H. M. 1975. *A First Course in Stochastic Processes, Second Edition*. San Francisco: Academic Press.
29. Allen, L. J. S. 2011. *An Introduction to Stochastic Processes with Applications to Biology, Second Edition*. Boca Raton, FL: CRC Press.
30. Insua, D. R., Ruggeri, F., and Wiper, M. P. 2012. *Bayesian Analysis of Stochastic Process Models*. New York: John Wiley & Sons.

3

Introduction to Stochastic Processes

3.1 Introduction

The first part of this chapter will review the basic terminology and notation involved with the presentation of materials about stochastic processes. This is to be followed by the definition of a stochastic process and the various types of stochastic processes. Next, the focus will be on defining the various types of states of a stochastic process. Lastly, several examples of discrete and continuous Markov processes will be discussed, while the last section will be devoted to the explanation of several types of normal processes. Our main emphasis is on the basic foundation of studying stochastic processes, which is necessary for implementing Bayesian inference. The Bayesian analysis of such processes will continue with Chapter 4, but for now, the main objective is to introduce the reader to the foundation of stochastic processes.

3.2 Basic Terminology and Notation

Recall Section 2.2, where a review of the probability theory was outlined. The following ideas about probability were included:

1. The definition of a probability space
2. The definition of a random variable X
3. The distribution function F of X (3.1)
4. The expectation and moments of X
5. The law of total probability
6. The probability mass function of a discrete random variable
7. The probability density function of a continuous random variable
8. The joint distribution function of several random variables
9. The joint probability mass function of several discrete random variables
10. The joint probability density function of several continuous random variables
11. The ideas of conditional probability, conditional distributions, and conditional expectation
12. Joint, marginal, and conditional distributions

13. Moments and characteristic functions of a random variable
14. Different types of convergence, including convergence with probability 1, convergence in probability, and convergence in quadratic mean

Also introduced in Chapter 1 were the well-known discrete and continuous distributions. With regard to the continuous type, the normal exponential, gamma, uniform, and beta were introduced, while for the discrete type, the binomial, Poisson, negative binomial, and geometric were defined. With regard to continuous multivariate distributions, the multivariate normal, the multivariate t, and Dirichlet were defined, and with respect to the multivariate discrete type of distributions, the multinomial and hypergeometric were explained. Of course, such basic concepts of probability and the distributions named earlier play an important role in the definition of stochastic processes.

3.3 Definition of a Stochastic Process

The study of stochastic processes involves investigating the probabilistic structure of sets of random variables. Let $X(t)$ be a random variable, where t is a parameter belonging to some index set T. A stochastic process is a family of random variables

$$\{X(t), t \in T\}, \tag{3.2}$$

where for each t, $X(t)$ is a random variable assuming real values. If the index set T is the set $T = \{0, 1, 2, \ldots\}$, then the stochastic process in Equation 3.2 represents values like those of a random walk. Of course, in order to investigate the process, what is required is the joint probability distribution of the set of random variables

$$X(t_1), X(t_2), \ldots, X(t_n) \tag{3.3}$$

for all $n = (1, 2, 3, \ldots)$ and all time points $0 \le t_1 < t_2 <, \ldots, < t_n$.

The values of $X(t)$ can be multidimensional, that is, one, two, or n dimensional. For example, suppose $X(t)$ represents the outcome of the tth toss of a die with possible outcomes $\{1, 2, 3, 4, 5, 6\}$, then a possible realization of the process for the first 10 time points (tosses) might be 4, 6, 1, 8, 3, 2, 1, 6, 6, 4; that is, on the first toss, the outcome was a 4, and on the tenth toss, the outcome was a 4. If the die is fair,

$$P[X(t) = i] = 1/6 \tag{3.4}$$

for $i = 1, 2, 3, 4, 5, 6$, and if the tosses are independent, one would know the joint distribution of the family $X(1), X(2), \ldots, X(10)$. Since the random variables of the process are independent, this is a very simple case of a stochastic process. Our interest will be mostly centered on the cases where the random variables of the process are correlated. The possible outcomes of this process are called the states of the process.

We will also study processes with index set $T = [0, \infty)$, and for now, be content with a brief introduction to the basic concepts of stochastic processes using two important examples.

3.3.1 Poisson Process

One of the most basic processes is the counting process called the Poisson process. Suppose $N(t)$ counts the number of times a well-defined event occurs over the interval $[0, t]$ and with index set $T = [0, \infty)$ and state space $S = [0, 1, 2, 3, 4, 5, \ldots]$; that is, the possible values of $X(t)$ are the nonnegative integers. Figure 3.1 displays a hypothetical realization of a Poisson process.

Note that the first event occurs at time 1; the second, at time 3; the third, at time 4; and the fifth, at time 5.

Typical events are the number of deaths over time of some well-defined place (e.g., the United States); the number of accidents in Seattle, Washington; the number of people in a queue; etc. The following description of a Poisson process is from pages 22–25 of Karlin and Taylor[1] and pages 200–204 of Allen.[2]

It is assumed that the number of events occurring in two disjoint time intervals is independent and that $N(t_0 + t) - N(t_0)$ depends only on t (the interval length) but not on t_0 or on the value of $N(t_0)$. The following with $h \to 0$ determines the probability characteristics of the process:

1. The probability of at least one event occurring in a time duration h is

$$p(h) = ah + o(h), \quad h \to 0,\ a > 0, \tag{3.5}$$

   where $o(h)$ is a quantity such that $\lim[o(t)/t] = 0, \quad t \to 0$.

2. The probability of two or more events occurring in an interval of length h (as $h \to 0$) is 0, thus ruling out the possibility of simultaneous events at any particular time.

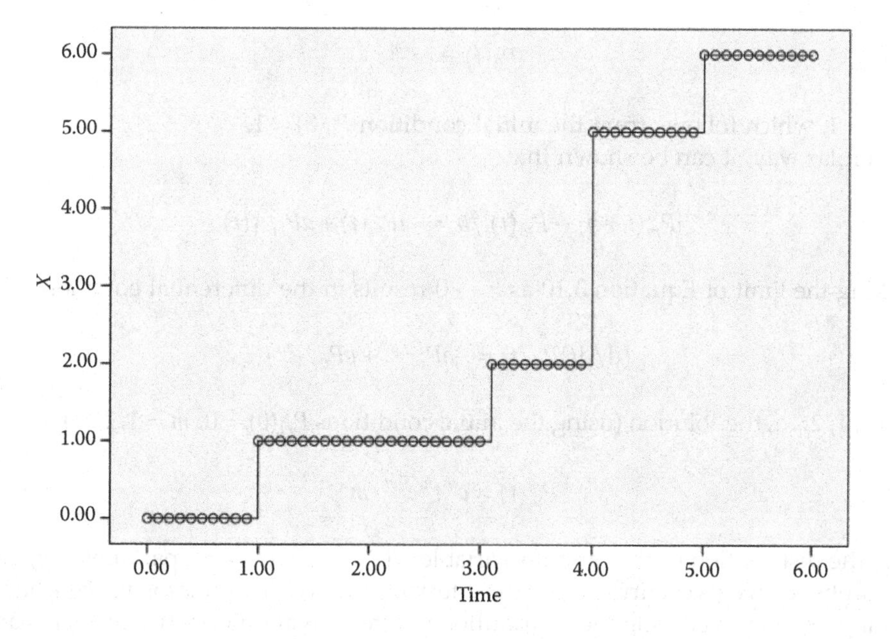

FIGURE 3.1
Realization of a Poisson process.

Let $P_m(t) = \Pr[N(t) = m]$ be the probability that m events occur over an interval from 0 to t, where $m = 0, 1, 2, \ldots$; then from the second assumption,

$$\sum_{m=2}^{\infty} P_m(h) = o(h), \tag{3.6}$$

which implies

$$p(h) = p_1(h) + p_2(h) + \ldots$$

Recall the assumption of independence of events over nonoverlapping intervals; then, it can be shown that

$$p_0(t+h) = p_0(t)p_0(h) = p_0(t)[1 - p(h)], \tag{3.7}$$

where $p(h)$ satisfies Equation 3.5. Now let us set up the difference quotient

$$[P_0(t+h) - P_0(t)]/h = -P_0(t)p(h)/h. \tag{3.8}$$

Then, taking the limit of Equation 3.8 as $h \to 0$ gives the differential equation

$$(\mathrm{d}/\mathrm{d}t)P_0(t) = -aP_0(t)$$

with solution

$$P_0(t) = ce^{-at}, \tag{3.9}$$

where $c = 1$, which follows from the initial condition $P_0(0) = 1$.

In a similar way, it can be shown that

$$[P_m(t+h) - P_m(t)]/h = -aP_m(t) + aP_{m-1}(t), \tag{3.10}$$

and taking the limit of Equation 3.10 as $h \to 0$ results in the differential equation

$$(\mathrm{d}/\mathrm{d}t)P_m(t) = -aP_m(t) + aP_{m-1}(t) \tag{3.11}$$

for $m = 0, 1, 2, \ldots$, the solution (using the initial conditions $P_m(0) = 0$, $m = 1, 2, \ldots$) of which is

$$P_m(t) = a^m t^m e^{-at}/m! \tag{3.12}$$

Thus, the distribution of the random variable $N(t)$ is Poisson with parameter at, and the mean number of events occurring over the interval 0 to t is at. The Poisson process has many interesting features, including the distribution of interarrival times between events and the waiting time until the nth event occurs. The Bayesian analysis of the Poisson process including the analysis of the waiting times (which have a gamma $X(t)$ distribution) was introduced in Chapter 2.

3.3.2 Wiener Process

An interesting stochastic process is $\{X(t): t > 0\}$ with state space $[0, \infty)$, where $X(t)$ represents the displacement of a particle undergoing Brownian motion. The displacement $X(t) - X(s)$ over the time period (s, t) is the result of a large number of small displacements of surrounding particles also undergoing Brownian motion. By using the central limit theorem, one can show that the displacement $X(t) - X(s)$ has a normal distribution. It is also assumed that the displacement $X(t + h) - X(s + h)$ over the interval from $s + h$ to $t + h$ $(h > 0)$ has the same distribution as the interval $X(t) - X(s)$ over (s, t), and that its displacement depends only on the time length $t - s$. The Brownian motion was described by R. Brown in 1827 and Einstein in 1905, who explained Brownian motion by asserting that the motions observed were the result of perpetual collisions with other molecules in the enveloping medium.

The Wiener process has the following probabilistic properties:

1. Let n be a positive integer and let the time points t_i, $i = 1, 2, \ldots, n$, be ordered as $t_1 < t_2 < \ldots < t_n$; then the increments $X(t_i) - X(t_{i-1})$, $i = 1, 2, \ldots, n$, are independent, and the process is characterized as having independent increments.
2. The distribution of $X(t) - X(s)$ depends only on the length $t - s$ and not on s, and the distribution of $X(t) - X(s)$ is normal with mean of 0 and variance of $\sigma^2(t - s)$. Thus, assuming $X(0) = 0$, the first two moments of the process are $E[X(t)] = 0$ and $\mathrm{Var}[X(t)] = \sigma^2 t$.

The Bayesian analysis of the Wiener process was performed in Chapter 2 and will be extended in Chapter 4. Both the Poisson and Wiener processes are examples of Markov processes.

3.4 Elements of a Stochastic Process

The main characteristics identifying stochastic processes are the state space S and the index set T and the correlation between the random variables making up the process.

3.4.1 State Space

The state space S is the set of all possible values $X(t)$ of the stochastic process. Suppose $S = \{0, 1, 2, \ldots, n\}$ or $S = \{0, 1, 2, \ldots\}$; then the process is referred to as a discrete state space process. In the first case, the possible value is finite, and in the latter, the process takes on an infinitely countable number of possible values. When S is the set of real numbers $(-\infty, \infty)$, the process is a real-valued process and is one dimensional, but it is possible that the possible values are elements of Euclidean K space, in which case the possible values of $X(t)$ are k-dimensional real vectors. It is very important that the state space be unambiguously described when making Bayesian inferences of a process.

3.4.2 Index Set

One must also know the index set T or the parameter space of the process. For example, if $T = \{0, 1, 2, \ldots, n\}$ or $T = \{0, 1, 2, \ldots\}$, the process $\{X(t), t \in T\}$ is referred to as a discrete-time stochastic process. In the former case, the process is observed over a finite set of times, while in the latter, the process is observed hypothetically, over an infinitely countable number of times. Lastly, when $T = [0, \infty)$, the process is called a continuous-time process. The Poisson process and the Wiener process are examples of continuous-time processes, where the Poisson state space is $S = \{0,1,2,...\}$ and infinitely countable, but for the latter Wiener process, the state space is $S = [0, \infty)$, the set of nonnegative reals. Of course, index set T can be multidimensional, such as the set of all ordered pairs in the plane.

3.4.3 Some Classical Types of Processes

The probabilistic relationships between the member of a stochastic process will be examined, where for convenience the index set $T = [0, \infty)$ and the state space $S = R$, the set of real numbers.

3.4.3.1 Processes with Independent Increments

Let n be a positive integer and consider the set of time points t_i ordered as $t_1 < t_2 < ... < t_n$; then, if the increments

$$X(t_2) - X(t_1),\ X(t_3) - X(t_2),\ ..,\ X(t_n) - X(t_{n-1}) \tag{3.13}$$

are independent, the process $\{X(t): t \in T\}$ is said to have independent increments. In addition, in the special case $T = \{0,1,2,...\}$, the process $\{Y(t),\ t \in T\}$ defined as $Y(0) = X(0)$, $Y(t) = X(t) - X(t-1)$, $t = 1,2,...$, is a sequence of independent random variables. Thus, it is easily seen that if one knows the distribution of the independent increments (Equation 3.13), the joint distribution of the original sequence $\{X(t): t \in T\}$ is known.

If, in addition, the distribution of $X(t_1 + h) - X(t_1)$ depends only on h and not on t_1, the process is said to have stationary increments. It should be noted that if a process has stationary independent increments, the distribution of $X(t_1 + h) - X(t_1)$ is the same as $X(t_2 + h) - X(t_2)$ for all h and values t_1 and t_2. Of special interest is to know that if a process has stationary independent increments and the index set $T = \{0,1,2,...\}$, the process has a finite mean:

$$E[X(t)] = m_0 + m_1 t, \tag{3.14}$$

where $t = 0, 1, 2, \ldots, m_0 = E[X(0)]$, and $m_1 = E[X(1)] - m_0$.

In a similar way, one may show that

$$\text{Var}[X(t)] = \text{Var}[X(0)] + \text{Var}[X(1)]t, \tag{3.15}$$

where

$$\text{Var}[X(0)] = E\left[(X(0) - m_0)^2\right]$$

and

$$\text{Var}[X(1)] = E\left[(X(1) - m_1)^2\right] - \text{Var}[X(0)].$$

It is left as an exercise to verify Equation 3.14. See Karlin and Taylor[1] for additional information on how to verify Equation 3.14.

3.4.3.2 Martingales

Consider a real-valued stochastic process $\{X(t): t \in T\}$, where T is discrete or continuous, and suppose that $E[|X(t)|] < \infty$, that is, the moments of each member of the process exits, then the process is a martingale if for all times such that $t_1 < t_2 < \ldots < t_n < t_{n+1}$ and n is a positive integer,

$$E[X(t_{n+1})|X(t_1) = c_1, X(t_2) = c_2, .., X(t_n) = c_n] = c_n, \tag{3.16}$$

for all time points and choices of the conditioning values $c_1, c_2, \ldots, c_n$. There is an interesting gambling interpretation of martingales in that the average amount a player will have at time t_{n+1}, given the player has an amount c_n at time t_n, is c_n regardless of their fortune (holdings) at times previous to t_n. Another example of a martingale can be expressed as sums of independent random variables $Y(i)$, namely,

$$X(n) = Y(1) + Y(2) + \ldots + Y(n) \tag{3.17}$$

for $n = 1, 2, \ldots$ and assuming $E[Y(i)] = 0$, for $i = 1, 2, 3, \ldots$.

Lastly, consider a process $\{X(t), t \geq 0\}$ with independent increments and with zero means; then it can be shown that the process is a martingale. This assertion is left as an exercise for the student.

3.4.3.3 Markov Processes

A stochastic process $\{X(t), t \in R\}$ is said to be Markov if

$$\Pr[X(t) \leq b|X(t_1) = x_1, \ldots, X(t_n) = x_n] = \Pr[X(t) \leq b|X(t_n) = x_n] \tag{3.18}$$

for all choices of times points $t_1 < t_2 < \ldots < t_n < t$ and all states $x_1 < x_2 < \ldots < x_n$.

What is the interpretation of such a process? A Markov process has the property that given the values of previous random variables $X(t_i)$, $i = 1, 2, \ldots, n$, the value of $X(t)$ depends on only the most recent value, namely, that of $X(t_n)$.

Consider the transition probability function

$$P(x, s; t, A) = \Pr[X(t) \in A|X(s) = x], \tag{3.19}$$

which is a quite useful subject in the study of Markov chains. Note that the Markov property can be represented as

$$P(x_n, t_n; t, A) = \Pr[X(t) \leq b|X(t_1) = x_1, .., X(t_n) = x_n], \tag{3.20}$$

where the event $A = \{t : t \leq b\}$, the set of all states that do not exceed b. Note that if the states of a Markov process can be finite or countable, the process is called a Markov chain, and this is the subject of Chapter 4.

3.4.3.4 Stationary Processes

Stationarity is a property of a stochastic process that induces a probabilistic stability among the member of the process $\{X(t),\ t \in T\}$, where T is $(-\infty, +\infty)$ or $[0, \infty)$ or $(\ldots, -2, -1, 0, 1, 2, \ldots)$ or $(1, 2, 3, \ldots)$. There are two types of stationarity: strict stationarity and covariance stationarity. Strict stationarity is defined as follows:

$$X(t_1 + h), X(t_2 + h), \ldots, X(t_n + h) \sim X(t_1), X(t_2), \ldots, X(t_n) \tag{3.21}$$

for all choices of n, $h > 0$, and ordered time points $t_1, t_2, \ldots, t_n$ of the index set T. Thus, for example, the mean of $X(t)$ is the same for $\forall\, t \in T$. Note that the distribution of the set of n random variables on the left of Equation 3.21 is the same as the set of n random variables on the right-hand side.

The second case of stationarity, called weak stationarity, assumes that the second moments of each member of the process exist and that $\mathrm{Cov}[X(t), X(t + h)]$ depends on h and only h for all $t \in T$. If a process is strictly stationary and has finite second moments, then the process is covariance stationary; however, the converse is not true. The preceding description of stationary processes is taken from Karlin and Taylor[1] and is similar to the description on pages 9–12 of Cox and Miller[3] and pages 69–72 of Parzen.[4] It may appear that stationarity is too strong a property to impose on a stochastic process, but it is an appropriate assumption for many applications, including communication theory, signal processing, astronomy, biology, and economics. Such applications will be presented in the following two sections of this chapter.

Recall that a Markov process is said to have stationary transition probabilities if $P(x, s; t, A)$ as defined by Equation 3.10 is only a function of $t - s$, but remember that $P(x, s; t, A)$ is a conditional distribution, given the present state. It can be shown that a Markov process which has stationary transition probabilities is not necessarily a strictly stationary process. The Poisson process and the Wiener process are not stationary. This is obvious for the Poisson process $\{N(T),\ t \geq 0\}$ with parameter $\rho > 0$, because the mean value $E[N(t)] = \rho t$ increases with time. It is left as an exercise to show that the Wiener process is not strictly stationary.

3.4.3.5 Renewal Processes

A renewal process is a sequence $T(i)$, $i = 1, 2, \ldots$, of independent and identically distributed random variables that represent lifetimes or the survival of a set of objects (say, people), where starting from 0, $T(1)$ is the survival time of the first patient; $T(2)$, that of the second; etc. Waiting times are defined as follows:

$$W(n) = T(1) + T(2) + \ldots + T(n). \tag{3.22}$$

That is, $W(n)$ is the time to the nth failure or, in the context of a survival study, the time of the nth death. Such processes are relevant to medical studies, to life table analysis in actuarial work, and to experiments in reliability. Thus, a renewal process is a counting process that keeps a record of the number of renewals $N(t)$ over time, and it takes on the value n according to

$$N(t) = n \Leftrightarrow W(n) \leq t < W(n + 1). \tag{3.23}$$

That is, the *n*th failure occurs at time *t* if and only if the waiting time to failure *n* does not exceed time *t* and is less than the waiting time for event $n + 1$. Thus, there are three processes associated with a renewal process: (1) the failure times $T(n)$, (2) the waiting times $W(n)$, and the counting process $N(t)$. Of course, the Poisson process is a renewal process with exponentially distributed interarrival times and gamma-distributed waiting times.

The Bayesian analysis of the Poisson process was introduced in Chapter 2, and renewal processes will be studied extensively in Chapter 4.

3.4.3.6 Point Processes

Let S be an n-dimensional space, and let Ξ be a family of subsets of S. A point process is a stochastic process indexed by the sets $A \in \Xi$, and suppose the state space of the process be the nonnegative integers. Suppose points are distributed over space S, and, in addition, suppose that $N(A)$ counts the number of points in set A; thus, $N(A)$ is a counting process that counts the number of points in set A. Certain consistency rules are required of $N(A)$:

1. $N(A_1 \cup A_2) = N(A_1) + N(A_2)$ for disjoint sets A_1 and A of Ξ.
2. If the empty set O is in Ξ, $N(O) = 0$.

   As an example, suppose S is a set of the plane and for every subset $A \subset S$ and let $N(A)$ be the area of A; then $\{N(A), A \subset S\}$ is a homogeneous Poisson point process with intensity ρ if the following apply:

   a. For each $A \subset S$, $N(A)$, the length of A has a Poisson distribution with parameter $\rho L(A)$, where $L(A)$ is the length of A.
   b. For every finite collection of disjoint subsets $\{A_1, A_2, \ldots, A_n\}$, the random variables $N(A_1), N(A_2), \ldots, N(A_n)$ are independent.

Poisson point processes occur in a large number of scientific fields. For example in astronomy, where $N(A)$ counts the number of galaxies in a region of space A, and in ecology, where $N(A)$ represents the number of plant types in the area A. For additional information about point processes, see the excellent monograph by Cox and Isham.[5] These authors give several examples of point processes, including emissions from a radioactive source, electrical energy in a nerve fiber, the time of accidents in coal mines in Wales, and the arrival times of customers in a queue. These examples will be employed to illustrate Bayesian inferences for stochastic processes.

3.4.4 Essential Factors for Working with a Stochastic Process

The most important aspect in studying stochastic processes is to know the relationship between the random variables of the process. We will consider a process $\{X(t), t \in T\}$ well defined if the state space, index space, and joint distribution of the following n random variables are known: $\{X(t_i) : i = 1, 2, \ldots, n\}$, where the time points are ordered as $t_1 < t_2 <, \ldots, < t_n$.

Special problems arise when index set T is continuous. To see this, consider two sequences $X(t)$ and $Y(t)$, where

$$
\begin{aligned}
X(t) &= 1, \quad U = t, \\
X(t) &= 0, \quad U \neq t,
\end{aligned}
\tag{3.24}
$$

where U is uniformly distributed over [0,1]. Also, let

$$Y(t) = 0, \quad t > 0. \tag{3.25}$$

It can be shown that

$$\Pr[X(t) \le 1/2] = 0$$

$$\Pr[Y(t) \le 1/2] = 1, 0 \le t \le 1.$$

This seeming paradox is due to the continuous nature of the index set, and for further information about resolving this issue, refer to pages 32 and 33 of Karlin and Taylor.[1]

3.5 States of a Markov Chain

3.5.1 Introduction

The behavior of a Markov chain is determined by the characteristics of the possible values. For example, some states are visited on a finite number of times, while others occur infinitely often. This section will classify the states of a Markov chain, and the idea of accessibility is the foundation of studying the behavior of a Markov chain. The following ideas will be explained: accessibility, a transient state, a recurrent state, a communicating class of states, and a periodic state of a given period.

3.5.2 Classification of the States of a Markov Chain

A state j is said to be accessible from state i if there is a positive probability of reaching state j from state i in a finite number of transitions, that is, when $P_{ij}^n > 0$, where the n-step transition probability is given by

$$P_{ij}^n = \Pr[X(n+m) = j \mid X(m) = i]. \tag{3.26}$$

If the states i and j are accessible from each other, the states are said to communicate, and this is designated by $i \leftrightarrow j$. On the other hand, if these two states do not communicate, then either

$$P_{ij}^n = 0, \quad \forall\, n \ge 0, \tag{3.27}$$

$$P_{ji}^n = 0, \quad \forall\, n \ge 0, \tag{3.28}$$

or both Equations 3.27 and 3.28 are true. The idea of communicating classes is an equivalence relation namely,

$$\begin{aligned}&1.\ i \leftrightarrow i,\\ &2.\ \text{If } i \leftrightarrow j \Rightarrow j \leftrightarrow i\\ &3.\ \text{If } i \leftrightarrow j \text{ and } j \leftrightarrow k \Rightarrow i \leftrightarrow k\end{aligned} \tag{3.29}$$

The third relation is called transitivity and is demonstrated as follows: since $i \leftrightarrow j$ and there exist times m and n such that $P_{ij}^{m} > 0$,

$$P_{ik}^{m+n} = \sum_{r=0}^{r=\infty} P_{ir}^{m} P_{rk}^{n} \geq P_{ij}^{m} P_{jk}^{n} > 0 \tag{3.30}$$

Of course, it can be shown that there exists a time l such that $P_{ki}^{l} > 0$; then using the same argument as Equation 3.30, transitivity is declared. Since communication $\leftrightarrow$ induces an equivalence class on the states of a Markov chain, the states of a chain can be partitioned into a set of communicating states. If all the states consist of one communication class, the chain is said to be irreducible, a very important concept for us to study. Let C_1 and C_2 be two communicating classes, and let i and j belong to the two, respectively, that is, $i \in C_1$ and $j \in C_2$; then it is possible that $i \to j$ (that is, j is accessible from i), but if that is the case, then it cannot be true that $j \to i$ (that is, i is accessible from j). As an example, consider the 5×5 transition matrix

$$P = \begin{pmatrix} P_1, 0 \\ 0, P_2 \end{pmatrix}, \tag{3.31}$$

where

$$P_1 = \begin{pmatrix} .5, .5 \\ .25, .75 \end{pmatrix} \tag{3.32}$$

and

$$P_2 = \begin{pmatrix} 0,1,0 \\ .5,0,.5 \\ 0,1,0 \end{pmatrix}. \tag{3.33}$$

Thus, it can be seen that the two states of P_1 do not communicate with the three states of P_2, and vice versa, and each of the two classes can be analyzed by itself.

Thus, if the state of $X(0)$ is in the first class, the state of the chain remains in the first class.

In order to perform a Bayesian analysis, it is important to know the communicating classes of a chain. Also, in order to generate realizations from a Markov chain, it is important to know the communicating classes. For example, for the preceding chain (Equation 3.31), one would generate observations from P_1 and P_2 separately, then it would be possible to find Bayesian inferences for all the transition probabilities.

The communicating classes of a chain can be quite complex. Consider the transition matrix of a random walk with integers for the state space, $S = \{0, 1, 2, ..., a-1, a\}$,

$$P = \begin{pmatrix} 1,0,0,0,......0,0,0 \\ q,0,p,0,.....0,0,0 \\ 0,q,0,p,,,,,0,0,0 \\ . \\ . \\ 0,..............q,0,p \\ 0,................0,0,1 \end{pmatrix}. \tag{3.34}$$

If the initial state is 0, the chain remains in 0, that is, 0 is an absorbing state, while if the initial state is 1, the process moves one unit to the right with probability p or one unit to the left with probability q and $p+q=1$. There are three communicating classes: $C_1 = \{0\}$, $C_2 = \{1,2,\ldots,a-1\}$, and $C_3 = \{a\}$, where a is a positive integer. This chain is similar to the gambler's ruin problem, to be presented in Section 3.6. The first and third classes can be reached from the second, but it is impossible to return to the second, from either the first or third!

3.5.3 Periodicity of a Markov Chain

The period of a state i, which is denoted by $d(i)$, is the greatest common divisor of all $n \geq 1$, where $P_{ii}^n > 0$, but if $P_{ii}^n = 0$ for all $n \geq 0$, $d(i)$ is defined as $d(i) = 0$. That is, the chain returns to state i at integer multiples of $d(i)$. For example, for a random walk with the transition matrix in Equation 3.34, each state has period 2. Now consider a chain with the $n \times n$ transition matrix

$$P = \begin{pmatrix} 0,1,0,0,....,0 \\ 0,0,1,0,.....,0 \\ . \\ . \\ 0,0,...........,1 \\ 1,0,0,........,0 \end{pmatrix}; \tag{3.35}$$

then each state has period n.

The following assertions about periodicity will be left for the student to prove:

1. If the states i and j communicate, then they have the same period or $i \leftrightarrow j \Rightarrow d(i) = d(j)$.
2. Suppose that the period of i is $d(i)$, then there exists an integer N, depending on i such that for all $n \geq N$, $P_{ii}^{nd(i)} > 0$.

3.5.4 Recurrent States

How often does a particular state occur while observing the outcomes of a Markov chain? A particular state might be visited only a finite number of times, infinitely often, or never. This section will define the ideas of a recurrent state and a transient state.

We state for a particular state i, the probability that the process will return to i for the first time $n \geq 1$; thus, let this probability be determined by the following equation:

$$f_{ii}^{n} = \Pr[X(n) = i, X(m) \neq i, m = 1,2,.,n-1|X(0) = i]. \tag{3.36}$$

It is easy to show that

$$P_{ii}^{n} = \sum_{k=0}^{k=n} f_{ii}^{k} P_{ii}^{n-k}. \tag{3.37}$$

That is to say, the probability that i will occur at time n (given initially that the process is in state i) is the probability that state i will occur first at time k, multiplied by the probability that the process will return to state i after $n - k$ time units. Equation 3.37 follows from the law of total probability. It thus follows that

$$P_{ij}^{n} = \sum_{k=0}^{k=n} f_{ij}^{k} P_{jj}^{n-k}, \tag{3.38}$$

where the first transition from state i to state j occurs at time n and is portrayed by

$$f_{ij}^{n} = \Pr[X(n) = j, X(m) \neq j, m = 1,2,..,n-1|X(0) = i]. \tag{3.39}$$

A state i is called recurrent if

$$\sum_{n=1}^{n=\infty} f_{ii}^{n} = 1. \tag{3.40}$$

That is, a state is recurrent if starting from state i, the probability of returning to state i in a finite time interval is 1, and additional information about recurrence is to follow.

For further information about recurrence, refer to pages 62–67 of Karlin and Taylor.[1] This information includes using probability-generating functions and ideas about Abel's type convergence to prove additional assertions about recurrent states, for example,

1. A state i is recurrent if and only if

$$\sum_{n=1}^{n=\infty} P_{ii}^{n} = \infty. \tag{3.41}$$

2. If i is recurrent and $i \leftrightarrow j$, then j is recurrent.

3. The expected number of returns to a given state i, given $X(0) = i$, is given by $\sum_{n=0}^{n=\infty} P_{ii}^{n}$.

Note that 1 implies that a state is recurrent if and only if it occurs infinitely often, while 3 implies that a state is recurrent if and only if the expected number of returns to that state is infinite. Also, it is easy to show that the assertion in Equation 3.41 implies Equation 3.40. If a state i is not recurrent, it is transient, that is, $\sum_{n=1}^{n=\infty} f_{ii}^{n} < 1$ or equivalently $\sum_{n=1}^{n=\infty} p_{ii}^{n} < \infty$. In addition, if a state i is recurrent and the mean recurrence time is positive, the state is called positive recurrent; however, if the mean recurrence time is 0, the state is called null recurrent.

A simple example of a recurrent Markov chain is the one-dimensional random walk, where at each transition, the probability of moving to the right by one unit is p and that of moving to the left by one unit is $1-p$; thus, $P_{00}^{2n+1} = 0$, $n = 0, 1, 2, \ldots$, and $P_{00}^{2n} = \binom{2n}{n} p^{n}(1-p)^{n}$, and the latter is approximated by using the Stirling formula for factorials, namely,

$$n! \simeq n^{n+1/2} e^{-n} \sqrt{2\pi}. \tag{3.42}$$

Thus,

$$P_{00}^{2n} \approx (p(1-p))^{n} 2^{2n} / \sqrt{n\pi} = (4p(1-p))^{n} / \sqrt{\pi n}, \tag{3.43}$$

and if $p = 1/2$, it can be thus shown that $\sum_{n=0}^{n=\infty} P_{00}^{2n} = \infty$. The conclusion is that state 0 is recurrent. If $p \neq 1/2$, it will be left to the student to show that 0 is not recurrent.

3.5.5 More on Recurrence

What is to follow tells one how to differentiate between recurrent and transient states. Let V_{ii} be the probability a particle starting in state i returns infinitely often and let V_{ii}^{n} be the probability that for a particle initially in state i, it returns infinitely often at least N times; then

$$V_{ii}^{n} = \sum_{k=1}^{k=\infty} f_{ii}^{k} V_{ii}^{N-1} = V_{ii}^{N-1} f_{ii}^{*}, \tag{3.44}$$

where

$$f_{ii}^{*} = \sum_{k=1}^{k=\infty} f_{ii}^{k}. \tag{3.45}$$

Thus, since $\lim V_{ii}^{N} = V_{ii}$ as $N \to \infty$, therefore V_{ii} is 1 or 0 if and only if $f_{ii}^{*} = 1$ or <1, that is, if and only if state i is recurrent or transient. Thus, if a state is recurrent, that state will occur infinitely often with a probability of 1, but on the other hand, if the state is transient, the probability is less than 1 that the state will be occupied infinitely often. Also, note that if i and j communicate and the class is recurrent, then

$$f_{ij}^* = \sum_{n=1}^{n=\infty} f_{ij}^n = 1. \tag{3.46}$$

Remember that in a finite-state Markov chain, not all states can be transient; that is, at least one state must be recurrent. Consider the example on pages 146 and 147 of Ross[6] of a Markov chain with states 0, 1, 2, and 3 and transition probability matrix

$$P = \begin{pmatrix} 0,0,.5,.5 \\ 1,0,0,0 \\ 0,1,0,0 \\ 0,1,0,0 \end{pmatrix}. \tag{3.47}$$

Since all states communicate, all states must be recurrent.

For a different situation with states 0, 1, 2, 3, and 4, consider a process with transition matrix

$$P = \begin{pmatrix} 1/2,1/2,0,0,0 \\ 1/2,1/2,0,0,0 \\ 0,0,1/2,1/2,0 \\ 0,0,1/2,1/2,0 \\ 1/4,1/4,0,0,1/2 \end{pmatrix}, \tag{3.48}$$

where there are three classes $\{0,1\}$, $\{2,3\}$, and $\{4\}$, with the first two recurrent but the third transient.

In the long run, what are the possible states of a Markov chain, and how do they depend on the initial state? Such problems come under the subject of limiting probabilities. The following shows how to find the limiting probabilities of certain types of Markov processes:

1. For an irreducible positive recurrent Markov chain, $\lim P_{ij}^n = \pi_j$, $n \to \infty$ and is independent of i.
2. π_j is the unique solution to the system of equations

$$\pi_j = \sum_{i=0}^{i=\infty} \pi_i P_{ij}, \quad j \geq 0, \tag{3.49}$$

and satisfying the constraint $\sum_{j=0}^{j=\infty} \pi_j = 1$.

This implies that the long-run proportion of times the process is in state j is the limiting probability π_j given by Equation 3.49.

An interesting example on pages 154 and 155 of Ross[6] is a problem of interest to sociologists that involves determining the proportion of society that has an upper class-type or lower class-type occupation. It is assumed that the transitions between the three classes, lower, middle, and upper classes, follow a Markov chain with matrix

$$P = \begin{pmatrix} .45, .45, .07 \\ .05, .70, .25 \\ .01, .50, .49 \end{pmatrix}. \tag{3.50}$$

For example, the child of a middle-class worker has a chance of .7 of remaining middle class and a 25% chance of being upper class. What are the limiting probabilities for this example? From Equation 3.49, the three equations are

$$\begin{aligned} \pi_0 &= .45\pi_o + .05\pi_1 + .01\pi_2, \\ \pi_1 &= .05\pi_o + .70\pi_1 + .50\pi_2, \\ \pi_2 &= .07\pi_o + .25\pi_1 + .49\pi_2, \end{aligned} \tag{3.51}$$

and subject to the constraint $\pi_o + \pi_1 + \pi_2 = 1$; it can be shown that the solution is $\pi_o = .07$, $\pi_1 = .62$, $\pi_2 = .31$. Thus, in the long run, the investigated population has 7% of the workers in lower class, 62% in the middle class, and 31% in the upper class. Of course, these limiting probabilities are just estimates because they are based on the estimated transition matrix in Equation 3.50, which was most likely determined by surveys conducted by the sociologists. This example will be studied in Chapter 4, where Bayesian inferences for the limiting probabilities will be developed.

An interesting question left to the student is: is it possible to find limiting probabilities for Markov chains that are not irreducibly positive recurrent?

3.6 Some Examples of Discrete Markov Chains

3.6.1 Introduction

This chapter presents the fundamental ideas that describe a Markov chain, a group of very important stochastic processes which have been shown to be relevant to many problems in science. Stochastic processes are a part of the probability theory, and this book will give the reader a good idea of how to construct Bayesian inferences for Markov chains. However, in this chapter, the main emphasis is not on inferences but on presenting the fundamental concepts of Markov chains; thus, we begin with the definition.

A discrete-time Markov chain (DTMC) $\{X(n), n = 0, 1, \ldots\}$ is a sequence of random variables with nonnegative integers as the index set and where n is interpreted as time. The state space is for now taken to be the nonnegative integers, and $X(n) = i$ is interpreted to be the state of the process at time n. The Markov chain is defined as a conditional probability that the future state of the chain is in state j, given that at the previous time, the state of the process is in state i, or

$$P_{ij}^{n,n+1} = \Pr[X(n+1) = j | X(n) = i], \tag{3.52}$$

where i and j belong to state space $S = \{0,1,2,...\}$ and time n is a nonnegative integer. Note the dependence of these transition probabilities not only on the initial and final states but also on the time n of the transition. If this one-step transition probability is independent of time variable, the Markov chain is said to have stationary transition probabilities, and this chapter will mainly emphasize such processes. In this case, $P_{ij}^{n,n+1}$ does not depend on n and the one-step transition probability is written as P_{ij}.

Note the probability of going from state i to j is the same for all times $n = 0, 1, 2, \ldots$. The transition probability matrix is shown as

$$P = \begin{pmatrix} P_{00}, P_{01}, P_{02}, \ldots.. \\ P_{10}, P_{11}, P_{12}, \ldots.. \\ \\ P_{i0}, P_{i1}, P_{i2}, \ldots.. \\ \\ \end{pmatrix}, \tag{3.53}$$

where the first row is the conditional probability of the states $0, 1, 2, \ldots$, given that the initial state is 0, while the second row is the same, but when the initial state of the process is 1, etc. Thus, each row is a conditional probability; hence, for all $i = 0, 1, 2, \ldots$,

$$\sum_{j \geq 0} P_{ij} = 1 \tag{3.54}$$

Of course, the number of states can be finite, in which case the transition matrix P is suitably modified.

The chain is completely specified when one knows the transition probability matrix P and the probabilities of the initial state $X(0)$. Also, for the process to be completely determined, one must be able to determine

$$\Pr[X(0) = i_0, X(1) = i_1, .., X(n) = i_n]. \tag{3.55}$$

Can this joint probability be computed using only the one-step transition probabilities? Consider Equation 3.55 expressed as

$$\begin{aligned} &\Pr[X(n) = i_n | X(0) = i_0, X(1) = i_1, \ldots X(n-1) = i_{n-1}], \\ &\Pr[X(0) = i_0, X(1) = i_1, \ldots X(n-1) = i_{n-1}]. \end{aligned} \tag{3.56}$$

Now the process is Markov; thus,

$$\begin{aligned}&\Pr[X(n) = i_n | X(0) = i_0, X(1) = i_1, ..X(n-1) = i_{n-1}] \\ &= \Pr[X(n) = i_n | X(n-1) = i_{n-1}] = P_{i_{n-1}, i_n},\end{aligned} \tag{3.57}$$

and it then follows that

$$\Pr[X(0) = i_0, X(1) = i_1, X(2) = i_2, ..X(n) = i_n] = P_{i_0} P_{i_0, i_1} ..P_{i_{n-1}, i_n},$$

and the joint distribution can be expressed as a product of one-step transition probabilities, as was to be shown.

3.6.2 Homogenous Markov Chain

Let $\eta_1, \eta_2, ..., \eta_m$ be independent and identically distributed as the discrete random variable η with probability mass function

$$\Pr[\eta = i] = a_i, \tag{3.58}$$

where $\sum_{i=1}^{i=\infty} a_i = 1$ and $a_i \geq 0$, $i = 1, 2, ...$

Consider the process $X(i) = \eta_i$; then the one-step transition probability matrix is

$$P = \begin{pmatrix} a_0, a_1, a_2, a_3, .. \\ a_0, a_1, a_2, a_3, .. \\ a_0, a_1, a_2, a_3, .. \\ . \\ . \\ \end{pmatrix}, \tag{3.59}$$

where each row is the same probability distribution, namely, that of the random variable η. This is a trivial example of a Markov chain since $X(i)$ is independent of $X(i+1)$. Another Markov process defined in terms of the random variables η is described in the following section.

Let $X(n) = \sum_{i=1}^{i=n} \eta_i$, $n = 1, 2, ...$; then this process is Markov with one-step transition probability

$$\begin{aligned}\Pr[X(n+1) = j | X(n) = i] &= a_{j-1}, \quad j \geq i \\ &= 0, \quad j < i.\end{aligned} \tag{3.60}$$

Thus, the transition probability matrix is

$$P = \begin{pmatrix} a_0, a_1, a_2, a_3, a_4, \ldots. \\ 0, a_0, a_1, a_2, a_3, \ldots.. \\ . \\ \\ . \\ \\ \end{pmatrix}. \tag{3.61}$$

One would have to modify this matrix if the index set is the set of all integers or if the index set is $\{0, 1, 2, \ldots, m\}$.

3.6.3 Embedded Markov Chain

A good example of a Markov chain is illustrated by the number of people in a queue waiting for service. Consider a theater with a single cashier and where the arrivals of customers occur according to a Poisson process. Also, assume that the service times of consecutive people in the queue are independent and identically distributed random events. For $n \geq 1$, let $X(n)$ be the number of people waiting in line for service when the nth person is to be served, then the sequence $\{X(n),\ n \geq 1\}$ is a Markov chain; that is, the conditional distribution of $X(n+1)$ given $X(1), X(2), \ldots, X(n)$ depends only on the most recent value of $X(n)$. Now let $W(n)$ be the number of customers joining the queue while the nth person is being served. Then it can be shown that

$$X(n+1) = X(n) - \delta(X(n)) + W(n+1), \tag{3.62}$$

where

$$\delta(x) = 1, \quad x = 1, \tag{3.63}$$

and $\delta(0) = 0$.

Equation 3.62 is interpreted as follows: The number of persons waiting for service when person $n+1$ leaves depends on whether that person was in line when the nth customer departed service. When $\delta(X(n)) = 0$, $X(n+1) = (n+1)$; otherwise, $X(n+1) = W(n+1) + X(n) - 1$.

Since $W(n+1)$ is independent of $X(1), X(2), \ldots, X(n)$, the implication is that given the value of $X(n)$, the values of the previous $n-1$ values of the process are not involved in determining the conditional distribution of $X(n+1)$. This chain is an example of an embedded chain in that it corresponds to the stochastic process $\{N(t),\ t \geq 0\}$, where $N(t)$ is the number of customers in line at time t and where the times $\{t_n\}$ are the corresponding arrival times, and the embedded chain can be written as $X(n) = N(t_n)$. See pages 190 and 191 of Parzen[4] for additional information about this queuing example.

3.6.4 Restricted Random Walk

This example is taken from pages 108 and 109 of Allen[2] and involves a process with state space $\{0, 1, 2, \ldots, N\}$, in which case there are two boundaries or with infinite and countable state space {0, 1, 2, ...} with boundary at 0. In such a model, the states are positions which are denoted by $X(n)$, where n is time. Let p be the probability of moving one unit to the right and $1 - p$ be the probability of one unit to the left; thus, the one-step transition probability is

$$\begin{aligned} \Pr[X(n+1) = j|X(n) = i] &= p, \quad j = i+1, \\ &= 1-p, \quad j = i-1 \end{aligned} \tag{3.64}$$

If the boundary at 0 is absorbing, $P_{00} = 1$. If the boundary 0 is reflecting,

$$P_{00} = 1 - p \tag{3.65}$$

and

$$P_{01} = p, \tag{3.66}$$

where $0 < p < 1$.

Thus, if 0 is an absorbing barrier, once the process reaches 0, it cannot leave 0. On the other hand, if the boundary 0 is reflecting, there is a positive probability of not staying at 0 and a positive probability of moving away from 0 by one unit. The boundary at 0 is elastic if

$$\begin{aligned} P_{12} &= p, \\ P_{11} &= sq, \\ p_{10} &= (1-s)q, \\ p+q = 1,& \quad P_{00} = 1, \\ 0 < p,& \quad s < 1 \end{aligned} \tag{3.67}$$

The elastic boundary is an intermediate situation between reflective and absorbing boundaries. If $s = 0$ and $p = 0$, then 1 is an absorbing boundary, but if $s = 1$, then 1 is a reflecting barrier, while finally, when $0 < s < 1$, an object moving toward the boundary from the position at 1 will reach 0 with probability $(1 - s)q$ or return to state 1 with probability $s(1 - p)$.

The random walk with absorbing boundaries is equivalent to the gambler's ruin on the state space $\{0,1,2,...,N\}$ and where $X(n) = x$ represents the capital of the gambler at the nth game. At the next game, the gambler can increase the capital by one unit to $x + 1$, or decrease it to $x - 1$, and when $X(n) = 0$, the gambler is ruined. The one-step transition matrix is

$$\begin{aligned} P_{ij} = \Pr[X(n+1) = j|X(n) = i] &= p, \quad j = i+1, \\ &= 1-p, \quad j = i-1, \end{aligned} \tag{3.68}$$

where is $0 < p < 1$ and where $i = 1, 2, ..., N - 1$. The two boundaries 0 and N are absorbing, that is, $P_{00} = 1$ and $P_{NN} = 1$.

Thus, the $N+1 \times N+1$ transition matrix is given by

$$P = \begin{pmatrix} 1,0,0,0,....0,0,0 \\ q,0,p,0,...0,0,0 \\ 0,q,0,p,0,....,0 \\ . \\ 0,0,0,0,..,0,p,0 \\ 0,0,0,0,..,q,0,p \\ 0,0,0,0,.....,0,1 \end{pmatrix} \tag{3.69}$$

and $q = 1 - p$. There are three communicating classes $\{0\}$, $\{1,2,...,N-1\}$, and $\{N\}$; that is, 0 communicates only with 0, N communicates only with N, and the remaining $N-1$ states communicate only with each other, but not with 0 or N.

There are many interesting aspects of the dynamics of the gambler's ruin problem, including (1) the probability the gambler either wins or loses all capital and (2) the average length of time in order to win or lose all capital. Suppose the gambler has to begin with total capital k; then the average duration of the game is

$$\mu = \mu_{ko} + \mu_{kN}, \tag{3.70}$$

where μ_{ko} is the average duration of the game until ruin, starting with capital k, and μ_{kN} is the average duration of the game until the gambler wins.

Let

$$a_{k0} = \Pr[X(n) = 0|X(0) = k] \tag{3.71}$$

be the probability of the gambler's loss of all capital given that initially, the capital of the gambler is k. In a similar fashion, let

$$b_{kN} = \Pr[X(n) = N|X(0) = k] \tag{3.72}$$

be the probability the gambler wins the game, given that they have initial capital k. Recall that if the state is 0, the gambler loses, but if the state is N, the gambler wins.

Note that the sequence $a_{kn} + b_{kn}$ $n = 0,1,2,...$ is the probability of absorption at trial n; hence,

$$\sum_{n=0}^{n=\infty} (a_{kn} + b_{kn}) = 1 \tag{3.73}$$

for $1 \le k \le N-1$.

The generating function technique is used to find the average duration until the gambler's worth is 0 or if the gambler wins. For further information about the generating function technique, see pages 6–9 of Bailey.[7]

Consider the two functions for the sequences a_{kn} and b_{kn}, $n = 0, 1, 2, \ldots$, namely,

$$A_k(t) = \sum_{n=0}^{n=\infty} a_{kn} t^n \tag{3.74}$$

and

$$B_k(t) = \sum_{n=0}^{n=\infty} b_{kn} t^n, \tag{3.75}$$

respectively, for $|t| \leq 1$. Thus, their sum

$$A_k(t) + B_k(t) = \sum_{n=0}^{n=\infty} (a_{kn} + b_{kn}) t^n \tag{3.76}$$

is the probability-generating function for the sequence $a_{kn} + b_{kn}$, $n = 0, 1, 2, \ldots$. Note that the nth term of this sequence is the probability of absorption at time n.

Let

$$a_k = A_k(1) \tag{3.77}$$

and

$$b_k = B_k(1); \tag{3.78}$$

then a_k is the probability of the gambler's ruin with beginning capital k, while b_k is the probability of winning all the pot, beginning with k as the capital.

Finally, let

$$\zeta_k = A_1'(\mathrm{t}) + A_2'(t) = \sum_{n=0}^{n=\infty} n(a_{kn} + b_{kN}) \tag{3.79}$$

be the mean duration of the games beginning with capital k, and let T_k be the random variable denoting the mean time to absorption; then obviously

$$\zeta_k = E(T_k) \tag{3.80}$$

for $1 \leq k \leq N - 1$. The primes in Equation 3.79 denote the derivatives with respect to t.

The main concern at this point is to determine the probability of absorption, and two approaches will be presented. The first is via a linear difference equation that relates a_{k-1}, a_k, and a_{k+1}, and the second is a numerical approach based on the transition matrix.

Initially, consider the difference equation for the probability of ruin when the gambler has capital k; they either win or lose the next game with probabilities p and q, respectively. When the gambler wins the next game, the capital is $k + 1$ with probability a_{k+1}, while if they lose, the resulting capital is $k - 1$ with probability a_{k-1}; hence, it follows that the fundamental difference equation is

$$a_k = pa_{k+1} + qa_{k-1}, \quad 1 \le k \le N - 1 \tag{3.81}$$

In order to solve this equation, the boundary conditions are $a_0 = 1$ and $a_N = 0$, thus if the capital is zero, the probability of ruin is 1, but if the gambler's capital is N, the probability of ruin is 0.

The method of characteristic equations is used to solve the difference equation in Equation 3.81, thus, let

$$a_k = \gamma^k \neq 0 \tag{3.82}$$

and substitute this value into Equation 3.81, which results in the characteristic equation

$$p\gamma^2 - \gamma + q = 0, \tag{3.83}$$

which has the two roots

$$\gamma_{1,2} = (1 \pm |p - q|)/2p. \tag{3.84}$$

Two cases are considered: (1) $p \neq q$ and (2) $p = q = 1/2$. In the first case, the roots of Equation 3.83 are $\gamma = 1$ and $\gamma = q/p$, and the general solution to the difference equation is

$$a_k = d_1 + d_2(q/p)^k, \tag{3.85}$$

where d_1 and d_2 are determined by imposing the boundary conditions; then it can be shown that the general solution is

$$a_k = \left[(q/p)^N - (q/p)^k\right] / \left[(q/p)^N - 1\right], \quad p \neq q. \tag{3.86}$$

It also follows that the general solution for the probability of winning with capital k is

$$b_k = \left[(q/p)^k - 1\right] / \left[(q/p)^N - 1\right], \quad p \neq q. \tag{3.87}$$

Finally, for the case $p = q = 1/2$, note that the characteristic equation in Equation 3.83 has one root 1 of order 2; thus, the solution to the difference equation for the probability of ruin (with initial capital k) is

$$a_k = (N - k)/N, \quad p = 1/2. \tag{3.88}$$

And the probability of winning the total capital N is

$$b_k = k/N, \quad p = 1/2. \tag{3.89}$$

TABLE 3.1

Gambler's Ruin with Capital $k = 50$ and Total Capital $N = 100$

Probability	a_{50}	b_{50}	ζ_{50}	$A'_{50}(1)$	$B'_{50}(1)$
$Q = .50$	0.5	0.5	2500	1250	1250
$q = .51$	.880825	.119175	1904	1677	227
$q = .55$	.999956	.000044	500	499.93	.07
$q = .60$	1.000	.0000	250	250	0

Source: Allen, L. J. S., *An Introduction to Stochastic Processes with Applications to Biology, Second Edition*, CRC Press, Boca Raton, FL, 2011, p. 113.

Consider the probability of ruin when the total capital is N and the initial capital is k, and suppose $N = \$100$ and $k = \$40$; then the probability of ruin is ½, which is the same as the probability of winning.

Refer to Equation 3.80 for the mean duration $\zeta_k = E(T_k)$; then we use the difference equation technique to compute the mean duration time. The difference equation is

$$\zeta_k = p(1 + \zeta_{k+1}) + q(1 + \zeta_{k-1}), \quad k = 1, 2, \ldots, N - 1, \tag{3.90}$$

but since $p + q = 1$ can be expressed as

$$p\zeta_{k+1} - \zeta_k + q\zeta_{k-1} = -1, \tag{3.91}$$

it can be shown that the general solution is

$$\zeta_k = (1/(|q - p|))[k - N[(1 - (q/p)^k)/(1 - (q/p)^N], \quad p \neq q, \tag{3.92}$$

and when $p = q = 1/2$,

$$\zeta_k = k(N - k). \tag{3.93}$$

Table 3.1 portrays the probability of ruin, the probability of winning, and the average duration of the game. Refer to Equations 3.86 and 3.87 for computing a_{50}, while for b_{50}, refer to Equations 3.45 and 3.48. For the average duration ζ_{50}, refer to Equations 3.51 and 3.52. The last two columns are the additive components of ζ_{50} given by Equations 3.35 and 3.36.

Of course, the statistical problem is to observe the gambler playing a game and estimating the probability p of winning a particular game. This will be presented in Chapter 4.

3.7 Continuous Markov Chains

3.7.1 Introduction

Continuous Markov chains have the parameter space $T = \{t : t \geq 0\}$, the nonnegative numbers, and the state space $S = \{0, 1, 2, \ldots\}$ and constitute an important class of stochastic processes. Several examples were introduced in Section 2.7.3. In the Poisson process, $N(t)$

represents the number of arrivals at time t recorded over the interval $[0, t]$ and evolves from state n to state $n + 1$, for $n \geq 0$ and $N(0) = 0$. This is an example of a pure birth process since the state of the system increases from n by one unit. This is a special case of the general birth and death process, which can either increase from n to $n + 1$ or decrease from n to $n - 1$, where the change from n to $n + 1$ is a birth and the transition from n to $n - 1$ is interpreted as a death. In what is to follow, continuous-time Markov processes are defined and the relation to discrete-type Markov chains is explained. This is followed by an illustration of general birth and death processes, and then continuous chains are determined by a system of differential equations that indeed characterize the probabilistic properties of a continuous stochastic process with the Markov property. As with discrete Markov chains, limiting long-run probabilities of the states of the system are described, and the section is finalized with some queuing examples.

Let $\{X(t), t \geq 0\}$ be a continuous-time stochastic process; then the process is said to be a continuous-time Markov chain (CTMC) if

$$\begin{aligned} &\Pr[X(t+s) = j | X(s) = i, X(v) = x(v), 0 \leq v < s] \\ &= \Pr[X(t+s) = j | X(s) = i] \end{aligned} \tag{3.94}$$

for all s and $t \geq 0$ and nonnegative integers $i, j, x(v), 0 \leq v < s$. How is Equation 3.94 interpreted? A stochastic process has a Markovian property if the conditional probability of the future state $X(t + s)$ given the present state $X(s)$ and past states $X(v)$, $0 \leq v < s$ depends only on the present state (and not on the past). Our study of these Markov chains will be confined to stationary processes, namely, those where $\Pr[X(t+s) = j | X(s) = i]$ is independent of s.

Associated with the continuous-time process is the random time between events thus, suppose at time 0, the process is observed for 15 minutes, at which time the first event i occurs. But since this is a Markov process, what is the probability that it remains in that state for the next 26 minutes before the next event occurs? Thus, let T_1 denote the amount of time the process stays in state i then

$$\Pr[T_1 > 26 | T_1 > 15] = P[T_1 > 11]. \tag{3.95}$$

Therefore, in general,

$$\Pr[T_1 > s + t | T_1 > s] = P[T_1 > t], \quad t \geq 0, s \geq 0, \tag{3.96}$$

and the interarrival time is said to be memoryless, and in fact, it can be shown that T_1 has an exponential distribution. Another way to define a CTMC is to specify the probabilistic properties of the process at which time it enters state i:

1. The amount of time it spends in state i before making a transition into a different state is exponentially distributed with parameter λ_i (with mean $1/\lambda_i$).
2. When the process leaves state i, it enters the next state j with some probability P_{ij}, where $P_{ii} = 0$ and $\sum_j P_{ij} = 1$.

That is to say, a CTMC is a stochastic process that moves from state to state in the manner of a DTMC; however, the time it spends in each state before entering the next state has an exponential distribution. Also, it can be seen that the interarrival times between events are independent random variables.

3.7.2 Birth and Death Processes

One of the most studied continuous-time processes is the birth and death process. See, for example, page 62 of Chiang,[8] who describes birth and death processes with application to medicine and biology. Suppose there are n items in the system at time t, denoted by $X(t) = n$ and suppose (1) new arrivals enter the system at an exponential rate γ_n and (2) items leave the system at an exponential rate δ_n; that is to say, when there are n items in the system, then the time until the next arrival is exponentially distributed with mean $1/\gamma_n$, and the time duration is independent of the time until the next departure, which has an exponential distribution with mean $1/\delta_n$. This describes a CTMC with state space $S = \{0, 1, 2, \ldots\}$, where transitions from state n goes either to $n + 1$ corresponding to a birth (or arrival) or to state $n - 1$ with parameters $\{\gamma_n : n = 0, \ldots\}$ and $\{\delta_n : n = 0, \ldots\}$ for the arrival and departure times, respectively. There are several important relations between the birth and death rates and the transition probabilities. Let

$$
\begin{aligned}
&v_0 = \gamma_0, \\
&v_i = \gamma_i + \delta_i, \quad i > 0, \\
&P_{01} = 1, \\
&P_{i,i+1} = \gamma_i/(\gamma_i + \delta_i), \quad i > 0, \\
&\text{and} \\
&P_{i,i-1} = \delta_i/(\gamma_i + \delta_i), \quad i > 0
\end{aligned}
\tag{3.97}
$$

This follows because if there are i items in the system, then the next state is $i + 1$ if a birth occurs before a death; and the probability that an exponential random variable with rate γ_i will happen earlier than an exponentially distributed random variable with rate δ_i is $\delta_i/(\gamma_i + \delta_i)$, and the time until one or the other occurs is $v_i = \gamma_i + \delta_i$. Of course, a special case of the birth and death process is the Poisson process with exponential rates $\gamma_n = 0$, $n \geq 0$, and $\delta_n = 0$, $n \geq 0$, which is the case where there are only arrivals but no departures. The Poisson process is also referred to as a pure birth process.

Another special case of a birth and death process is a pure birth process with a linear birth rate (sometimes called the Yule process) with $\gamma_n = n\gamma$. Still another example of the same time of process is a linear growth model with immigration. The following description is from pages 259 and 260 of Ross.[6] For such a situation, the immigration model is defined as a birth and death process where $\gamma_n = n\gamma$ and $\delta_n = n\delta + \phi$, $n \geq 0$. Such processes are employed to describe biological reproduction and population growth, where each individual in the population gives birth at an exponential rate δ, and in addition, there is an additive contribution due to immigration occurring at an exponential rate ϕ; thus, the total birth rate is $n\delta + \phi$. Departures occur at an exponential rate γ for each item in the population. It is interesting to derive the average size at time t; thus, let the mean be

$$\mu(t) = E[X(t)], \tag{3.98}$$

where $X(t)$ is the population size at time t. A differential equation will be developed to derive an expression for $\mu(t)$ that obviously depends on γ, δ, and ϕ. To do this, let $h > 0$ and consider

$$\begin{aligned}\mu(t+h) &= E[X(t+h)] \\ &= EE[X(t+h)|X(t)],\end{aligned} \tag{3.99}$$

where the outer expectation is with respect to the distribution of $X(t)$ and the inner is with respect to the conditional distribution of $X(t+h)$ given $X(t)$. Given the population size at time t, the population at time $t+h$ will either increase by one item if an arrival occurs or an immigration occurs in $(t, t+h)$. Also, if a death happens, the population will decrease by one in the interval $(t, t+h)$. Also, the population size does not change if neither of the two preceding possibilities occurs. To describe this in symbols, it is stated as

$$\begin{aligned}&X(t+h) \\ &= X(t)+1 \quad \text{with probability } [\phi + X(t)\gamma]h + o(h), \\ &= X(t)-1 \quad \text{with probability } X(t)\delta h + o(h), \\ &= X(t) \quad \text{with probability } [1-\phi + X(t)\gamma + X(t)\delta]h + o(h).\end{aligned} \tag{3.100}$$

From Equation 3.100, it follows that

$$E[X(t+h)|X(t)] = X(t) + [\phi + X(t)\gamma + X(t)\delta]]h + o(h).$$

The preceding expectation with respect to the distribution of $X(t)$ is

$$\mu(t+h) = \mu(t) + [\gamma - \delta]\mu(t)h + \phi h + o(h). \tag{3.101}$$

This will be put into the form of a derivative as

$$[\mu(t+h) - \mu(t)]/h = [\gamma - \delta]\mu(t) + \phi + o(h)/h,$$

and taking the limit as $h \to 0$ yields the differential equation

$$(d/dt)\mu(t) = [\gamma - \delta]\mu(t) + \phi, \tag{3.102}$$

and from this it can be shown that the solution is

$$\mu(t) = [\phi/(\gamma - \delta)]\left[e^{(\gamma-\delta)} - 1\right] + ie(\gamma-\delta)t, \quad \gamma \neq \delta, \tag{3.103}$$

and when $\gamma = \delta$, the solution is

$$\mu(t) = \phi t + i. \tag{3.104}$$

3.7.3 Kolmogorov Equations

Consider two states i and j of a continuous Markov chain, and let P_{ij} be the corresponding transition probability and suppose ω_i is the rate of the process when it makes a transition while in state i; then

$$q_{ij} = \omega_i P_{ij} \tag{3.105}$$

is the rate of the process when making the transition from i to j and is called the instantaneous transition rate. Our goal is to express the probabilistic properties of the process in terms of the instantaneous transition rates and the transition probabilities.

Consider the probability that a chain presently in state i will be in the future in state j, namely,

$$P_{ij}(t) = \Pr[X(t+s) = j|X(s) = i], \tag{3.106}$$

then the main question is, how does one express Equation 3.106 in terms of the quantities involving Equation 3.105? To answer this question, consider the following three conditions about the transition probabilities:

1. $\lim[1 - P_{ij}(h)]/h = \omega_i$ as $h \to 0$.
2. $\lim P_{ij}(h)/h = q_{ij}$ as $h \to 0$. (3.107)
3. For all $s,t \geq 0$, $P_{ij}(t+s) = \sum_{k=0}^{k=\infty} P_{ik}(t)P_{kj}(s)$.

These limits are needed to set up the differential equations that will evolve into the Kolmogorov differential equations. For the proof of Equation 3.107, see pages 266 and 267 of Ross.[6] The set of equations found in part 3 of Equation 3.107 are called the Chapman–Kolmogorov equations, and from them, it follows that

$$\begin{aligned} P_{ij}(t+h) - P_{ij}(t) &= \sum_{k=0}^{k=\infty} P_{ik}(h)P_{kj}(t) - P_{ij}(t) \\ &= \sum_{k=0}^{k=\infty} P_{ik}(h)P_{kj}(t) - [1 - P_{ii}(h)]P_{ij}(t); \end{aligned}$$

dividing by h and setting up the difference quotient gives

$$\lim\left[P_{ij}(t+h) - P_{ij}(t)\right]/h$$

$$= \lim\left\{\sum_{k \neq i} [P_{ik}(h)/h]\right\} P_{kj}(t) - [[1 - P_{ik}(h)]/h]P_{ij}(t),$$

and taking the limit as $h \to 0$ results in the differential equations (which holds for all states i and j and times ≥ 0) called the Kolmogorov backward equations:

$$(d/dt)P_{ij}(t) = \sum_{k \neq j} q_{ik}P_{kj}(t) - \omega_i P_{ij}(t). \tag{3.108}$$

What are the Kolmogorov equations for the pure birth process and the birth and death process?

For a birth and death process, the Kolmogorov equations in Equation 3.108 reduce to

$$(d/dt)P_{0j}(t) = \gamma_0 P_{1j}(t) - \gamma_0 P_{oj}(t)$$

and

$$(d/dt)P_{ij}(t)$$
$$= (\gamma_i + \delta_i)\left[\gamma_i/(\gamma_i + \delta_i)P_{i+1,j}(t) + \delta_i/(\gamma_i + \delta_i)P_{i-1,j}(t)\right] - (\gamma_i + \delta_i)P_{ij}(t),$$

and these two equations are equivalent to

$$(d/dt)P_{oj}(t) = \gamma_0\left[P_{1j}(t) - P_{0j}(t)\right]$$

and (3.109)

$$(d/dt)P_{ij}(t) = \gamma_i P_{i+1,j}(t) + \delta_i P_{i-1,j}(t) - (\gamma_i + \delta_i)P_{ij}(t)$$

for $i > 0$.

These equations will be applied to a special case of the birth and death process with two states.

Consider a continuous-time birth and death process with two states. A machine works for an exponential time with mean $1/\gamma$ before failing; then, it takes a random amount of time, with an exponential distribution with mean of $1/\delta$, to be repaired. If the machine is working, what is the chance the machine will be working at time $t = 20$? Let the two states be 0 and 1, where if it is in state 0, it is working, while if the state is 1, the machine is being repaired. That is, we want to determine $P_{00}(20)$. The relevant parameters are $\gamma_0 = \gamma$, $\delta_1 = \delta$, $\gamma_i = 0\,(i \neq 0)$, and $\delta_i = 0\,(i \neq 1)$. Using the differential equations in Equation 3.109 for the general birth and death process, an expression for $P_{00}(20)$ will be derived.

It follows that

$$(d/dt)P_{00}(t) = \gamma[P_{10}(t) - P_{00}(t)]$$

and (3.110)

$$(d/dt)P_{10}(t) = \delta[P_{00}(t) - P_{10}(t)].$$

Using the initial condition $P_{00}(0) = 1$, one may show that the solutions to the Kolmogorov equations (Equation 3.110) are

$$P_{00}(t) = [\gamma/(\gamma+\delta)]\exp-(\gamma+\delta)t + \delta/(\gamma+\delta)$$

and (3.111)

$$P_{10}(t) = \delta/(\delta+\gamma) - \delta/(\delta+\gamma)\exp-(\delta+\gamma)t.$$

Thus, to find $P_{00}(20)$, one must have value for the rates of the machine working and the repair times of the machine. Of course, from a statistical point of view, by observing the working and repair times of the machine, one can estimate the rates and the probabilities of Equation 3.111.

3.7.4 Limiting Probabilities

This last section on CTMCs emphasizes the limiting probabilities

$$P_j = \lim P_{ij}(t), \quad t \to \infty,$$

assuming the limit exists and is independent of the state i. Similar to the discrete-time process, we are interested in determining the solutions P_j to a set of equations. The appropriate set of equations will be based on the Chapman–Kolmogorov equations of the previous section. These equations are

$$(d/dt)P_{ij}(t) = \sum_{k\neq j} q_{kj}P_{ik}(t) - \omega_j P_{ij}(t), \tag{3.112}$$

and taking the limit of both sides of Equation 3.112, namely,

$$\lim(d/dt)P_{ij}(t) = \lim\left[\sum_{k\neq j} q_{kj}P_{ik}(t) - \omega_j P_{ij}(t)\right], \quad t\to\infty,$$
$$= \sum_{k\neq j} q_{kj}P_k - \omega_j P_j.$$

The limits in Equation 3.113 exist because the transition probabilities are $P_{ij}(t)$ and are bounded over [0,1].

Thus, the relevant set of equations is

$$\omega_j P_j = \sum_{k\neq j} q_{kj}P_k, \quad \forall j \tag{3.113}$$

and are solved for the limiting probabilities P_j, subject to the constraint $\sum_j P_j = 1$.

Consider the limiting probabilities of the birth and death process of Section 3.7.2 and the defining quantities given by Equation 3.97. We consider the special case by

TABLE 3.2

Birth and Death Process: Limiting Probabilities

State	Rate of Leaving = Entering Rate
0	$\gamma_0 P_0 = \delta_1 P_1$
1	$(\gamma_1 + \delta_1)P_1 = \delta_2 P_2 + \gamma_0 P_0$
2	$(\gamma_2 + \delta_2)P_2 = \delta_3 P_3 + \gamma_1 P_1$
$n \geq 1$	$(\gamma_n + \delta_n)P_n = \delta_{n+1} P_{n+1} + \gamma_{n-1} P_{n-1}$

equating the rate at which the chain leaves a state with the rate at which it enters a state (Table 3.2).

Add to each equation of Table 3.2 the preceding equation, which yields the system

$$\begin{aligned} \gamma_0 P_0 &= \delta_1 P_1, \\ \gamma_1 P_1 &= \delta_2 P_2, \\ &\vdots \\ \gamma_n P_n &= \delta_{n+1} P_{n+1}, \quad n \geq 0. \end{aligned} \tag{3.114}$$

This system is equivalent to the following system:

$$P_1 = (\gamma_0/\delta_0)P_0,$$

$$P_2 = (\gamma_1/\delta_1)P_1 = (\gamma_0\gamma_1/\delta_0\delta_1)P_0,$$

and, in general,

$$P_n = (\gamma_0\gamma_1 \ldots \gamma_{n-1}/\delta_0\delta_1 \ldots \delta_{n-1})P_0,$$

and using the constraint $\sum_{j=0}^{j=\infty} P_j = 1$ implies that the solution is

$$P_0 = \left[1\Big/\left[1 + \sum_{n=1}^{n=\infty}(\gamma_0\gamma_1 \ldots \gamma_{n-1})/(\delta_0\delta_1 \ldots \delta_{n-1})\right]\right],$$

and in general, for $n \geq 1$, the solution for the limiting probabilities is

$$P_n = [\gamma_0\gamma_1 .. \gamma_{n-1}]\Big/\left\{(\delta_1\delta_2 .. \delta_n)\left[1 + \sum_{n=1}^{n=\infty}(\gamma_0\gamma_1 \ldots \gamma_{n-1})/(\delta_0\delta_1 .. \delta_{n-1})\right]\right\}. \tag{3.115}$$

Thus, if the entry and departure rates are known, the preceding equations determine the limiting probabilities.

3.8 Normal Processes

3.8.1 Introduction

This section takes us to the final class of Markov processes where the state and index sets are both continuous and the emphasis will be on investigating the Brownian motion. The Brownian motion is a normal process and plays an important role in the study of physics, economics, communication theory, biology, and other scientific endeavors. This process has been briefly introduced in Section 2.7.4, where Bayesian inference was employed to estimate the variance parameter of the Wiener process and, because of its importance, will be the focus of more advanced Bayesian techniques in Chapter 6. As a biological mechanism, the Brownian motion was discovered by Brown in 1827, and later a mathematical description of the process was presented by Einstein in 1905. Since then, the Brownian motion has been extensively studied by many, including Wiener[9] in his 1918 dissertation.

3.8.2 Brownian Motion

As stated earlier, the Brownian motion is an example of a Markov process with continuous state and index sets. Let $X(t)$ be the x-component of a particle undergoing Brownian motion, and let x_0 be the position of the particle at time t_0, that is, $X(t_0) = x_0$. Suppose $p(x, t|x_0)$ denotes the conditional probability density of $X(t + t_0)$ given $X(t_0) = x_0$; thus, since p is a density, it follows that $p(x, t|x_0) \geq 0$ and

$$\int_{-\infty}^{\infty} p(x, t|x_o)dx = 1. \tag{3.116}$$

It will be shown that the transition probabilities in going from one state to the other are stationary. Also, we will assume that for small h, $X(h + t_0)$ is close to $X(t_0) = x$, that is,

$$\lim p(x, h|x_o) = 0, \quad x \neq x_0, h \to 0 \tag{3.117}$$

It was demonstrated by Einstein that the density $p(x, t|x_o)$ is a solution to the partial differential equation

$$(\partial / \partial t)p = D(\partial^2 / \partial^2 x^2)p, \tag{3.118}$$

the so-called diffusion equation, with diffusion coefficient D, and

$$D = 2RT/Nf, \tag{3.119}$$

where R is the gas constant, T is the temperature, N is Avogadro's number, and f is the coefficient of friction. It can be shown that $D = 1/2$, if the proper units are chosen for the terms in the diffusion equation (Equation 3.119). According to pages 341–343 of Karlin and Taylor,[1] the solution to the partial differential equation (Equation 3.118) is

$$p(x,t|x_0) = \left(1/\sqrt{2\pi}\right)\exp-(1/2t)(x-x_0), \tag{3.120}$$

which is a normal density with mean x_0 and variance $2t$.

Another way to derive the normal density for the conditional probability (Equation 3.120) is as an approximation based on the symmetric random walk for a discrete Markov chain. See page 342 of Karlin and Taylor[1] for details. The definition for the Brownian motion is given by the following three properties:

1. Every increment $X(t+s)-X(s)$ has a normal distribution with mean 0 and variance $\sigma^2 t$.
2. For each pair of disjoint intervals $[t_1,t_2]$ and $[t_3,t_4]$, the increments $X(t_2)-X(t_1)$ and $X(t_4)-X(t_3)$ are independent.
3. $X(0)=0$ and $X(t)$ are continuous.

From this, one may show that for n and time points satisfying $t > t_0 > t_1 > \ldots > t_n$,

$$\begin{aligned} &\Pr[X(t) \le x|X(t_0)=x_0,\ldots,X(t_n)=x_n] \\ &= \Pr[X(t) \le x|X(t_0)=x_0]. \end{aligned} \tag{3.121}$$

Thus, the three postulates of the definition of the Brownian motion imply that the process $\{X(t),\ t \ge 0\}$ satisfies the Markov property in Equation 3.121. Under the condition that $X(t)=0$, the variance of $X(t)$ is $\sigma^2 t$, and σ^2 is called the variance parameter of the process. Note that the joint distribution of $X(t_1), X(t_2), \ldots, X(t_n)$ can be found via transformation of the n increments

$$X(t_n)-X(t_{n-1}), X(t_{n-1})-X(t_{n-2}), \ldots, X(t_2)-X(t_1), X(t_1).$$

But we know that the distribution of each increment is normal with mean 0 and variance $\sigma(t_i - t_{n-1})$, and $X(t)=0$, which is enough information to find the joint distribution of $X(t_i)$, $i=1,2,\ldots,n$. When $\sigma^2=1$, the process is referred to as the standard Brownian motion.

There are several interesting generalizations of the Brownian motion, including the Brownian motion with drift, defined as follows:

1. $X(t+s)-X(s) \sim N(\mu t, \sigma^2 t)$, where μ and σ^2 are constants.
2. For each pair of disjoint intervals $[t_1,t_2]$and $[t_3,t_4]$, the increments $X(t_2)-X(t_1)$ and $X(t_4)-X(t_3)$ are independent.
3. $X(0)=0$ and $X(t)$ are continuous.

Obviously, the Brownian motion with drift satisfies the Markov property. From a statistical viewpoint, realizations of this process would be used to make inferences about the mean μ and σ^2. For our purposes, Bayesian estimates and tests of hypotheses about these parameters are the subjects of Chapter 6. Another interesting generalization of Brownian motion is the geometric Brownian motion. Let $\{X(t),\ t \ge 0\}$ be the Brownian motion with drift μ and diffusion coefficient σ^2; then the process

$$Z(t) = e^{X(t)},\ t \geq 0 \tag{3.122}$$

is referred to as the geometric Brownian motion with state space $(0, \infty)$. It can be shown that the mean and variance of $Z(t) = Z(0)e^{X(t)-X(0)}$ are

$$E[Z(t)|Z(0) = z_o] = z_0 \exp\{t(\mu + \sigma^2/2)\}$$

and (3.123)

$$\text{Var}[Z(t)|Z(t_0) = z_0] = z_0^2 \exp[2t(\mu + \sigma^2/2)][\exp(t\sigma^2) - 1],$$

respectively.

An interesting version of the Brownian motion is the Ornstein–Uhlenbeck[10] process, given by a normal process with mean 0 and covariance

$$\text{Cov}[X(s), X(t)] = \alpha \exp -\beta|s - t|, \tag{3.124}$$

where α and β are unknown parameters.

3.8.3 Normal Processes and Covariance Stationary Processes

When working with normal stochastic processes, it is important to know that the process is completely determined by the mean function and covariance kernel of the process. Recall from the distribution theory that a finite realization from a normal process has a multivariate normal distribution, with a given mean vector and variance–covariance matrix.

Let $\{X(t), t \in T\}$ be a stochastic process with finite second moments; then the mean value function is

$$m(t) = E[X(t)], \quad t \in T, \tag{3.125}$$

and the covariance kernel is

$$K(s,t) = \text{Cov}[X(t), X(s)], \quad s,t \in T. \tag{3.126}$$

As our first example, consider the process defined by

$$X(t) = X_0 + Vt, \tag{3.127}$$

where $X(t)$ is the position of a body in motion, with X_0 and V representing the initial position and velocity, respectively. What are the mean value function and covariance kernel of this process? They are

$$m(t) = E[X_0] + tE[V]$$

and

$$K(s,t) = \text{Var}[X_0] + (s + t)\text{Cov}[X_0, V] + st\text{Var}[V].$$

Recall for the Brownian motion that the mean of $X(t)$ is 0 and the variance is $\sigma^2 t$; but what is its covariance kernel? It can be shown to be

$$K(s,t) = \text{Cov}[X(s), X(t)] = \text{Cov}[X(s), X(t) + X(s) - X(s)],$$

but because of independent increments,

$$\begin{aligned} K(s,t) &= \text{Cov}[X(s), X(t) - X(s)] + \text{Cov}[X(s), X(s)] \\ &= \text{Var}[X(s)] = \sigma^2 s. \end{aligned} \tag{3.128}$$

For a Poisson process $\{N(t), t \geq 0\}$ with parameter λ, the mean value function and variance are $m(t) = \lambda t$ and $\text{Var}[X(t)] = \lambda t$, respectively. What is the covariance kernel? It can be shown that

$$K(s,t) = \text{Cov}[N(s), N(t)] = \lambda \min(s,t). \tag{3.129}$$

3.8.4 Stationary and Evolutionary Processes

The concept of a stationary process was introduced for DTMCs and will be generalized to Markov processes with continuous state and index sets. When a process is stationary, it conveys the stability of the probabilistic properties of the process over time. On the other hand, with an evolutionary process, the probabilistic characteristics (the joint distribution) change over time. The Poisson process is an example of an evolutionary process because its mean value function increases over time. In order to define *stationary*, the index set T is usually restricted to be linear, that is, when it is closed under addition. Thus, the index set $T = \{1, 2, ...\}$ is linear since the sum of positive integers is a positive integer, as is index set $T = \{t : t \geq 0\}$. A stochastic process with a linear index set is said to be stationary of order m if for all k time points $t_i \in T$, $i = 1, 2, ..., m$,

$$X(t_1), X(t_2), ..., X(t_m) \sim X(t_1 + h), X(t_2 + h), ..., X(t_m + h). \tag{3.130}$$

Thus, the joint distribution of the m random variables remains the same when the time is shifted by an amount h for all h.

If for all m = 1, 2, ... the process is stationary of order m, then the process is said to be strictly stationary. Often, it is difficult to verify that a given process is strictly stationary, because stationarity has to be verified for all positive integers.

A weaker type of stationarity is called covariance stationarity and is described as follows: It is assumed that the process has finite second moments; then it is called covariance stationary if the covariance kernel $K(s,t)$ is a function only of $|s - t|$, $s, t \in T$, where T is a linear index set. It thus follows that there exists a function R (called the covariance function) such that for all s and t, $K(s,t) = R(s - t)$ or, equivalently,

$$\text{Cov}[X(s), X(s + h)] = R(h), \tag{3.131}$$

where s and h belong to the index set T of the process. The idea of stationarity is important because for such processes, ergodic theorems were first proved. Ergodic theorems show that the sample means of realizations from stationary processes have desirable frequency properties and, hence, properties that imply good sampling properties for Bayesian estimates of certain unknown parameters of a stationary stochastic process.

Suppose $\{X(t), t \geq 0\}$ is a stochastic process with mean value function $m(t) = E[X(t)]$ and covariance function $K(s,t) = \text{Cov}[X(s), X(t)]$, and suppose $[X(t), 0 \leq t \leq c]$ is a finite record or realization. Under what conditions do these observations provide optimal properties for estimating, say, the mean value function $m(t)$? Birkhoff[11] and Von Neumann[12] were among the first to provide satisfactory proofs for ergodic theorems (about ensemble averages). In the proof by Birkhoff, it was assumed that the process $\{X(t), t = 1, 2, \ldots\}$ is strictly stationary; then for any function g where the ensemble average

$$E[g(X(t))] \tag{3.132}$$

exists, the corresponding sample average is

$$(1/T)\sum_{i=1}^{i=T} g[X(i)], \tag{3.133}$$

where $T \to \infty$ with a probability of 1, and $\{X(t), t = 1, 2, \ldots, T\}$ is a realization of size T observed of the process. The ergodic theorems based on strictly stationary processes is beyond the scope of this book; however, according to Parzen,[4] a process need not be strictly stationary in order to provide sample statistics that have optimal frequency properties.

Suppose

$$M_T = (1/T)\sum_{i=1}^{i=T} X(i) \tag{3.134}$$

is a sequence of sample means observed from the discrete parameter process

$$\{X(t), t = 1, 2, \ldots\}; \tag{3.135}$$

then the sequence of sample means is said to be ergodic if

$$\lim \text{Var}[M_T] = 0, T \to \infty. \tag{3.136}$$

How is this version of ergodic interpreted? Equation 3.136 implies that the successive sample means in Equation 3.134, formed from a sample function of the process, have variances which tend to be 0 as the sample size T becomes larger. In a sense, this implies that the sample mean is approximately the same as the ensemble mean. The following theorem from pages 74 and 75 of Parzen[4] provides necessary and sufficient conditions for the sample means to be ergodic:

Let $\{X(t), t = 1, 2, \ldots\}$ be a stochastic process with covariance kernel $K(s,t)$ where

$$\text{Var}[X(t)] = K(t,t) \leq c_0, \quad t = 1, 2, \ldots,$$

and let

$$c(t) = \mathrm{Cov}[X(t), M_t]$$

be the covariance between the sample mean and the tth observation; then in order for

$$\lim \mathrm{Var}[M_T] = 0, \quad T \to \infty,$$

it is necessary and sufficient that

$$\lim c(t) = 0, \quad t \to \infty.$$

Thus, the sample means are ergodic if and only if the correlation between the sample mean and the last observation becomes smaller and smaller as the sample size increases. For additional details about this ergodic theorem, see pages 74 and 75 of Parzen.[4]

3.8.5 Stochastic Calculus

This is the last section of the chapter, but it is imperative that stochastic integration and differentiation be described because they are important ways to transform a stochastic process to another process. For example, consider a process that models the random position of a particle over time; then the corresponding integrated process would describe the velocity of the particle over time, etc. A good example of this is the Brownian motion, which was described in the preceding section.

Suppose a continuous process $\{X(t),\ t \geq 0\}$ is continuously observed over the interval $[0, T]$; then the sample mean is

$$M_T = (1/T)\int_0^T X(t)dt. \tag{3.137}$$

However, this integral needs to be defined! Similar to the definition of the ordinary Riemann integral, a natural way of defining $\int_a^b X(t)dt$ as a limit of approximating sums $\sum_{i=1}^{i=n} X(t_i)(t_i - t_{i-1})$, where the limit is taken over partitions of the interval $(a, b]$ into subintervals by the points $a = t_0 < t_1 < \ldots < t_n = b$ in such a way that the maximum length among the intervals $(t_i - t_{i-1})$, $i = 1, 2, \ldots, n$. For the purpose of this book, the type of convergence is taken to be convergence in mean square, which is defined as follows:

A sequence of random variables Y_i, $i = 1, 2, \ldots$ is said to converge in mean square to the random variable Y if

$$\lim E\left[|Y_n - Y|^2\right] = 0,\ n \to \infty. \tag{3.138}$$

A necessary and sufficient condition for convergence in mean square is based on the product moment $E[X(s), X(t)]$ considered as a function of s and t over the region $S = [a,b]x[a,b]$. If the product moment has a Riemann integral over S, then it can be shown that the stochastic integral $\int_a^b X(t)dt$ exists in the sense of convergence in mean square of the approximating sums described earlier. Suppose that the process has a mean value function $m(t)$ and covariance kernel $K(s,t)$; then it can be shown that

$$
\begin{aligned}
&1.\ E\left[\int_a^b X(t)dt\right] = \int_a^b m(t)dt \\
&2.\ E\left[\left|\int_a^b X(t)dt\right|^2\right] = \int_a^b\int_a^b E[X(s)X(t)]ds\,dt \\
&3.\ \mathrm{Var}\left[\int_a^b X(t)dt\right] = \int_a^b\int_a^b K(s,t)ds\,dt \\
&4.\ \mathrm{Cov}\left[\int_a^b X(s)ds, \int_c^d X(t)dt\right] = \int_a^b ds \int_c^d dt K(s,t).
\end{aligned}
\tag{3.139}
$$

Thus, the moments of stochastic integrals are Riemann integrals of the appropriate function. Of course, the mean value function and covariance kernel of the process have to exist and be known. In order to illustrate the stochastic integrals of Equation 3.139, an example with the displacement of a particle in free Brownian motion is considered.

Suppose a body is moving in a straight line (a simplification) because the body is colliding with other bodies and suppose $N(t)$ is the number of hits on the body and that $N(t)$ is Poisson with parameter κ; that is, the average number of hits per unit time is κ. As a result of each hit, the body reverses velocity either from d to $-d$ or from $-d$ to d; thus, the velocity of the body is given by the stochastic process

$$v(t) = v(1)(-1)^{N(t)}.$$

Thus, the process $\{v(t), t \geq 0\}$ has mean value function $E[v(t)] = 0$ and covariance function $E[v(s)v(t)] = d^2 e^{-2d|s-t|}$.

The displacement of the body is given by

$$X(t) = \int_0^t v(x)dx; \tag{3.140}$$

thus,

$$E\left[\left|\int_0^t v(x)dx\right|^2\right] = E\left[\int_0^t\int_0^t v(x_1)v(x_2)\mathrm{d}x_1\mathrm{d}x_2\right],$$
$$= \int_0^t\int_0^t E[v(x_1)v(x_2)]\mathrm{d}x_1\mathrm{d}x_2. \tag{3.141}$$

Since $= E[v(s)v(t)]$ is continuous in s and t, the integral in Equation 3.140 exists; therefore, the mean square displacement is

$$E[|X(t)|^2] = (2d^2/\beta^2)\left(e^{-\beta t} - 1 + \beta t\right), \tag{3.142}$$

where $\beta = 2d$.

The stochastic integral and stochastic derivatives are equally important, and the derivative process of the stochastic process $\{X(t), t \geq 0\}$ is defined as

$$(d/dt)X(t) = \lim[X(t+h) - X(t)]/h, \quad h \to \infty, \tag{3.143}$$

where the limit is mean square convergence. Under what conditions does the limit exist? It can be shown that a necessary and sufficient condition for the limit to exist is that the following two limits exist as $h \to \infty$ and $k \to \infty$:

$$\lim E[X(t+h) - X(t)]/h$$
$$\text{and} \tag{3.144}$$
$$\lim \mathrm{Cov}\{[X(t+h) - X(t)]/h, [X(t+k) - X(t)]/k\}.$$

Thus, convergence in mean square depicted by the stochastic derivative is possible if ordinary convergence of the sequences in Equation 3.144 occurs. Another way to determine the possibility of convergence in the mean square of the stochastic derivative is in terms of the differentiability of the mean value and covariance kernel of the process. The following is a sufficient condition for the stochastic integral to be well defined as a limit in the sense of convergence in mean square:

1. The mean value function $m(t)$ is differentiable with respect to t.
2. The mixed partial second derivative $(\partial^2/\partial s\,\partial t)K(s,t)$ exists and is continuous.

Of main importance is the mean value function and covariance kernel of the derivative process $\{X'(t), t \geq 0\}$ given by the stochastic derivative in Equation 3.143. It can be shown that

$$E[X'(t)] = E[(d/dt)X(t)] = (d/dt)m(t) \tag{3.145}$$

and

$$\text{Cov}[X'(s), X'(t)] = (d^2/ds\,dt)K(s,t). \tag{3.146}$$

Let $\{X(t), t \geq 0\}$ be a covariance stationary process, and let $\{X'(t), t \geq 0\}$ be the corresponding derivative process. Then, is the derivative process also covariance stationary? The answer is yes, and the student should verify the assertion.

3.9 Summary and Conclusions

This chapter presents a brief review of stochastic processes and gives the reader the necessary material in order to understand the background essential for making Bayesian inferences for the unknown parameters of a stochastic process. This chapter begins by introducing the basic terminology and notation for the study of stochastic processes, and this is followed by the formal definition of a stochastic process. Probabilistic characteristics of the two most important processes are described, namely, the Poisson and Wiener processes, where the former is a continuous parameter and discrete state model, but the latter is a continuous index set and continuous state space. Next to be described are the essential components when working with stochastic processes, namely, the state space and the index set, and explained are the various types of processes, such as martingales, Markov chains, stationary processes, renewal, and point processes.

Of course, when working with Markov chains, it is important to know the various types of states the chain can assume. At this point, transient states, recurrent states, and periodic states are defined and several examples provided. Next to be presented are the concepts of homogeneous and embedded Markov chains and Markov processes with continuous index sets. Of fundamental importance are the Kolmogorov equations, which are used to derive and define various stochastic processes. The last class of Markov processes to be described is those with a continuous state space, such as the Brownian motion. Concluding the chapter are explanations of the stationary and evolutionary processes and the introduction of stochastic calculus, including the integral and derivative of a stochastic process.

The material for this chapter is based on well-known textbooks. Thus, it is recommended that the reader refer to some of the references, including Karlin and Taylor,[1] Allen,[2] Cox and Miller,[3] Parzen,[4] Cox and Isham,[5] Ross,[6] Bailey,[7] and Chiang.[8]

3.10 Exercises

1. a. Define the probability mass function of a discrete random variable.
   b. Define the probability density function of a random vector with continuous random variables as components.
   c. Let X and Y be continuous random variables. Define the conditional expectation of Y given $X = x$.
2. Refer to Section 3.3 and define a discrete parameter stochastic process with state space $S = \{0, 1, 2, \ldots\}$.
3. Define the Poisson process with parameter $\lambda > 0$ and calculate $\Pr[N(7) > 2]$ if $\lambda = 1$.
4. Describe a Wiener process with diffusion coefficient σ^2 and calculate $\Pr[X(3) > 1]$ when $\sigma^2 = 2$.
5. What are the index set and state space of a stochastic process? What are the index set and state space of a Poisson process and of a Wiener process?
6. Describe a stochastic process with independent process. Does the Wiener process have independent increments? For a Wiener process with parameter σ^2, what is the joint distribution of $X(t_1), X(t_2), \ldots, X(t_n)$ for $t_1 < t_2 < \ldots < t_n$ and n is a positive integer?
7. Explain the Markov property for a Markov chain with index set $[0, \infty)$.
8. Explain the difference between a process which is stationary of order k and a process with the strict stationary property.
9. For a DTMC with state space $S = \{0, 1, 2, \ldots\}$, define a recurrent state, a transient state, and a state with period d.
10. Consider a Markov chain with a countable state space. Explain how the communicating relation $\leftrightarrow$ induces a set of equivalence classes among the states of the chain.

11. Consider the transition matrix

$$P = \begin{pmatrix} P_1, 0 \\ 0, P_2 \end{pmatrix}$$

where

$$P_1 = \begin{pmatrix} .5, .5 \\ .25, .75 \end{pmatrix}$$

and

$$P_2 = \begin{pmatrix} 0, 1, 0 \\ .5, 0, .5 \\ 0, 1, 0 \end{pmatrix}.$$

Do the states of P_1 communicate with those of P_2? Explain your answer.

12. What are the three communicating the classes of the chain with transition matrix

$$P = \begin{pmatrix} 1, 0, 0, 0, \ldots\ldots 0, 0, 0 \\ q, 0, p, 0, \ldots\ldots 0, 0, 0 \\ 0, q, 0, p, ,,,, 0, 0, 0 \\ . \\ . \\ 0, \ldots\ldots\ldots\ldots\ldots q, 0, p \\ 0, \ldots\ldots\ldots\ldots\ldots 0, 0, 1 \end{pmatrix}$$

and states 0, 1, 2, ..., $a - 1$, and a. Is the state 0 an absorbing boundary?

13. Consider a Markov chain with states 0, 1, 2, and 3 and transition probability matrix

$$P = \begin{pmatrix} 0, 0, .5, .5 \\ 1, 0, 0, 0 \\ 0, 1, 0, 0 \\ 0, 1, 0, 0 \end{pmatrix}.$$

Show that all the states communicate and all are recurrent.

14. For a different situation with states 0, 1, 2, 3, and 4, consider a process with transition matrix

$$P = \begin{pmatrix} 1/2, 1/2, 0, 0, 0 \\ 1/2, 1/2, 0, 0, 0 \\ 0, 0, 1/2, 1/2, 0 \\ 0, 0, 1/2, 1/2, 0 \\ 1/4, 1/4, 0, 0, 1/2 \end{pmatrix},$$

where there are three classes $\{0,1\}$, $\{2,3\}$, and $\{4\}$. Show that the first two are recurrent but the third is transient.

15. Suppose $X(i)$, $i = 1,2,...,n$, is a random sample of size n from a discrete population with probability mass function $\Pr[X = i] = a_i$, $i = 1,2,...,$ and consider the stochastic process $\{X(i), i = 1,2,...\}$. What is the transition probability matrix for this process? See Equation 3.59. Now consider the process $\{Y(n) = \sum_{i=1}^{i=n} X(i)$, $i = 1,2,...\}$; thus, $\Pr[Y(n+1) = k|Y(n) = j] = a_{k-j}$. What is the transition probability matrix of this process? See Equation 3.61.

16. This example is taken from pages 108 and 109 of Allen[2] and involves a process with state space $\{0,1,2,..,N\}$, in which case there are two boundaries or with infinite and countable state space {0, 1, 2, ...} with boundary at 0. In such a model, the states are positions which are denoted by $X(n)$, where n is time. Let p be the probability of moving one unit to the right and $1 - p$ be the probability of moving one unit to the left; thus, the one-step transition probability is

$$\Pr[X(n+1) = j|X(n) = i] = p, \quad j = i + 1,$$
$$= 1 - p, \quad j = i - 1$$

If the boundary at 0 is absorbing, $P_{00} = 1$. Write down the $(N + 1) \times (N + 1)$ transition matrix of this random walk process.

17. Refer to Section 3.7.2 and define the birth and death process with state space $S = \{0,1,2,...\}$ and verify the transition probabilities in Equation 3.97.

18. Derive the Kolmogorov backward equations in Equation 3.108.

19. Derive the general equations for solving for the limiting probabilities P_j, and verify Equation 3.113 for the limiting probabilities of a birth and death process.

20. Derive Equation 3.120, which is the solution to the diffusion equation (Equation 3.118). The solution is a normal density which is the conditional probability $\Pr[X(t + t_0)|X(t_0) = x_0]$, where $\{X(t), t \geq 0\}$ is the Brownian motion with parameter σ^2, and $X(t)$ is the position of a particle undergoing the Brownian motion.

21. For a geometric Brownian motion process (Equation 3.122), verify the conditional mean and variance of the process given by Equation 3.123.

22. Define the stochastic integral $\int_a^b X(t)dt$ as the limit of a sequence of partial sums where convergence is in the sense of mean square, and verify the four equations of Equation 3.139 involving the first and second moments of stochastic integrals.
23. Suppose that the body is colliding with other bodies and suppose that $N(t)$ is the number of hits on the body and that $N(t)$ is Poisson with parameter κ; that is, the average number of hits per unit time is κ. As a result of each hit, the body reverses velocity either from d to $-d$ or from $-d$ to d; thus, the velocity of the body is given by the stochastic process

$$v(t) = v(1)(-1)^{N(t)}.$$

Show that the process $\{v(t), t \geq 0\}$ has mean value function $E[v(t)] = 0$ and covariance function $E[v(s)v(t)] = d^2 e^{-2d|s-t|}$.

The displacement of the body is given by

$$X(t) = \int_0^t v(x)\, dx;$$

thus,

$$E\left[\left|\int_0^t v(x)dx\right|^2\right] = E\left[\int_0^t \int_0^t v(x_1)v(x_2)\, dx_1\, dx_2\right]$$

$$= \int_0^t \int_0^t E[v(x_1)v(x_2)]\, dx_1\, dx_2.$$

Since $= E[v(s)v(t)]$ is continuous in s and t, the integral in Equation 140 exists. Show that the mean square displacement is

$$E[|X(t)|^2] = (2d^2/\beta^2)\left(e^{-\beta t} - 1 + \beta t\right),$$

where $\beta = 2d$.

24. Suppose $\{(d/dt)X(t), t \geq 0\}$ is the derivative process of the process $\{X(t), t \geq 0\}$, which has mean value function $m(t)$ and covariance kernel $K(s, t)$. Show

$$E[X'(t)] = E[(d/dt)X(t)] = (d/dt)m(t)$$

and

$$\text{Cov}[X'(s), X'(t)] = (d^2/ds\,dt)K(s, t).$$

References

1. Karlin, S., and Taylor H. M. 1975. *A First Course in Stochastic Processes, Second Edition*. San Francisco: Academic Press.
2. Allen, L. J. S. 2011. *An Introduction to Stochastic Processes with Applications to Biology, Second Edition*. Boca Raton, FL: CRC Press.
3. Cox, D. R., and Miller, H. D. 1965. *The Theory of Stochastic Processes*. London: Chapman and Hall.
4. Parzen, E. 1962. *Stochastic Processes*. San Francisco: Holden Day Wilson.
5. Cox, D. R., and Isham, V. 1980. *Point Processes*. New York: Chapman and Hall.
6. Ross, S. M. 1972. *Probability Models, Fifth Edition*. San Diego, CA: Academic Press.
7. Bailey, N. T. J. 1964. *The Elements of Stochastic Processes*. New York: John Wiley & Sons.
8. Chiang, C. L. 1968. *Introduction to Stochastic Processes in Biostatistics*. New York: John Wiley & Sons.
9. Wiener, N. 1923. Differential space. *Journal of Mathematics and Physics/Massachusetts Institute of Technology* 2:131–174.
10. Uhlenbeck, G. E., and Ornstein, L. S. 1930. On the theory of Brownian motion. *Physical Review* 36:823–841.
11. Birkhoff, G. D. 1931. Proof of the Ergodic theorem. *Proceedings of the National Academy of Sciences of the United States of America* 17:656–660.
12. Von Neumann, J. 1932. Proof of the quasi-ergodic hypothesis. *Proceedings of the National Academy of Sciences of the United States of America* 18:70.

4

Bayesian Inference for Discrete Markov Chains

4.1 Introduction

This chapter begins the formal approach to using Bayesian methods of making inferences for stochastic processes, in particular, those processes with a countable number of states and an index over the set of nonnegative integers. Bayesian methods of inference consist of estimation, testing hypotheses, and making predictions (of future observations), and these methods were introduced in Chapter 2 for several well-known processes.

In general, Bayesian inferences will be provided for each case of a Markov chain introduced in the following sections of the chapter. The chapter begins with a brief review of the definition of a Markov chain, followed by the presentation of many examples, including the example given by Andreyevich Markov,[1] the gambler's ruin problem, the Wright–Fisher model in genetics, a random walk on a graph, cycle and complete graphs, a birth and death process with a countable number of states, the idea of weighted directed graph with associated examples and transition matrices, and an interesting example by Dobrow[2] about cancer metastasis.

This is followed by an explanation of how to compute the n-step transition probabilities illustrated by examples including the gambler's ruin problem and an example involving a chain that explains the status of the weather. In order to determine the long-term behavior of a chain, the integer powers of the one-step transition matrix P are needed, and it is at this point the R Code for computing such powers of P is presented. The long-term behavior using R to compute the powers of P is illustrated with several examples, including the risk of fire in Ontario, example from Diaconis.[3] Another interesting and an example by Dobrow[2] of long-term behavior is given by a random walk on a cycle with 26 states, and another example (a random walk on a cycle graph with six vertices) of where the long-term behavior exhibits an alternating sequence of distributions, depending on the initial state of the chain, is also described. For each example, Bayesian inferences are provided by simulating the states of the chain, estimating the associated transition matrix P, testing hypotheses about the entries of the transition matrix, and generating future observations from the chain based on the estimated transition matrix P.

Next to be considered is the limiting behavior of a Markov chain; thus, the idea of a limiting distribution and a stationary distribution are explained. It is shown that the long-term probability of a particular state is the same as the proportion of the time the chain is in that state. Both analytical and computation methods using R illustrate ways of determining the limiting probabilities for some interesting cases of Markov chains, including the general two-state chain. Closely associated with the liming distribution of a chain is the stationary distribution of the chain. The stationary distribution of a chain is defined, and it is demonstrated that the limiting distribution of a chain in a stationary distribution π is one that satisfies

$$\pi = \pi P, \tag{4.1}$$

where P is the transition matrix of the chain.

The chapter continues with the definition of a regular transition matrix and finding the stationary distribution of such chains. Computational methods for determining the stationary distribution of regular chains are introduced and involve using R. The details of computing the stationary distribution with R are made clear by referring to the following examples: The Ehrenfest dog-flea model, a random walk on a graph, a random walk on a weighed directed graph, and a random walk on a hypercube.

In order to fully understand the long-term behavior of a chain, one must know how often the states of a chain are visited, and this in turn relates to the idea of a communicating class, the ideas of recurrent and transient states, and, finally, the idea of an irreducible chain. In order to understand the idea of a recurrent and transient state, the associated graph showing the transition probabilities of the chain provides an invaluable tool. The definition of recurrence and transience depends on the long-run behavior of the chain: A state is recurrent if the probability is one if it is eventually revisited, but the state is transient if the probability is less than one if it is eventually revisited. Many examples of chains with recurrent and/or transient states are given including one by Tsai of earthquakes in Taiwan, which is described by a four-state chain where all the states are recurrent, and its stationary distribution is found along with the expected number of returns to each state. Also presented is the R Code, which computes the expected return to the states of a Markov chain. The period of a state of a chain is also a class property as is recurrence. That is, consider a class of states that all communicate, then if one state of the class is recurrent all are recurrent, and if one in the class has period 2, they all have period 2. For some of these examples, simulations provide realizations that will illustrate the period of a state and the recurrence and transience of a state, and then Bayesian inferences for the unknown transition matrices are determined. In general, the communicating classes of a chain will be determined. Then, for each class, the states of that class will be examined as to their nature; that is, are they recurrent or transient? And what are their periods? This can be a challenge if all one has is the number of pairwise transitions among the states of the chain, and it is consequently a challenge in providing Bayesian inferences for the chain. A chain is called periodic if it is irreducible (all the states communicate) and all states have period greater than one; if not, the chain is said to be aperiodic, that is, if the states have period 1.

The last important idea considered in this chapter is ergodic Markov chains, namely, those chains that are irreducible and aperiodic. For such chains, it is known that they have a unique stationary distribution, which is its limiting distribution. For example, the Ehrenfest dog-flea model is discrete and ergodic and can be shown to have a limiting distribution that is binomial with parameters N and ½. For example, if there are $N = 6$ fleas, the limiting distribution is easily found. See page 110 of Dobrow.[2] The stationary command in R easily computes the stationary distribution of an ergodic chain. From the perspective of the Bayesian, realizations will be generated from the transition matrix, and then the transition probabilities of the chain estimated from the posterior distribution.

Absorbing chains is the last topic to be discussed in this chapter. A state is called absorbing if the probability of remaining in the state over one time period is 1. Of course, a good example is the gambler's ruin chain with transition matrix

$$P = \begin{pmatrix} 0,.6,0,0,.4,0 \\ .4,0,.6,0,0,0 \\ 0,.4,0,.6,0,0 \\ 0,0,.4,0,0,.6 \\ 0,0,0,0,1,0 \\ 0,0,0,0,0,1 \end{pmatrix}. \tag{4.2}$$

Assuming the gambler starts with \$2 with a chance of winning \$1 on each round of .6 and either gains \$5 or loses, and then the absorbing states are 0 and 5. When P is arranged into the canonical form, one is able to compute the probability of the eventual ruin using R. From a Bayesian approach, I will generate realizations from the chain with transition matrix P and then test the hypotheses that the probability of winning one dollar is .6.

Lastly, Chapter 6 is concluded with a section for comments and conclusions, which will summarize the Bayesian inferences presented in the chapter.

4.2 Examples of Markov Chains

4.2.1 Biased Coins

Recall the definition of a Markov chain: Consider a sequence of random variables $X(n)$, $n = 0,1,2,...$; then this sequence is said to be a Markov chain if for positive integers and states $i, j, i_{n-1}, ..., i_0$

$$\begin{aligned} &P[X(n+1) = j|X(n) = i, X(n-1) = i_{n-1}, ..., X(0) = i_0] \\ &= P[X(n+1)] = j|X(n) = i. \end{aligned} \tag{4.3}$$

The states are assumed to be nonnegative integers, and the first example is taken from Diaconis,[3] who studied the results of a large number of coin tosses resulting in the one-step transition matrix

$$P = \begin{pmatrix} .51,.49 \\ .49,.51 \end{pmatrix}. \tag{4.4}$$

This shows evidence of a slight bias, where if the previous toss results in heads, the current toss results in heads with probability .51, and if the previous toss resulted in a head, the current toss occurs with a tail with probability .49. Also, if the previous toss resulted in a tail, the current toss results in a head with probability .49. Suppose we generate a realization from this chain with transition matrix P using the following R Code given by Dobrow.[2] Note that this is a Markov chain where $X(n) = 1$ denotes a head and $X(n) = 2$ signifies a tail.

R Code 4.1

```
> markov <- function(init,mat,n,labels) {
+ if (missing(labels)) labels <- 1:length(init)
+ simlist <- numeric(n+1)
+ states <- 1:length(init)h
+ simlist[1] <- sample(states,1,prob=init)
+ for (i in 2:(n+1))
+ { simlist[i] <- sample(states,1,prob=mat[simlist[i-1],]) }
+ labels[simlist]
+ }
> P<-matrix(c(.51,.49,.49,.51),nrow=2,ncol=2,byrow=TRUE)
> init<=c(1,0)
```

Markov is a function with inputs init P and n. init is initial distribution where c(1,0) denotes the initial toss is head, while c(0,1) denotes the initial toss is a tail.

The following realization was based on the following inputs for the Markov function given by R Code 4.1: init=c(1,0), P given by 4.3, and n=100:

```
1 2 1 2 1 2 2 2 2 1 1 1 2 1 2 2 2 2 2 2 2 1 2 1 1 1 2 2 2 2 2 2 1 1 1 2 1
2 1 2 2 1 2 2 1 1 1 2 2 1 2 1 2 2 1 1 2 1 1 2 2 1 2 2 1 2 1 1 2 2 1 1 1 2
2 1 1 2 1 2 1 1 2 2 2 2 2 2 1 1 2 2 2 2 2 1 1 1 1 1 2.
```

Thus, the usual estimated values of the transition probabilities for P_{11} and P_{12} are .4545 and .5454, respectively and are .4363 and .5636 for P_{21} and P_{22}, respectively. Consider the Bayesian approach to estimating P_{11}, then one must place a prior distribution on this parameter, which, in turn, depends on the count n_{11}, which has a binomial distribution with parameters P_{11} and $n_1 = n_{11} + n_{12}$. I am assuming noninformative prior information about P_{11} with the improper prior density

$$\pi(P_{11}) \propto P_{11}^{-1}(1 - P_{11})^{-1}, \quad 0 < P_{11} < 1. \tag{4.5}$$

Thus, the posterior density

$$\pi(P_{11}|n_1) \propto P_{11}^{n_{11}-1}(1 - P_{11})^{n_{12}-1}, \quad 0 < P_{11} < 1, \tag{4.6}$$

which is a beta distribution with parameters n_{11} and n_{12}. If one employs the posterior mean to estimate P_{11}, then the estimator is $E(P_{11} \mid n_{1.}) = n_{11} / (n_{11} + n_{12}) = 20 / (20 + 24) = .4545$, which is the usual maximum likelihood estimator of P_{11}. It should be noted that if one uses the uniform prior for P_{11}, the posterior mean of this parameter is $E(P_{11} \mid n_{1.}) = (n_{11} + 1) / (n_{11} + n_{12} + 2) = .4565$, which is not much different from the estimate based on the prior distribution (Equation 4.6). Of course, it is of interest to know the 95% credible interval for P_{11}, which is left as an exercise for the reader.

4.2.2 Rainy Day

Suppose that whether or not it rains today is a function of what happened the previous two days. In particular, suppose that if it rained for the past two days, then it will rain tomorrow with probability of .7, and if it rained today but not yesterday, then it will rain tomorrow with probability of .5. In addition, if it rained yesterday but not day, it will rain the following day with probability of .4, and lastly, if it has not rained yesterday and the day before yesterday, it will rain tomorrow with probability of .2. This is enough information to define a four-state Markov chain as follows:

- State 1: It rained today and yesterday.
- State 2: It rained today but not yesterday.
- State 3: It rained yesterday but not today.
- State 4: It did not rain yesterday or today.

The resulting Markov chain has transition matrix

$$P = \begin{pmatrix} .7, .0, .3, .0 \\ .5, .0, .5, .0 \\ .0, .4, .0, .6 \\ .0, .2, .0, .8 \end{pmatrix}. \tag{4.7}$$

Let us now generate a realization from this process using R Code 4.1., presented in the previous section. Assuming the initial state is 1 (it rained both today and yesterday), 100 generated values are as follows:

```
1 1 3 4 4 4 4 4 2 1 1 1 3 2 3 4 4 4 2 1 1 1 3 2 1 3 4 2 3 4 4 2 3 4 4 4 4
4 4 4 4 4 2 1 3 4 4 4 2 3 4 4 4 4 4 4 2 1 1 1 1 1 3 4 4 4 4 4 4 4 2 3 4 4
4 4 4 4 2 1 3 4 4 4 4 4 4 4 4 4 2 3 4 4 4 4 4 4 4 4 4.
```

Conditioning on the initial state of 1, state 1 was visited nine times; and state 3, six times; and of course, states 2 and 4 were visited zero times. Thus, as in the previous example, the distribution of n_{11} is binomial with parameters P_{11} and $n_{11} + n_{13} = 15$. And, using the uniform prior for this parameter, the posterior distribution of P_{11} is beta with parameters $n_{11} + 1 = 10$ and $n_{13} + 1 = 7$, yielding a posterior mean of $E(P_{11}|n_{11} = 9, n_{13} = 6) = 10/17 = .5882$.

I will use a completely different method of performing the posterior analysis by employing WinBUGS to generate observations for the transitions based on the transition matrix P given by Equation 4.7. Note that given P_{11}, the distribution of the count n_{11} is binomial with distribution P_{11} and $n_{11} + n_{13} = n_{1.} = 50$. Using WinBUGS Monte Carlo Markov chain (MCMC) techniques allows one to construct credible intervals for the transition probabilities. I begin with the first row of P which conditions on the first state of the process where $P_{11} = .7$ and $P_{13} = .3$, and the other two transition probabilities of the first row are zero; that is, it is impossible to go from state 1 to either states 2 or 4. It is also assumed that the initial state is 1, that it rained both today and yesterday. I am also assuming 50 transitions with initial state 1. The following code was used to perform the Bayesian analysis where the data are given in the list statement.

WinBUGS Code 4.1

```
model {
p11~dbeta(1,1)
     for (i in 1 : 50) {
            n11[i] ~ dbin(p11,50)         }
}
list(
n11 = c(
37.0,34.0,39.0,27.0,32.0,
36.0,35.0,37.0,42.0,28.0,
29.0,35.0,32.0,34.0,33.0,
34.0,38.0,33.0,43.0,37.0,
31.0,37.0,30.0,36.0,27.0,
37.0,29.0,34.0,33.0,38.0,
36.0,39.0,34.0,33.0,36.0,
35.0,38.0,38.0,36.0,37.0,
36.0,36.0,31.0,36.0,33.0,
45.0,34.0,32.0,39.0,30.0))
```

The Bayesian analysis assumes a binomial distribution for n_{11}, which can vary from 0 to 50 and where P_{11} is given a uniform prior distribution.

The 50 observations given in the list statement where generated from a binomial distribution with $\theta = .7$ using the following code:

WinBUGS Code 4.2

```
model {
            for (i in 1 : 100) {
                   n11[i] ~ dbin(.7,50)
            }
      }
```

When executed with 35,000 observations with a burn-in of 5,000, the result of the posterior analysis is presented in Table 4.1.

Thus, the posterior mean of P_{11} is .6963 with a 95% credible interval of (.6783, .7142), and Figure 4.1 portrays the posterior density. The posterior mean of .6963 differs very little from the $P_{11} = .7$ used to generate the observations used for the posterior analysis; thus, a Bayesian test of the null hypothesis H: $P_{11} = .7$ versus the alternative A: $P_{11} \neq .70$ will be performed.

The reader is referred to Section 2.5.3 to review the fundamental aspects of testing hypotheses from a Bayesian perspective.

The posterior probabilities of the null and alternative hypotheses p_0 and p_1 are required where that of p_0 is given by Equation 4.9; thus, it follows that

$$p_0 = \pi_0 \xi(n_{11} | P_{11} = .70) / \xi(n_{11}), \tag{4.8}$$

TABLE 4.1
Posterior Analysis for P_{11}

Parameter	Mean	SD	Error	2 1/2	Median	97 1/2
P_{11}	.6963	.009167	.0000521	.6783	.6964	.7142

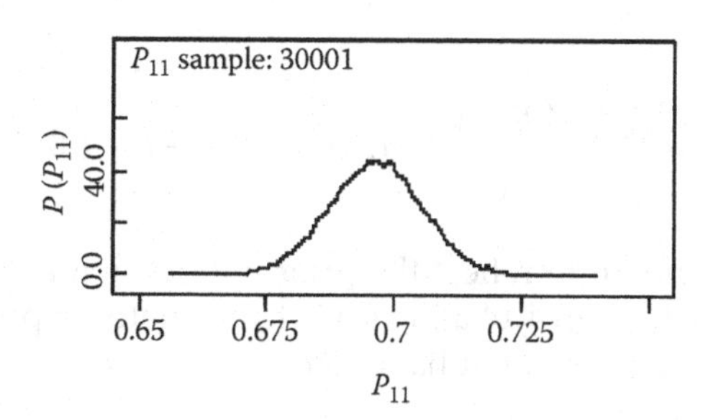

FIGURE 4.1
Posterior density of P_{11}.

where

$$\xi(n_{11}) = \pi_0\xi(n_{11}|P_{11} = .7) + \pi_1\xi_1(n_{11}). \tag{4.9}$$

π_0 is the prior probability of the null hypothesis, and the probability mass function of n_{11} (given the null hypothesis) is

$$\xi(n_{11}|P_{11} = .70) = \binom{n_{1.}}{n_{11}}(.7)^{n_{11}}(.3)^{n1.-n_{11}}, \tag{4.10}$$

where $n_{11} = 1, 2, \ldots, n_{1.}$.

Let ξ be the generic symbol for density and mass functions and recall that n_{11} is the number of transitions corresponding to the transition probability P_{11}.

In addition

$$\xi_1(n_{11}) = \pi_1\int\xi_1(P_{11})\xi(n_{11}|P_{11})\mathrm{d}P_{11} \tag{4.11}$$

and

$$\xi_1(P_{11}) = [\Gamma(\alpha+\beta)/\Gamma(\alpha)\Gamma(\beta)]{P_{11}}^{\alpha-1}(1-P_{11})^{\beta-1}, \tag{4.12}$$

where π_1 is the prior probability of the alternative hypothesis and $\pi_0 + \pi_1 = 1$.

Note that in the marginal mass function of n_{11} given by Equation 4.10 that $\xi_1(P_{11})$ is the prior density of P_{11} over interval [0,1], that is, over the values specified by the alternative hypothesis. It is convenient to choose

$$\xi_1(P_{11}) = [\Gamma(\alpha+\beta)/\Gamma(\alpha)\Gamma(\beta)]{P_{11}}^{\alpha-1}(1-P_{11})^{\beta-1}, \quad (4.13)$$

where $0 \le P_{11} \le 1$ with α and β as positive parameters; thus, the prior distribution of P_{11} is a beta with parameters α and β. Now it can be shown that the marginal distribution of n_{11} is given by

$$\xi_1(n_{11}) = [\Gamma(\alpha+\beta)/\Gamma(\alpha)\Gamma(\beta)]\binom{n_{1.}}{n_{11}}[\Gamma(n_{11}+\alpha)\Gamma(n_{1.}-n_{11}+\beta)/\Gamma(n_{1.}+\alpha+\beta)], \quad (4.14)$$

where α and β must be chosen to reflect the prior information about the null hypothesis. Combining Equations 4.8 through 4.14 allows one to evaluate the posterior probability p_0 of the null hypothesis. For the problem at hand, let

$$\begin{aligned} \pi_1 = \pi_0 &= 1/2; \\ \alpha &= 7; \\ \beta &= 3;. \end{aligned} \quad (4.15)$$

$n_{11} = 37$, the first value generated from the appropriate binomial (see WinBUGS Code 3.1); and $n_{1.} = 50$ (the total number of transitions with an initial state of 1).

One can show that the posterior probability of the null hypothesis is p_0 =.99513. Note that these posterior probabilities can vary according to the values of the number of transitions n_{11} as well as to the specifications of the prior distributions of the null and alternative hypotheses. For example, if the observed transition is 27, then one can show that $p_0 = .9998$, but if $n_{11} = 45$, then it can be demonstrated that $p_0 = .01800$. The latter calculation shows that there is little evidence indicating that the null hypothesis is true, but when $n_{11} = 37$, there is strong evidence that the null hypothesis is true as indicated by $p_0 = .99513$.

The next phase for Bayesian inference for the rainy day example is forecasting future observations. The reader is referred to Section 2.6.2, which presents the essentials for predicting future observations from a binomial population.

Suppose the binomial case is again considered, where the posterior density of the binomial parameter P_{11} is

$$\xi(P_{11}|x) = [\Gamma(\alpha+\beta)\Gamma(n+1)/\Gamma(\alpha)\Gamma(\beta)\Gamma(x+1)\Gamma(n-x+1)]{P_{11}}^{\alpha+x-1}(1-P_{11})^{\beta+n-x}, \quad (4.16)$$

a beta with parameters $\alpha + x$ and $n - x + \beta$, and x is the sum of the set of n observations. The population mass function of a future observation Z is $f(z|\theta) = \theta^z(1-\theta)^{1-z}$; thus, the predictive mass function of Z, called the beta-binomial, is

$$\begin{aligned} g(z|x) &= \Gamma(\alpha+\beta)\Gamma(n+1)\Gamma\left(\alpha+\sum_{i=1}^{i=n}x_i+z\right)\Gamma(1+n+\beta-x-z) \\ &\div \Gamma(\alpha)\Gamma(\beta)\Gamma(n-x+1)\Gamma(x+1)\Gamma(n+1+\alpha+\beta), \end{aligned} \quad (4.17)$$

where $z = 0, 1$. If $z = 1$, there is a transition from state 1 to state 1, and if $z = 0$, there is a transition from state 1 to state 3. Also, note that this function does not depend on the unknown parameter P_{11}. Assuming that the sum of the number of transitions from state 1 to state 1 is $n_{11} = x = 37$, that the total number of transitions beginning with state 1 is $n = 50$, and that $\alpha = \beta = 1$ (a uniform prior for P_{11}), the following WinBUGS Code 4.3 will generate observations from the predictive beta binomial distribution (Equation 4.17).

The following list statement gives the 50 predicted transitions for the rainy day example:

WinBUGS Code 4.3

```
model;

{
# the parameters of the beta
# α + n11 = 1 + 37 = 38
# β + n-n11 = 1 + 50-37 = 14
theta~dbeta(38,14)

for (i in 1:50){
y[i]~dbern(theta)

}}

list(
theta = 0.7682,
y = c(
1.0,1.0,1.0,1.0,0.0,
0.0,1.0,1.0,1.0,1.0,
1.0,1.0,0.0,1.0,1.0,
0.0,1.0,1.0,1.0,1.0,
1.0,1.0,1.0,1.0,0.0,
1.0,1.0,1.0,1.0,1.0,
1.0,1.0,1.0,1.0,0.0,
1.0,1.0,0.0,0.0,1.0,
0.0,1.0,1.0,1.0,0.0,
1.0,1.0,0.0,0.0,1.0))
```

The total number of ones in the preceding list statement is the predicted number of transitions from state 1 to state 1, while the total number of zeroes is the number of transitions from state 1 to state 3. There are 38 out of 50 predicted transitions from state 1 to state 1. Recall that the ratio $38/50 = .768$ should be compared to the transition probability $P_{11} = .70$ in the rainy day example.

4.3 Fundamental Computations

This section will explore how to compute the n-step transition probabilities of a Markov process, the joint distribution of such a process at arbitrary time points, and the marginal distribution of the process at a given time point. Also presented are the R routines that will be used to compute the n-step transition probabilities and the associated Bayesian inference procedures. For example, the estimation of the n-step transition probabilities of a Markov process is a primary goal, but other inference procedures such as testing hypotheses and prediction procedures will also be described. This will entail the simulation of realizations for interesting examples from biology.

Let us now consider the n-step transition matrix with ij-th element:

$$P_{ij}^n = P[X(n) = j|X(0) = i], \tag{4.18}$$

where $P_{ij}^0 = P[X(0) = j|X(0) = i] = 1$ if $i = j$ and 0 if otherwise. We now show that the n-step transition matrix is the nth power of the one-step transition matrix P. Note in Equation 4.18 that P_{ij}^n is not the nth power of P_{ij}, and that the nth power of P is denoted by $(P_{ij})^n$.

Consider

$$\begin{aligned}
P[X(n) = j|X(0) = i] &= \sum_k P[X(n) = j|X(n-1) = k, X(0) = i]P[X(n-1) = k|X(0) = i] \\
&= \sum_k P[X(n) = j|X(n-1) = k]P[X(n-1) = k|X(0) = i] \\
&= \sum_k P_{kj} P[X(n-1) = k|X(0) = i],
\end{aligned} \tag{4.19}$$

which is valid because of the Markov property and the fact that the process is time homogenous; thus, for $n = 3$,

$$P[X(3) = j|X(0) = i] = \sum_k P_{kj} P[X(2) = k|X(0) = i] = \sum_k P_{kj} P_{ik}^2 = \left(P^3\right)_{ij}, \tag{4.20}$$

which is the ijth element of the third power of the first step transition matrix P, as was to be shown.

Consider the following example of a random walk on a cycle graph consisting of five vertices labeled 0, 1, 2, 3, and 4, then the one-step transition matrix is

$$P = \begin{pmatrix} 0,.5,0,0,.5 \\ .5,0,.5,0,0 \\ 0,.5,0,.5,0 \\ 0,0,.5,0,.5 \\ .5,0,0,.5,0 \end{pmatrix}. \tag{4.21}$$

That is, starting at vertex zero (the initial state), the probability of remaining in that state is 0, but is ½ of moving to the right is ½. Note that the probability of remaining in the initial state is always 0. What is the transition matrix after six moves? It can be shown to be

$$P^6 = \begin{pmatrix} .312500 & .109375 & .234375 & .234375 & .109375 \\ .109375 & .312500 & .109375 & .234375 & .234375 \\ .234375 & .109375 & .312500 & .109375 & .234375 \\ .234375 & .234375 & .109375 & .312500 & .109375 \\ .109375 & .234375 & .234375 & .109375 & .312500 \end{pmatrix}. \tag{4.22}$$

Note the pattern of the six-step transitions is as follows: the probability of a return to the initial position is .312500 and the probability of moving one vertex to the right is .109375 as is the probability of moving one vertex to the left, etc. Also, the probability of moving two units to the right or left is .234375.

The following code is used to generate the n-step transition probabilities, where the function matrixpower has two arguments, the matrix labeled "mat" of one-step transition probabilities given by Equation 4.20 and the desired power k, which refers to the k-step transition probabilities, where $k = 6$. This example is from Dobrow.[2] See the following R Code 4.2:

R Code 4.2

```
matrixpower <- function(mat,k) {
      if (k == 0) return (diag(dim(mat)[1]))
      if (k == 1) return(mat)
      if (k > 1) return( mat %*% matrixpower(mat, k-1))
```

Of interest to the Bayesian is to estimate the probabilities of the one-step transition matrix (Equation 4.22), and then use those estimates to estimate the six-step transition probabilities (Equation 4.23) and compare them to the entries of the matrix in Equation 4.23. Consider the first row of Equation 4.22, which is the conditional distribution of the five states (vertex number) given the initial state 0, then I will assume that the distribution of the number of transitions is multinomial with probabilities 0, .5, 0, 0, .5, and 0 and will generate four realizations assuming a total of 50 transitions with the following R statement for generating observations from the multinomial distribution.

```
> rmultinom(4,50,prob)
     [1] [2] [3] [4]
[0]   0   0   0   0
[1]  22  32  23  18
[2]   0   0   0   0
[3]   0   0   0   0
[4]  28  18  27  32
```

Thus, the first realization generates 22 transitions from state 0 to state 1 and 28 one-step transitions from state 0 to state 4, while there are zero transitions from 0 to the other three states 0, 2, and 3. The obvious estimates for the transition probabilities are $22/50 = .44$ and $28/50 = .56$ for P_{01} and P_{04}, respectively. Based on the first realization, what are the Bayesian estimates for these two transition probabilities? To perform the Bayesian analysis, I assumed that there were 50 transitions that the conditional distribution of observed transitions had a Bernoulli distribution with probability $P_{01} = .5$. Fifty observations were generated using WinBUGS, then using those fifty simulated observations, P_{01} was estimated based on WinBUGS Code 4.4. Note that the 50 simulated observations are included in the list statement and that the prior distribution of P_{01} is uniform. The Bayesian analysis is executed with 35,000 observations generated from the posterior distribution and initially with 5,000 observations.

WinBUGS 4.4

```
model {
        p01~dbeta(1,1)
            for (i in 1 : 50) {
                  y[i] ~ dbern(p01)
            }
        }
list(
y = c(
1.0,0.0,0.0,1.0,1.0,
0.0,0.0,1.0,1.0,1.0,
1.0,1.0,1.0,0.0,0.0,
1.0,0.0,1.0,0.0,1.0,
1.0,1.0,1.0,0.0,1.0,
1.0,0.0,0.0,1.0,0.0,
0.0,0.0,1.0,0.0,0.0,
0.0,1.0,1.0,1.0,0.0,
1.0,0.0,1.0,1.0,1.0,
1.0,1.0,0.0,0.0,1.0))
```

Table 4.2 presents the posterior distribution of P_{01}.

Based on the posterior mean and median, the estimate of P_{01} is .5769 with a posterior standard deviation of .06797, and the MCMC error is quite small implying that the estimate of .5769 is within .000385 units of the actual posterior mean. The 95% credible interval of P_{01} varies from .4433 to .7066. The posterior mean of .5769 is different for the value .50 of P_{01} used to generate the 50 transitions given in the list statement of WinBUGS Code 4.4. Recall that the goal of this section is to provide estimates of the six-step transition matrix (Equation 4.22). I will use R Code 4.2 to generate the sixth power of

$$\hat{P} = \begin{pmatrix} 0, .5769, 0, 0, .4321 \\ .5769, 0, .4321, 0, 0 \\ 0, .5769, 0, .4321, 0 \\ 0, 0, .5769, 0, .4321 \\ .5769, 0, 0, .4321, 0 \end{pmatrix}. \tag{4.23}$$

TABLE 4.2

Posterior Distribution of P_{01}

Parameter	Mean	SD	Error	2 1/2	Median	97 1/2
P_{01}	.5769	.06797	.000385	.4423	.5766	.7066

Note that I used $\hat{P_{01}} = \hat{P_{10}} = \hat{P_{21}} = \hat{P_{32}} = \hat{P_{40}} = .5769$ for estimates of the one-step transition probabilities in the matrix in (Equation 4.23). As a consequence of using R Code 4.2, the estimated six-step transition probabilities are given by

$$\hat{P}^6 = \begin{pmatrix} .40965410 & .1149245 & .2626511 & .1819211 & .08607882 \\ .09515764 & .4294210 & .1058457 & .1819211 & .24288429 \\ .35066753 & .1149245 & .3216377 & .1058457 & .16215428 \\ .24288429 & .3242767 & .1149245 & .2409077 & .13223654 \\ .14131536 & .3242767 & .2164934 & .1058457 & .26729857 \end{pmatrix}. \tag{4.24}$$

One should compare the estimated six-step transition probabilities given by Equation 4.24 with the corresponding entries of the matrix in Equation 4.21, the matrix of "actual" six-step transition probabilities.

A computation of interest in this section is that of the marginal distribution of $X(n)$. It is easy to show that the marginal distribution is

$$P[X(n) = j] = (\alpha P^n)_j, \quad n \geq 0, \tag{4.25}$$

namely, the jth component of the vector αP^n, where α is the row vector denoting the initial distribution of the process; that is, the jth component of α is $P[X(0) = j]$ and P^n is the nth power of the one-step transition matrix P. In order to explain this idea, suppose that $P[X(6) = 1]$ is to be estimated assuming that $\alpha = (1,0,0,0,0)$; that is, that the initial state is 0. Note that the estimated value of P^6 is given by Equation (4.24;, thus, $P[X(6) = 1]$ is estimated by the first component of the 1×5 vector:

$$\hat{P}[X(6) = 0] = (1,0,0,0,0) \begin{pmatrix} .40965410 & .1149245 & .2626511 & .1819211 & .08607882 \\ .09515764 & .4294210 & .1058457 & .1819211 & .24288429 \\ .35066753 & .1149245 & .3216377 & .1058457 & .16215428 \\ .24288429 & .3242767 & .1149245 & .2409077 & .13223654 \\ .14131536 & .3242767 & .2164934 & .1058457 & .26729857 \end{pmatrix},$$

which is the first component of (.4096, .1149, .2626, .1819, .0860), namely, .4096. Thus, at time 6, the estimated probability that the process returns to the first state (the starting vertex of the cycle graph) is .4096. The student will be asked to estimate $P[X(6) = 0]$ given other initial distributions α. Refer to the exercises at the end of the chapter.

Last to be considered is the Bayesian estimation of the joint distribution of the process at an arbitrary number of time points, but for now, I consider the following example based on the cycle graph with estimated one-step transition matrix (Equation 4.25). For example, it can be shown that

$$P[X(5) = i, X(6) = j, X(9) = k, X(17) = l] = P^8_{kl} P^3_{jk} P_{ij} \left(\alpha P^5\right)_i \tag{4.26}$$

for states i, j, k, and $l = 0, 1, 2, 3, 4$. How does one estimate this joint probability from a Bayesian viewpoint? Note that this probability depends on the one-step transition matrix P and its powers of order 8, 3, and 5. This probability will be estimated by

$$\hat{P}[X(5) = i, X(6) = j, X(9) = k, X(17) = l] = \hat{P}_{kl}^{8}\hat{P}_{jk}^{3}\hat{P}_{ij}\left(\alpha\hat{P}^{5}\right)_{i} \tag{4.27}$$

where $\hat{P}$ is given by Equation 4.23 and i, j, k, and $l = 0, 1, 2, 3, 4$. Recall that the entries of $\hat{P}$ are Bayesian estimates. Consider a special case of Equation 4.28, namely,

$$\hat{P}[X(5) = 0, X(6) = 1, X(9) = 2, X(17) = 3]; \tag{4.28}$$

that is, the probability that the process begins at vertex 0 and is at 0 at time 5, moves to the right at the next step, at time 9, is at vertex 2, and, at time 17, is at vertex 3. Suppose it is assumed that the initial state is at vertex 0; thus, let $\alpha = (1, 0, 0, 0, 0)$. Using the matrix power function of R, it can be shown that Bayesian estimates of the powers of $\hat{P}$ are

$$\hat{P}^{3} = \begin{pmatrix} .0000000, & .4796177, & .1077132, & .1077132, & .33219956 \\ .4796177, & .0000000, & .3592352, & .1077132, & .08067757 \\ .1077132, & .4796177, & .0000000, & .2961040, & .14380876 \\ .1920002, & .1438088, & .3953308, & .0000000, & .29610404 \\ .4435222, & .1438088, & .1438088, & .2961040, & .00000000 \end{pmatrix}, \tag{4.29}$$

$$\hat{P}^{5} = \begin{pmatrix} .06269902, & .42559064, & .13651139, & .13651139, & .28450488 \\ .42559064, & .06269902, & .31876879, & .13651139, & .10224748 \\ .13651139, & .42559064, & .06269902, & .23875887, & .18225740 \\ .24333324, & .18225740, & .31876879, & .06269902, & .23875887 \\ .37984463, & .18225740, & .18225740, & .23875887, & .06269902 \end{pmatrix}, \tag{4.30}$$

and

$$\hat{P}^{8} = \begin{pmatrix} .3712193, & .1560907, & .2545221, & .1772599, & .1152173 \\ .1359664, & .3913436, & .1370367, & .1772599, & .2327027 \\ .3375513, & .1560907, & .2881902, & .1330784, & .1593988 \\ .2396826, & .3129464, & .1508060, & .2109279, & .1599465 \\ .1806956, & .3129464, & .2097930, & .1330784, & .2377960 \end{pmatrix}. \tag{4.31}$$

This is sufficient information to estimate the desired probability (Equation 4.27), namely,

$$\hat{P}_{23}^{8}\hat{P}_{12}^{3}\hat{P}_{01}\left(\alpha\hat{P}^{5}\right)_{0} = (.2387)(.3592)(.5769)(.0626) = .00309 \tag{4.32}$$

Thus, the joint probability of the three events (going from the initial vertex 0 to the final vertex 3, at times 5, 6, 7, and 19) is .00309.

Of course, different initial distributions α could have been used, resulting in different probabilities of the event. Referring to the exercises at the end of the chapter, the student is invited to explore the use of various initial distributions and their effect on the primary event of interest described earlier. Also, left as an exercise is to develop a Bayesian test of the hypothesis that $P_{01} = .5$ versus the alternative using the information in Section 2.5.3, and in addition, to be left as an exercise is the prediction of future transitions from the cycle graph process using information from Section 2.6.2.

4.4 Limiting Distributions

It is very important to know the long-term behavior of a stochastic process. In the long run, what are the possible states of a Markov chain, and how do they depend on the initial state? Such problems come under the subject of limiting probabilities. R Code 4.2 (matrixpower) will be employed to determine the long-term behavior of a Markov chain. An example, based on a study by Martell,[4] reveals the long-term behavior of the Canadian Forest Fire Weather Index. A five-state transition matrix is based on data taken over 26 years at 15 weather stations. The time unit is a day, and the following matrix, taken from one location in the early summer, gives the probability of daily changes in the fire index (the states of the chain). The fire index has five values: nil, low, moderate, high, and extreme.

$$P = \begin{pmatrix} .575, .118, .172, .109, .026 \\ .453, .243, .148, .123, .033 \\ .104, .343, .367, .167, .019 \\ .015, .066, .381, .505, .096 \\ .000, .060, .149, .567, .224 \end{pmatrix}, \tag{4.33}$$

Thus, the probability of going from nil risk to low risk (in a one-day period) is .118, and the probability of remaining at a nil risk (over one day) is .575, etc. On the other hand, the probability of a daily change from a nil risk to an extreme risk is only .026. It is interesting to note that the probability of a daily change from an extreme risk to a nil risk is essentially 0, to three decimal places. Of interest to the forest service is the long-term behavior of the daily risk index; that is, what is the long-term chance of risk on a typical day in late summer?

This will be answered by using R Code 4.2 to compute powers of the one-step transition matrix P (Equation 4.33). Thus, consider powers 3, 10, 17, and 18:

$$P^3 = \begin{pmatrix} .3317973 & .1762260 & .2353411 & .2111096 & .04552595 \\ .3263579 & .1753868 & .2352439 & .2160688 & .04694271 \\ .2830784 & .1922351 & .2466504 & .2293466 & .04868947 \\ .1579034 & .1832159 & .2798370 & .3123858 & .06665790 \\ .1177433 & .1654309 & .2858074 & .3532858 & .07773251 \end{pmatrix}, \tag{4.34}$$

$$P^{10} = \begin{pmatrix} .2643504 & .1812413 & .2518115 & .2491008 & .05349592 \\ .2642635 & .1812455 & .2518332 & .2491513 & .05350655 \\ .2640283 & .1812567 & .2518919 & .2492878 & .05353532 \\ .2625915 & .1813257 & .2522504 & .2501214 & .05371100 \\ .2618765 & .1813600 & .2524288 & .2505362 & .05379840 \end{pmatrix}, \tag{4.35}$$

$$P^{17} = \begin{pmatrix} .2636889 & .1812730 & .2519766 & .2494847 & .05357682 \\ .2636880 & .1812731 & .2519768 & .2494852 & .05357692 \\ .2636856 & .1812732 & .2519774 & .2494866 & .05357722 \\ .2636711 & .1812739 & .2519810 & .2494950 & .05357899 \\ .2636639 & .1812742 & .2519828 & .2494992 & .05357987 \end{pmatrix}, \tag{4.36}$$

and

$$P^{18} = \begin{pmatrix} .2636856 & .1812732 & .2519774 & .2494866 & .05357721 \\ .2636852 & .1812732 & .2519775 & .2494869 & .05357727 \\ .2636839 & .1812733 & .2519778 & .2494876 & .05357742 \\ .2636764 & .1812736 & .2519797 & .2494919 & .05357834 \\ .2636727 & .1812738 & .2519806 & .2494941 & .05357880 \end{pmatrix}. \tag{4.37}$$

This demonstrates that the day 17 and day 18 probabilities of a change agree to at least four decimal places and implies that the long-run probability of risk in late summer is as follows: nil, .2636; low, .18127; moderate, .25197; high, .24984; and extreme, .05257.

Of course, the three long-run probabilities of Table 4.3 are somewhat misleading because the one-step transition matrix P (Equation 4.34) gives only an estimate of the transition probabilities. What should be remembered is that these probabilities are based on the number of observed transitions from one state to the other, which is not available. Thus, I will generate transition counts corresponding to the transition probabilities of Equation 4.33 with the R command:

```
> rmultinom(5,100,prob)
```

where prob = (.575., 118, .172, .109, .026) is the first row of the one-step transition matrix P (Table 4.4).

TABLE 4.3

Long-Term Behavior of Risk of Forest Fire

Nil	Low	Moderate	High	Extreme
.2636	.1812	.2519	.2494	.0535

TABLE 4.4

Five Realizations of Forest Fire Index Risk

Transition	R1	R2	R3	R4	R5
n_{11}	60	64	65	61	57
n_{12}	9	12	7	6	17
n_{13}	16	12	11	20	17
n_{14}	12	10	15	9	8
n_{15}	3	2	2	4	1

Thus, for the first realization, there were 60 transitions over one day from a nil to a nil risk, 9 daily changes from a nil to a low risk, 16 daily changes from a nil to a moderate risk, 12 from a nil to a high risk, and 3 from a low to an extreme risk. One can see that the transition counts do indeed follow the transition probabilities given by the first row of Equation 4.33.

The multinomial mass function for the transition counts is

$$f\left(n_{11}, n_{12}, n_{13}, n_{14}, n_{15} \mid p_{11}, p_{12}, p_{13}, p_{14}, p_{15}\right) = \left[n! / \prod_{j=1}^{j=5} n_{1j}!\right] \prod_{j=1}^{j=5} p_{1j}^{n_{1j}}, \tag{4.38}$$

where $n = \sum_{j=1}^{j=5} n_{1j}$ is the total number of transition counts with initial fire index nil; the transition probabilities are unknown; and $\sum_{j=1}^{j=5} p_{1j} = 1$. Assuming the improper prior density

$$\xi(p_{11}, p_{12}, p_{13}, p_{14}, p_{15}) \propto \left[1 / \prod_{j=1}^{j=5} p_{1j}\right] \tag{4.39}$$

for $0 < p_{1j} < 1$, $j = 1,2,3,4,5$ and $\sum_{j=1}^{j=5} p_{1j} = 1$, it is seen that the posterior distribution of the five transition probabilities is Dirichlet$(n_{11}, n_{12}, n_{13}, n_{14}, n_{15})$ = Dirichlet(60,9,16,12,3).

Thus, the various posterior means are

$$\begin{aligned} E(P_{11} \mid \text{data}) &= 60/100 = .6, \\ E(P_{12} \mid \text{data}) &= 9/100 = .09, \\ E(P_{13} \mid \text{data}) &= 16/100 = .16, \\ E(P_{14} \mid \text{data}) &= 12/100 = .12, \end{aligned} \tag{4.40}$$

and

$$E(P_{15} \mid \text{data}) = 3/100 = .03.$$

The posterior means should be compared to the corresponding transition probabilities .575, .118, .172, .109, and .026, respectively, used to generate the multinomial realizations of

Table 4.3. It is seen that the agreement is quite good. How do we construct credible intervals for these parameters?

For P_{11}, it can be shown that (5027, .6934) is a 95% credible interval, where .5049 is the 2½ percentile and .694 is the 97½ percentile. In a similar manner (.04241, .15327) is a 95% credible interval for P_{12}. The exercises at the end of this chapter will involve finding credible intervals for the other transition probabilities for the evolution of the forest fire index.

The following R command was used to compute the pth 100 percentile of the beta distribution with parameters alpha=shape1 and beta=shape2:

```
qbeta(p, shape1, shape2, ncp = 0, lower.tail = TRUE, log.p = FALSE)
```

In particular for the posterior distribution of P_{11}, the following command was employed to find the 97½ percentile, which gives an answer of .6934.

```
qbeta(.975,60,40, = 0, lower.tail = TRUE, log.p = FALSE).
```

Also, one needs the posterior variance of the transition probabilities; thus, recall that if a random variable has a beta distribution with parameters α and β, then its variance is $\alpha\beta/[(\alpha+\beta)^2(\alpha+\beta+1)]$. Therefore,

$$\text{VAR}(P_{11}|\text{data}) = (60)(40)/\left[(60+40)^2(60+40+1)\right] = .002376237. \tag{4.41}$$

We now develop a Bayesian method of predicting future transitions for the forest fire index model with transition matrix P given by Equation 4.34. The Bayesian predictive density is defined as follows:

Let m_{1j}, $j = 1,2,3,4,5$, be the future transitions counts corresponding to the first row the transition matrix P of Equation 4.33 and assume that the transition counts follow a multinomial distribution with density

$$f(m_{11}, m_{12}, m_{13}, m_{14}, m_{15}|p) = [m!/m_{11},!m_{12},!m_{13},!m_{14},!m_{15}!]\prod_{j=}^{j=5} p_{1j}^{m_{1j}}. \tag{4.42}$$

That is, a multinomial mass function with parameters $p = (p_{11}, p_{12}, p_{13}, p_{14}, p_{15})$ and $m = \sum_{j-1}^{j=5} m_{1j}$. Therefore, the posterior density of the transition probabilities P_{1j}, $j = 1,2,3,4,5$, is Dirichlet $(n_{11}, n_{12}, n_{13}, n_{14}, n_{15})$ with density

$$\xi(p|n) = \left[\Gamma(n)/\prod_{j=1}^{j=5}\Gamma\left(n_{1j}\right)\right]\prod_{j=1}^{j=5} p_{1j}^{n_{1j}-1}, \tag{4.43}$$

where $n = \sum_{j=1}^{j=5} n_{1j}$ and $1 = \sum_{j=1}^{j=5} p_{1j}$.

The Bayesian predictive mass function of the m_{1j}, $j = 1,2,3,4,5$, is

$$g(m|n) = E[f(m|p)], \tag{4.44}$$

where $m = \sum_{j-1}^{j=5} m_{1j}$ is the total number of transitions, m_{1j} is the number of transitions from state 1 to state j, and $j = 1, 2, 3, 4, 5$.

Here, E of Equation 4.44 denotes the expectation of the conditional mass function (Equation 4.43) of the future transitions (given the transition probabilities) with respect to the posterior distribution of the transition probabilities with density (Equation 4.43). It can be shown that the predictive mass function (Equation 4.44) reduces to

$$g(m_{11}, m_{12}, m_{13}, m_{14}, m_{15} | n_{11}, n_{12}, n_{13}, n_{14}, n_{15}) = \left[m!\Gamma(n)\prod_{j=1}^{j=5}\Gamma\left(n_{1j} + m_{1j}\right)\right] / \left[\prod_{j=1}^{j=5} m_{1j}! \prod_{j=1}^{j=5}\Gamma\left(n_{1j}\right)\Gamma(m+n)\right], \tag{4.45}$$

where $m = \sum_{j-1}^{j=5} m_{1j}$ is the total number of transitions, m_{1j} is the number of transitions from state 1 to state j, and $j = 1, 2, 3, 4, 5$.

Note that Equation 4.45 is the conditional mass function of the future transition counts given the past transition counts. WinBUGS Code 4.5 generates 1000 observations from the predictive mass function (Equation 4.45), where the posterior distribution of the transition probabilities is the Dirichlet with parameters (60,9,16,12,3). This assumes that the prior density of the transition probabilities is the improper prior given by Equation 4.39. The 100 transition counts are given by the list statement with matrix y. Note that the first vector of predicted transition counts is (59,16,9,14,2), that is, the number of transitions from state 1 (nil) to state 1(nil) is 59, the number of predicted transitions from state 1(nil) to state 5 (extreme) is 2, etc. Also note the variation across the 100 prediction vectors of the counts of the fire index.

WinBUGS Code 4.5

```
{

alpha[1]<-60
alpha[2]<-9
 alpha[3]<-16
 alpha[4]<-12
  alpha[5]<-3

for( i in 1:100){

y[i,1:5]~dmulti(p[i,1:5],100)
p[i,1:5]~ddirch(alpha[1:5])

}}

List(y = structure(.Data = c(
58.0,16.0,9.0,14.0,3.0,
69.0,8.0,7.0,15.0,1.0,
60.0,3.0,22.0,13.0,2.0,
```

```
77.0,4.0,9.0,8.0,2.0,
53.0,8.0,18.0,17.0,4.0,
48.0,8.0,26.0,16.0,2.0,
63.0,5.0,11.0,16.0,5.0,
64.0,8.0,21.0,7.0,0.0,
62.0,8.0,7.0,22.0,1.0,
70.0,6.0,9.0,13.0,2.0,
56.0,10.0,16.0,13.0,5.0,
52.0,13.0,17.0,15.0,3.0,
63.0,12.0,16.0,9.0,0.0,
58.0,3.0,20.0,15.0,4.0,
64.0,6.0,14.0,16.0,0.0,
64.0,9.0,19.0,7.0,1.0,
59.0,5.0,20.0,14.0,2.0,
64.0,7.0,14.0,13.0,2.0,
70.0,12.0,11.0,6.0,1.0,
59.0,7.0,3.0,27.0,4.0,
61.0,9.0,14.0,16.0,0.0,
72.0,7.0,15.0,6.0,0.0,
53.0,11.0,13.0,20.0,3.0,
63.0,8.0,14.0,11.0,4.0,
57.0,9.0,18.0,13.0,3.0,
68.0,13.0,11.0,6.0,2.0,
70.0,7.0,14.0,7.0,2.0,
47.0,18.0,12.0,14.0,9.0,
60.0,12.0,16.0,12.0,0.0,
58.0,8.0,12.0,14.0,8.0,
57.0,6.0,26.0,10.0,1.0,
71.0,10.0,6.0,11.0,2.0,
44.0,21.0,24.0,5.0,6.0,
63.0,10.0,16.0,4.0,7.0,
60.0,8.0,17.0,9.0,6.0,
65.0,6.0,12.0,15.0,2.0,
63.0,6.0,20.0,9.0,2.0,
52.0,12.0,28.0,7.0,1.0,
65.0,5.0,20.0,9.0,1.0,
47.0,11.0,25.0,15.0,2.0,
50.0,13.0,21.0,14.0,2.0,
61.0,2.0,16.0,15.0,6.0,
61.0,9.0,12.0,12.0,6.0,
71.0,2.0,15.0,9.0,3.0,
66.0,2.0,9.0,16.0,7.0,
69.0,10.0,13.0,6.0,2.0,
55.0,2.0,16.0,24.0,3.0,
61.0,14.0,14.0,11.0,0.0,
65.0,7.0,14.0,13.0,1.0,
53.0,2.0,27.0,9.0,9.0,
74.0,8.0,5.0,12.0,1.0,
63.0,10.0,13.0,14.0,0.0,
```

```
49.0,8.0,15.0,23.0,5.0,
80.0,5.0,4.0,6.0,5.0,
55.0,5.0,28.0,7.0,5.0,
58.0,10.0,16.0,14.0,2.0,
53.0,10.0,19.0,16.0,2.0,
69.0,7.0,13.0,7.0,4.0,
62.0,9.0,14.0,15.0,0.0,
67.0,5.0,16.0,11.0,1.0,
55.0,20.0,14.0,10.0,1.0,
58.0,6.0,19.0,16.0,1.0,
60.0,5.0,15.0,15.0,5.0,
61.0,10.0,16.0,10.0,3.0,
68.0,11.0,6.0,14.0,1.0,
57.0,10.0,21.0,9.0,3.0,
71.0,6.0,11.0,10.0,2.0,
55.0,15.0,14.0,8.0,8.0,
43.0,12.0,21.0,21.0,3.0,
70.0,3.0,13.0,14.0,0.0,
60.0,10.0,12.0,11.0,7.0,
45.0,14.0,10.0,21.0,10.0,
49.0,10.0,28.0,13.0,0.0,
51.0,8.0,22.0,15.0,4.0,
38.0,8.0,28.0,22.0,4.0,
68.0,5.0,14.0,12.0,1.0,
51.0,8.0,30.0,8.0,3.0,
62.0,8.0,21.0,9.0,0.0,
46.0,9.0,30.0,14.0,1.0,
63.0,9.0,19.0,8.0,1.0,
67.0,11.0,10.0,9.0,3.0,
67.0,8.0,12.0,11.0,2.0,
45.0,23.0,17.0,12.0,3.0,
54.0,18.0,12.0,9.0,7.0,
56.0,18.0,14.0,9.0,3.0,
63.0,7.0,14.0,10.0,6.0,
52.0,14.0,16.0,14.0,4.0,
55.0,8.0,10.0,19.0,8.0,
65.0,12.0,14.0,8.0,1.0,
67.0,3.0,12.0,15.0,3.0,
62.0,8.0,16.0,12.0,2.0,
60.0,12.0,20.0,7.0,1.0,
70.0,8.0,8.0,13.0,1.0,
54.0,8.0,19.0,14.0,5.0,
70.0,4.0,12.0,10.0,4.0,
50.0,9.0,15.0,23.0,3.0,
58.0,19.0,10.0,12.0,1.0,
54.0,8.0,8.0,27.0,3.0,
49.0,13.0,18.0,14.0,6.0,
56.0,6.0,17.0,18.0,3.0),
.Dim = c(100,5)))
```

The student will be asked to verify the list of predicted transition counts given by matrix y of the preceding list statement in WinBUGS Code 4.5.

Consider the following test of hypotheses concerning the first row of the transition matrix of the fire index sample:

$$H_0: P_{11} = .575,\ P_{12} = .118,\ P_{13} = .172,\ P_{14} = .109,\ P_{15} = .026 \tag{4.46}$$

versus

$$H_1: H_0 \text{ is not true.} \tag{4.47}$$

How does one assign prior information to this case? A reasonable approach is to assign a positive probability π_0 for the null hypothesis and, for the alternative assign, a prior density $\pi_1\zeta_1(P)$, where

$$\int_{P:H_1} \zeta_1(P)dP = 1. \tag{4.48}$$

Thus, $\pi_0 + \pi_1 = 1$, and it is seen that the prior probability of the alternative is π_1, and for values of the alternative, ζ_1 is the density of the continuous random vector P that expresses the prior knowledge one has for the alternative hypothesis. Note that $P = (P_{11}, P_{12}, P_{13}, P_{14}, P_{15})$ is the first row of the transition matrix and $P_0 = (.575, .118, .172, .109, .026)$ is the hypothesized value under the null hypothesis.

Let

$$\zeta(n_{obs}) = \pi_0\zeta(n_{obs}|P_0) + \pi_1 \int \zeta_1(P)p(n_{obs}|P)dP, \tag{4.49}$$

where $n_{obs} = (n_{11}, n_{12}, n_{13}, n_{14}, n_{15})$ is the vector of observations with conditional mass function $\zeta(n_{obs}|P)$ and where $\zeta(n_{obs})$ is the marginal mass function of the observations.

By letting

$$\zeta_1(n_{obs}) = \int_{P \neq P_0} \zeta_1(P)\zeta(n_{obs}|P)dP, \tag{4.50}$$

the marginal density (Equation 4.50) can be expressed as

$$\zeta(n_{obs}) = \pi_0\zeta(n_{obs}|P_0) + \pi_1\zeta_1(n_{obs}), \tag{4.51}$$

and the posterior probabilities of the null and alternative hypotheses can be expressed as

$$p_0 = \pi_0\zeta(n_{obs}|P_0)/\zeta(n_{obs}). \tag{4.52}$$

In a similar manner, the posterior probability of the alternative hypothesis is

$$p_1 = \pi_1\zeta_1(n_{obs})/\zeta(n_{obs}). \tag{4.53}$$

In order to compute the probability of the null and alternative hypotheses, the following distributions are relevant.

First, the probability mass function of the observations given the unknown parameters is multinomial

$$\zeta(n_{\text{obs}}|P) = \left[n!/\prod_{j=1}^{j=5} n_{1j}!\right]\prod_{j=1}^{j=5} P_{1j}^{n_{1j}}, \tag{4.54}$$

where $\sum_{j=1}^{j=5} P_{1j} = 1$ and $\sum_{j=1}^{j=5} n_{1j} = n$. Also, the prior density of unknown parameters under the alternative is Dirichlet, namely,

$$\zeta_1(P) = \left[\Gamma\left(\sum_{j=1}^{j=5}\alpha_{1j}\right)/\prod_{j=1}^{j=5}\Gamma(\alpha_{1j})\right]\prod_{j=1}^{j=5} P_{1j}^{\alpha_{1j}-1}, \tag{4.55}$$

where $\sum_{j=1}^{j=5} P_{1j} = 1$. To compute the posterior probability (Equation 4.52) of the null hypothesis, relevant information required is $n_{\text{obs}} = (60,9,16,12,3)$, and for the parameters of the prior I used $\alpha = (23,4.72,6.88,4.36,1.04)$. Now one can show that

$$\begin{aligned}\zeta(n_{\text{obs}}|P_0) &= [100!/60!9!16!12!3!](.575)^{60}(.118)^{9}(.172)^{16}(.109)^{12}(.026)^{3} \\ &= .00025145477\end{aligned} \tag{4.56}$$

and

$$\zeta_1(n_{\text{obs}}) = .001114.$$

Therefore, the probability of the null hypothesis (Equation 4.52) is

$$p_0 = .184599; \tag{4.57}$$

thus, the evidence suggests the null hypothesis is not true. The hypothesized values were those used to generate various realizations (depicted in Table 4.4) from the fire index example with transition matrix (Equation 4.33). The student will be asked to repeat this hypothesis testing example using the second realization of Table 4.4; consequently, one would expect a different (different from .18459) posterior probability of the null hypothesis.

Of course, the limiting probabilities of a stochastic process are related to the stationary distribution of the process, and this will be explored in the next section.

4.5 Stationary Distributions

The objective of this section is to provide Bayesian inferences for the stationary distribution of a stochastic process. It is interesting to consider what will happen if the limiting distribution of a Markov chain $\{X(n), n = 0,1,...\}$ is assigned as the initial distribution of the chain.

Consider the two-state chain with transition matrix

$$P = \begin{pmatrix} 1-p, p \\ q, 1-q \end{pmatrix}, \tag{4.58}$$

where $0 < p < 1$ and $0 < q < 1$, then, it can be shown that the limiting distribution is given by the vector

$$\theta = (q/(p+q), p/(p+q)). \tag{4.59}$$

Also, it can be shown that the distribution of $X(1)$ is given by $\theta P = \theta$. A vector π that satisfied the set of equations $\pi P = \pi$ sets the stage for the concept of a stationary distribution of a Markov chain. The definition of a stationary distribution is as follows:

For a Markov chain with transition matrix P, the stationary distribution of the chain is given by the vector π that satisfies

$$\pi = \pi P,$$

which is the system of equations

$$\pi_j = \sum_i \pi_i P_{ij}, \quad \forall j. \tag{4.60}$$

Bayesian inferences for the stationary probability vector π will be the principal topic of this section. Consider the following example of a Markov chain with transition matrix

$$P = \begin{pmatrix} .45, .48, .07 \\ .05, .70, .25 \\ .01, .50, .49 \end{pmatrix}, \tag{4.61}$$

where the three states represent the social class of a person, namely, 1 denotes lower class, 2 signifies middle class, 3 represents the upper class. The transition probabilities denote the class mobility of a family member. Thus .48 is the probability a person with lower-class parents will be a member of the middle class, while the probability is .07 that the person will have a higher class occupation. See page 154 of Ross[5] for additional details. The stationary distribution π satisfies the following system of equations:

$$\begin{aligned} \pi_1 &= .45\pi_1 + .05\pi_2 + .01\pi_3, \\ \pi_2 &= .48\pi_1 + .70\pi_2 + .50\pi_3, \\ \pi_3 &= .07\pi_1 + .25\pi_2 + .49\pi_3, \\ \pi_1 &+ \pi_2 + \pi_3 = 1 \end{aligned} \tag{4.62}$$

It can be shown that the solution is $\pi_1 = .07$, $\pi_2 = .62$, and $\pi_3 = .31$; thus, in the long term, 7% will be in the lower class, 62% in the middle, and 31% in higher-class occupations. The following R Code computes the stationary distribution of a Markov chain with a given transition matrix mat:

R Code 4.3

```
stationary <- function(mat) {
x = eigen(t(mat))$vectors[,1]
as.double(x/sum(x))
}

mat<-matrix(c(.45,.48,.07,.05,.70,.25,.01,.50,.49),
nrow=3,ncol=3,byrow=TRUE)
```

π_1 = .06238859, π_2 = .62344029, and π_3 = .31417112 is the solution given by R.

Remember that in practice, what one knows are the transition counts of the chain, from which the transition probabilities are computed; thus, in reality, the transition probabilities (Equation 4.61) are only estimates.

In order to make Bayesian inferences, I will generate several realizations from the chain which will provide one with transition counts, then using those counts as the sample information, Bayesian inferences are possible. Table 4.5 portrays five realizations from a multinomial distribution with three classes and probabilities (.05, .70, .25) for a total of $n = 200$ outcomes. The following R Code was used to generate the five realizations from the second row of the transition matrix (Equation 4.61) of the social mobility example. Note that this routine generates samples from the appropriate multinomial distribution.

```
< rmultinom(5,100,prob)
```

where prob = (.05, .70, .250) is the second row of the one-step transition matrix P (Equation 4.62).

Consider the second row of the transition matrix, then using the first realization, there are 14 transitions from a middle to a lower-class occupation, 139 people with middle-class occupations and whose parents have middle-class occupations, and 47 people with higher-class occupations and whose parents are middle class (Table 4.5).

Thus, for the first realization, there are 14 transitions from a middle- to a lower-class occupation, 139 people with middle-class occupations and with parents that had middle-class occupations, and 47 people with higher-class occupations and whose parents are

TABLE 4.5

Five Realizations for Social Mobility Study

Transition Count	1	2	3	4	5
n_{21}	14	11	10	12	8
n_{22}	139	128	148	141	157
n_{23}	47	61	42	47	4
n_{11}	98	82	86	86	81
n_{12}	84	105	106	99	107
n_{13}	18	13	8	15	13
n_{31}	4	3	6	3	3
n_{32}	99	102	101	99	109
n_{33}	97	95	93	98	88

middle class. Notice the similarity from realization to realization as well as the variation. For the first realization, one would estimate P_{21} by $14/200 = .07$, P_{22} by $139/200 = .695$, and P_{23} by $47/200 = .235$. I arbitrarily set $n = 200$, which should be a large enough sample size to efficiently estimate the transition probabilities. How should the Bayesian estimate the stationary distribution of the social mobility example? Obviously, one needs estimates for all the nine transition probabilities. Two sources of information are needed for the Bayesian analysis, the information prior to the study (the prior density for the transition probabilities) and the information from the sample, the transition counts of Table 4.5, expressed as a multinomial mass function for the transition counts given the vector of transition probabilities. Using an improper prior for the transition probabilities for the first row of P, the prior density is

$$\xi(P_{11}, P_{12}, P_{13}) \propto \prod_{j=1}^{j=3} P_{1j}^{-1}, \tag{4.63}$$

where for the first row of transition probabilities, $\sum_{j=1}^{j=3} P_{1j} = 1$.

Assume that the transition counts for the first row follow a multinomial distribution with mass function

$$g(n_{11}, n_{12}, n_{13} | P_{11}, P_{12}, P_{13}) = [200!/98!84!18!] P_{11}^{98} P_{12}^{84} P_{13}^{18}, \tag{4.64}$$

where $\sum_{j=1}^{j=3} P_{1j} = 1$. Thus, by Bayes theorem, the posterior distribution of the transition probabilities of the first row is Dirichlet(98,84,18). In a similar manner, the posterior distribution of the transition probabilities of the second row is Dirichlet(14,139,47), and for the third is Dirichlet(4,99,97). This is sufficient information to provide estimates of the transition probabilities. See page 91 of Degroot[6] for the formulas for the moments of the Dirichlet distribution.

Consider first estimating the transition probabilities and stationary distribution via Bayes theorem. Based on the first realization for the social mobility example, I will generate samples from the posterior distribution of the nine transition probabilities as well as the posterior distribution of the stationary distribution. Refer to Equation 4.62, and then in general, the solution (π_1, π_2, π_3) is the stationary distribution of the mobility example.

The constraint is imposed by solving an associated system of equations:

$$\begin{aligned} P_{11} + P_{21}x_2 + P_{31}x_3 &= 1, \\ P_{12} + P_{22}x_2 + P_{32}x_3 &= x_2 \end{aligned} \tag{4.65}$$

with solution $x = (1, x_2, x_3)$, where

$$x_2 = [P_{32}(1 - P_{11}) + P_{12}P_{31}]/[P_{32}P_{21} - P_{31}(P_{22} - 1)], \tag{4.66}$$

$$x_3 = [1 - P_{11} - P_{21}x_2]/P_{31}.$$

Let $T = 1 + x_2 + x_3$ be the sum of the components of x; then the stationary distribution is

$$\pi = (1/T)(1, x_2, x_3)\,. \tag{4.67}$$

WinBUGS Code 4.6 generates the posterior distribution of the transition probabilities, and the stationary distribution and the code statements are similar to those expressed by Equations 4.66 and 4.67. Thirty-five thousand observations are generated from the posterior distribution with a burn-in of 500.

WinBUGS Code 4.6

```
model;

{
# transition probabilities social mobility
p11~dbeta(98,102)
p12~dbeta(84,116)
p13~dbeta(18,182)
p21~dbeta(14,186)
p22~dbeta(139,61)
p23~dbeta(47,153)
p31~dbeta(4,196)
p32~dbeta(99,101)
p33~dbeta(97,103)

x2<-(p32*(1-p11)+p12*p31)/(p32*p21-p31*(p22-1))
x3<-(1-p11-p21*x2)/p31
tot<-1+x2+x3

# stationary distribution
pi1<-(1/tot)
pi2<-x2*(1/tot)
pi3<-x3*(1/tot)

}
```

The Bayesian analysis is reported in Table 4.6.

Table 4.6 reveals a lot of information for estimating the stationary distribution of a Markov chain. Thus, in the long run, one would expect that 9.5% of the participants would be in a lower-class occupation with a 95% credible interval of (.0565, .1446), from 5.5% to 14.4%. The Bayesian estimates of the stationary distribution should be compared to the stationary distribution (.0623, .6234, .3144), which was based on the reported transition probabilities of Equation 4.61. On the other hand, the Bayesian stationary distribution is based on one realization generated from the transition matrix (Equation 4.61). The Bayesian medians of the stationary distribution are quite close: .0565 compared to .0955, .6103 compared to .6234, and .2942 compared to .3144.

Appearing in Figure 4.2 is a plot of the posterior density for π_2, which portrays a distribution which is symmetric about the mean .0955hh.

TABLE 4.6

Posterior Distributions for the Social Mobility Example

Parameter	Mean	SD	Error	2 1/2	Median	97 1/2
P_{11}	.4903	.0353	.0002103	.4219	.4901	.5589
P_{12}	.4199	.0347	.0002319	.353	.4194	.4886
P_{13}	.0897	.0201	.000127	.0544	.0883	.1329
P_{21}	.0699	.0179	.0001153	.0386	.0684	.1089
P_{22}	.6953	.0326	.0001996	.6296	.6958	.7577
P_{23}	.2349	.0299	.000188	.1788	.2342	.2956
P_{31}	.0200	.0098	.0000624	.0055	.0184	.0134
P_{32}	.4947	.0353	.000249	.4254	.495	.5636
P_{33}	.4885	.0355	.000211	.4162	.4854	.5555
π_1	.0955	.0224	.000144	.0565	.0939	.1446
π_2	.6103	.0282	.000180	.5557	.6102	.6661
π_3	.2942	.0368	.000237	.2212	.295	.3645

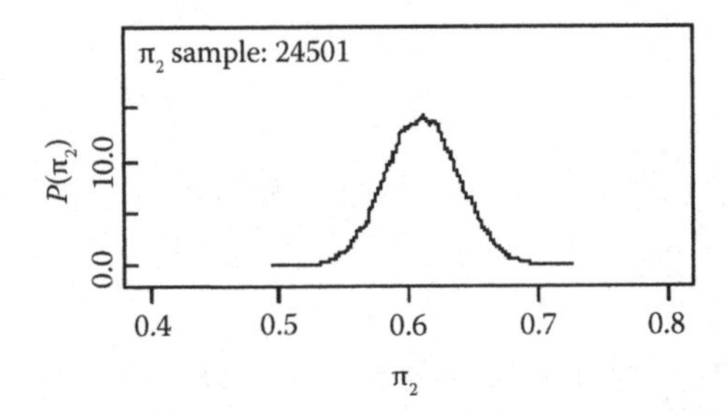

FIGURE 4.2
Posterior density of π_2.

4.6 Where Is That Particular State?

4.6.1 Introduction

How is a Bayesian analysis developed for a Markov chain with several communicating classes? The long-term behavior of a Markov chain involves many topics of interest, including knowing how often a particular state is visited, and this all depends on the concepts of the accessibility of a state from the other states of the chain. Recall the key concept of accessibility, which is defined as follows: state j is accessible from state j if there exists a $n \geq 0 \ni P_{ij}^{n} > 0$. Refer to Section 3.4.2 for additional information about accessible states. Accessibility leads to the idea of communication between states, namely, two states communicate if j is accessible from i and i is accessible from j, and this relation is designated by $i \leftrightarrow j$. Remember that the communication relation $i \leftrightarrow j$ is reflexive, symmetric, and transitive and partitions the state space into various communication classes. The member of a class communicates with itself and all other states in that class, but not with states outside of that class. The following example demonstrates how the communication relation partitions a chain with six states into three classes. Let the 6×6 transition matrix with states 1, 2, 3, 4, 5, and 6 be

$$P = \begin{pmatrix} 1/6, 1/3, 0, 0, 1/2, 0 \\ 0, 1, 0, 0, 0, 0, \\ 0, 0, 0, 0, 3/4, 1/4 \\ 1, 0, 0, 0, 0, 0, \\ 4/5, 0, 0, 1/5, 0, 0 \\ 0, 0, 1/2, 0, 1/2 \end{pmatrix}. \tag{4.68}$$

Thus, the probability that state 1 communicates with itself is $1/6$, while the probability of going from state 3 to state 5 is $3/4$, etc.

The R Code 4.4 develops a transition graph that shows the partitioning of the six states into three communication classes (Figure 4.3).

R Code 4.4

```
library(igraph)
P<-matrix(c(1/6,1/3,0,0,1/2,0,
+ +        0,1,0,0,0,0,
+ +        0,0,0,0,3/4,1/4,
+ +        1,0,0,0,0,0,
+ +        4/5,0,0,1/5,0,0,
+ +        0,0,1/2,0,1/2,0),nrow=6,ncol=6,byrow=TRUE)
> g<-graph.adjacency(P,weighted=TRUE)
> plot(g)
```

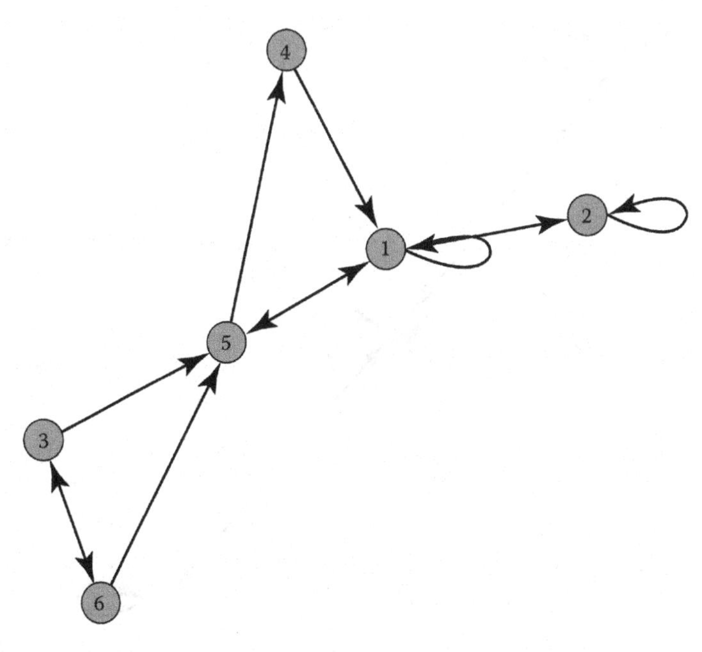

FIGURE 4.3
Transitions into three classes.

For example, for this chain, the probability of a transition from state 1 to state 2 is 1/3, but once the state is in state 2, it stays there; thus, 2 is an absorbing state. When all the states of a chain form one communicating class, the chain is called irreducible. A communicating class is closed if it is impossible to transition to a state outside of the communicating class.

4.6.2 Irreducible Chains

An example of an irreducible chain with states 1, 2, and 3 is one with transition matrix

$$P = \begin{pmatrix} 1/2, 1/2, 0 \\ 1/2, 1/4, 1/4 \\ 0, 1/3, 2/3 \end{pmatrix}, \tag{4.69}$$

where one can easily show that each state communicates with each of the other two. The corresponding transition graph is represented by Figure 4.4 and demonstrates that the chain is irreducible. The graph was executed using a suitable modification of R Code 4.4.

One thousand observations were generated from the chain with transition matrix P (Equation 4.80). Using only those simulations starting with the 3 state, I found the following number of transitions from the conditional distribution of the third row of P. There were 0 transitions from 3 to 1, 76 transitions from 3 to 2, and 175 transitions from 3 to state 3; thus, the fraction of 3 to 2 transitions is $76/251 = .302$, and for 3 to 3, the fraction is $175/251 = .697$. Note that based on the third row of P, $P_{32} = 1/3$ (compared to .302) and $P_{33} = 2/3$ (compared to .697), and it appears that the Markov chain simulation is believable. Another way to generate observations from the conditional distribution is to assume a multinomial distribution with parameters 0, 1/3, 2/3 and use the R Code

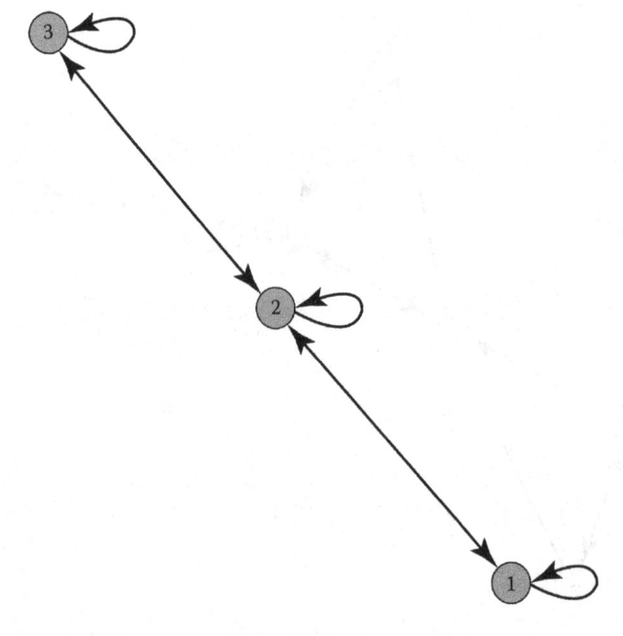

FIGURE 4.4
Transitions for irreducible chain.

```
prob<-c(0,1/3,2/3)
> rmultinom(3,100,prob)
```

which produces three multinomial realizations with parameter vector (P_{31}, P_{32}, P_{33}) (Table 4.7).

Bayesian inferences are made the usual way: (1) assume that the distribution of the transition counts of Table 4.7 is a multinomial with unknown parameter vector (P_{31}, P_{32}, P_{33}), (2) assign a uniform prior distribution to these three unknown parameters, and (3) determine the parameters of the posterior Dirichelt distribution of (P_{31}, P_{32}, P_{33}). Using the first realization of Table 4.7 and assigning the uniform prior to the unknown parameters result in a Dirichlet posterior with parameter vector (1, 31,71). Thus, the marginal posterior mean vector of (P_{31}, P_{32}, P_{33}) is $(1/103, 31/103, 71/103) = (.0097, .3009, .6893)$. Are these reasonable estimates? The following WinBUGS Code pertains to estimating the transition probabilities for the third row of the transition probability matrix and is executed with 35,000 observations for the simulation and 5,000 for the burn-in.

```
model;

{

# transition probabilities irreducible chain
p31~dbeta(1,102)
p32~dbeta(31,72)
p33~dbeta(71,32)

}
```

The posterior distributions of P_{32} and P_{33} appear to be symmetric about the posterior mean and the MCMC errors imply the simulation was successful for estimating the posterior means (Table 4.8). Note that estimation of P_{31} could have been ignored because it is

TABLE 4.7

Three Multinomial Realizations

	R1	R2	R3
P_{31}	0	0	0
P_{32}	30	27	28
P_{33}	70	70	72

TABLE 4.8

Posterior Distribution of Transition Probabilities of Irreducible Chain

Probability	Mean	SD	Error	2 1/2	Median	97 1/2
P_{31}	.00972	.009502	.00004776	.000251	.006879	.006879
P_{32}	.3012	.04501	.0002719	.2176	.2998	.3932
P_{33}	.6895	.04561	.000249	.5955	.691	.7745

known to be zero. Was the information in the first realization sufficient to accurately estimate the transition probabilities P_{32} and P_{33}? It will be left as an exercise to execute the Bayesian estimation of these two transition probabilities based on the other two realizations of Table 4.7.

As a last example, consider the random walk with state space {1, 2, ..., N} and transition matrix

$$P = \begin{pmatrix} 1,0,0,0.........0 \\ q,0,p,0,......0 \\ 0,q,0,p,0,....0 \\ . \\ . \\ . \\ 0,0,0,......0,p,0 \\ 0,0,0,.....q,0,p \\ 0,0,0,......0,0,1 \end{pmatrix}, \tag{4.70}$$

where $0 \le p \le 1$, $0 \le q \le 1$, $p + q = 1$.

Notice that the walker starts at state x, where $x = 2, 3, 4, 5, 6, 7$, or 8, and the walker stops walking when $x = 1$ or when $x = 9$.

The probability of moving to the right one step is $p = .5$, and the probability of moving to the left is also .5. The states of 1 and 9 represent absorbing states, that is, the walker stops moving when $x = 1$ or $x = 9$. It is seen from the graph and the transition matrix that there are three communicating classes: {1}, {2,3,4,5,6,7,8}, and {9}.

4.6.3 Bayesian Analysis of Transient and Recurrent States

It is well known that a Markov chain has two types of states, transient and recurrent. To demonstrate this, consider the chain with transition matrix

$$P = \begin{pmatrix} .34,.66,0.0 \\ 1.0,0.0,0.0 \\ .25,.50,.25 \end{pmatrix}, \tag{4.71}$$

where from 1, the chain either returns to state 1 in one step, or first moves to 2 and then returns to 1 at the second step. From 1, the chain revisits 1 with certainty.

On the other hand, for the chain that begins with 2, the chain first moves to 1, and it may continue to revisit 1 for many steps, but finally will return to 2, because the probability that it will remain at 1 forever is the chance that it repeatedly transitions from 1 to 1, which is the probability of the limit as $n \to \infty$,

$$\lim (P_{11})^n = \lim (.34)^n = 0$$

Therefore, it follows that from 2, the chain revisits 2 with probability 1.

Now consider the case where the process starts in state 3; then the chain may revisit 3 in successive steps but with positive probability will eventually be in state 1 or 2; thus, from state 3, there is a positive chance that the chain that starts in 3 will never revisit 3, and this probability is ¾ = 1 − (1/4).

It is easily seen that 1 is a recurrent state; that is, it will occur an infinite number of times with certainty. For example, consider a simulation of 200 transitions, starting with state 1 (Table 4.9).

How well does the simulation follow the first row of the transition matrix, where one would expect 1/3 of 200 transitions to be from state 1 to state 1, 2/3 from state 1 to state 2, and 0 from state 1 to state 3? How many times does the chain revisit state 1? It is revisited 120 times out of 200. Of course, we know that 1 is a recurrent state. Also, of course, states 1 and 2 form a communicating class. A similar situation exists with state 2, which is also recurrent; on the other hand, state 3 is transient; thus, of interest is a simulation of 200 transitions starting with state 3 (Table 4.10).

It is seen that state 3 is revisited only once, since once the transition to the state 2 occurs, it is impossible the chain will return to state 3. Recall that a state is transient if there is a positive probability the chain that starts with state 3, never returns to 3.

For the Bayesian, it would be of interest to test the hypothesis H: $P_{31} = 1/4$, $P_{32} = 1/2$, $P_{33} = 1/4$ versus the alternative that H is not true.

The data will be based on simulating realizations from the multinomial with parameter vector (1/4, 1/2, 1/4); thus, one is interested in determining if the generated values actually came from the appropriate transition probabilities for the third row of P (Equation 4.72). This corresponds to the conditional distribution of the chain with initial value $X(0) = 3$. This is left as an exercise for the student. Refer to Section 2.5.3 for relevant information on testing hypotheses from a Bayesian viewpoint.

One last topic to consider is a Bayesian estimator of the average return time (to a particular state) of an irreducible Markov chain. The analysis is based on the relationship

TABLE 4.9

Initial Value 1 with 200 Transitions

1 2 1 2 1 1 2 1 2 1 2 1 2 1 2 1 2 1 1 2 1 1 2 1 2 1 2 1 2 1 2 1 2 1 1 2 1
1 1 2 1 2 1 2 1 1 1 1 1 1 1 1 2 1 1 1 2 1 2 1 1 1 2 1 2 1 2 1 2 1 1 2 1 2
1 2 1 2 1 1 2 1 2 1 2 1 2 1 2 1 1 2 1 2 1 1 1 2 1 1 2 1 2 1 2 1 1 2 1 2 1
2 1 1 1 2 1 2 1 2 1 1 2 1 1 1 1 2 1 2 1 1 2 1 2 1 2 1 2 1 2 1 2 1 2 1 1 2
1 2 1 2 1 2 1 2 1 2 1 1 2 1 1 2 1 2 1 2 1 1 2 1 1 2 1 1 1 2 1 2 1 1 2 1 2
1 1 1 2 1 2 1 2 1 2 1 1 2 1 2 1

TABLE 4.10

Initial Value 3:200 Transitions

3 3 2 1 2 1 1 2 1 2 1 2 1 2 1 1 1 1 2 1 1 1 2 1 2 1 1 1 1 2 1 1 1 2 1 2 1
2 1 2 1 1 2 1 1 2 1 2 1 2 1 2 1 1 2 1 1 1 2 1 2 1 2 1 2 1 1 2 1 2 1 2 1 2
1 2 1 2 1 2 1 2 1 2 1 2 1 1 2 1 2 1 2 1 1 1 1 2 1 1 2 1 2 1 2 1 1 2 1 2 1
2 1 1 1 2 1 2 1 2 1 1 1 2 1 2 1 2 1 2 1 2 1 2 1 2 1 1 1 2 1 1 2 1 2 1 2 1
2 1 1 2 1 2 1 2 1 1 2 1 2 1 2 1 2 1 2 1 2 1 2 1 2 1 1 2 1 2 1 1 2 1 2 1 1
2 1 2 1 1 1 2 1 1 1 1 1 2 1 1 2

between the stationary probability π_j (of the stationary distribution π of the chain) to the average number of steps between visits to state j. To be more precise, let

$$T_j = \min\{n : n > 0, X(n) = j\} \tag{4.72}$$

and let

$$\mu_j = E\left(T_j \middle| X(0) = j\right) \tag{4.73}$$

be the expected return time to state j, and then it can be shown that μ_j is finite and that

$$\pi_j = 1/\mu_j \tag{4.74}$$

and

$$\pi_j = \lim(1/n) \sum_{m=0}^{m=n-1} P_{ij}^{m} . \tag{4.75}$$

Recall that the stationary distribution of the chain is determined by solving a linear system of equations involving the transition probabilities P_{ij} of the chain; thus, the Bayesian analysis is easily executed once one knows the posterior distribution of these transition probabilities. See page 103 of Dobrow[2] for the details on the verification of Equation 4.84.

Recall the example of determining the stationary distribution of an irreducible chain involving the transition matrix

$$P = \begin{pmatrix} .45, .48, .07 \\ .05, .70, .25 \\ .01, .50, .49 \end{pmatrix}, \tag{4.61}$$

with three states 1, 2, and 3, where state 1 denotes lower class; 2, middle class; and 3, upper class. This is an example of social mobility, where .48 is the probability of moving from the lower to the middle class in one generation. The Bayesian estimation of the stationary distribution was based on Equations 4.66 and 4.67, which expressed the stationary probabilities in terms of the one-step transition matrix (Equation 4.61). Data for this example were generated via the multinomial distribution (see Table 5.4) and WinBUGS Code 4.6, which is used to execute the Bayesian analysis reported in Table 5.5. In order to estimate the first return times of the three states of the example of social mobility, the following code was amended to WinBUGS Code 4.6:

```
mui1<-1/pi1
mui2<-1/pi2
mui3<-1/pi3
```

with Table 4.11 reporting the Bayesian estimation of the first return times. An improper prior was used for the prior distribution of the transition probabilities. See Equation 4.61.

TABLE 4.11

First Return Times for Social Mobility Example

Parameter	Mean	SD	Error	2 1/2	Median	97 1/2
μ_1	11.1	2.766	0.01619	6.91	10.68	17.69
μ_2	1.642	0.0762	0.0004363	1.501	1.639	1.799
μ_3	3.454	0.4639	0.00269	2.744	3.387	4.518

Therefore, the posterior mean for the average return time to the lower class is 11.1 generations, 1.642 for the middle class, and 3.454 generations for the upper class. This appears plausible and is reflected in the transition matrix (Equation 4.61). It should be remembered that for each row of the transition matrix, the data were generated assuming a multinomial distribution, and then assuming an improper prior, the posterior distribution was Dirichlet for each row and, hence, for each cell a beta posterior density.

The average first return times are estimated with the following R program:

R Code 4.5

```
 > markov <- function(init,mat,n,labels) {
+ if (missing(labels)) labels <- 1:length(init)
+ simlist <- numeric(n+1)
+ states <- 1:length(init)
+ simlist[1] <- sample(states,1,prob=init)
+ for (i in 2:(n+1))
+ { simlist[i] <- sample(states,1,prob=mat[simlist[i-1],]) }
+ labels[simlist]
 + }
> P<-matrix(c(0,1,0,
+          .5,0,.5,
+          .333,.333,.333),nrow=3,ncol=3,byrow=TRUE)
> init<-c(1,0,0)
> markov(init,P,25)
[1] 1 2 3 3 1 2 3 3 3 1 2 3 2 3 3 3 2 3 1 2 1 2 1 2 1 2
> trials<-10000
> simlist<-numeric(trials)
 > for ( i in 1:trials){path<-markov(init,P,25)
+ returntime<-which(path=="1")-1   + simlist<-returntime}
> mean(simlist)
[1] 11.85714
```

And the average return time to state 1 is computed as 11.85714, which is the usual average based on a run of 10,000 simulations. Note how much more informative the Bayesian analysis is as reported in Table 4.11. For example, in addition to the posterior mean (one's estimate of the return time), the posterior standard deviation, posterior median, and the 95% credible interval are given. See page 106 of Dobrow[2] for additional information about using R to estimate the average return times of the states of a Markov chain.

This section is concluded noting that recurrence and transience are class properties; that is, for a communicating class, all the states are recurrent or all are transient. For additional information about recurrence, see Section 3.5.4. Also, it is important to remember that for an irreducible finite chain, all the states are recurrent.

4.7 Period of a Markov Chain

The goal of this section is to provide Bayesian inferences about the period of the states of an irreducible Markov chain. In particular, this will involve testing hypotheses using the Bayesian approach where the main interest is to test the hypothesis that a particular state has a stated period. Recall that the period of state i is defined as

$$d(i) = \gcd\{n > 0 : P_{ii}^{n} > 0\}, \tag{4.76}$$

where P_{ii}^{n} is the n-step transition probability that the chain will return to state i in n time units. Thus, it is possible for the chain to return to state i in multiples of $d(i)$. Consider the chain with states 1, 2, 3, and 4 and transition matrix

$$P = \begin{pmatrix} 0, .5, 0, .5 \\ .5, 0, .5, 0 \\ 0, .5, 0, .5 \\ .5, 0, .5, 0 \end{pmatrix}, \tag{4.77}$$

then each state has period 2. Consider state 1, then the chain can return to 1 by first going to state 2, then returning to 1, or it can first go to 2, then 3, then 4, and then return to 1, for a total of four transitions. The greatest common divisor of 2 and 4 is 2. Of course, there are other paths of returning to state 1, but they are multiples of 2. The following R statements generate the graph in Figure 4.5. Using the R matrix power code, it can be shown that all powers of P (Equation 4.61) are P, that is,

$$P^{n} = P, \quad n = 1, 2, 3, \ldots$$

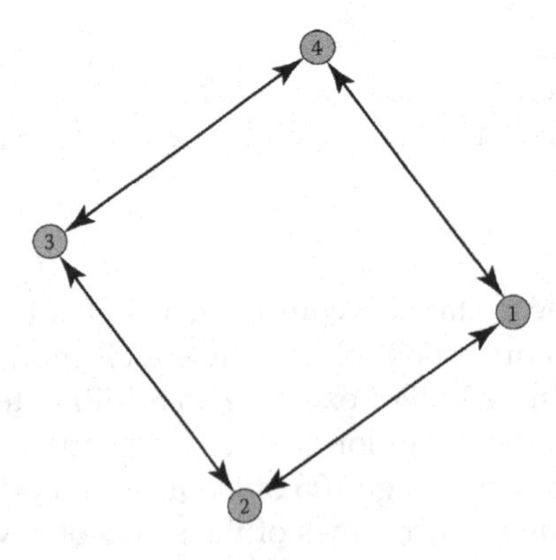

FIGURE 4.5
Transition graph of chain with period 2.

Therefore, in particular

$$P_{11}^{n} = P_{11} = .5 > 0, \quad n = 2,4,6 \tag{4.78}$$

This further confirms that the period of state 1 is 2, and of course, the period of the other three states is also 2.

R Code 4.6

```
P<-matrix(c(0,.5,0,.5,
            .5,0,.5,0,
            0,.5,0,.5,
            .5,0,.5,0),nrow=4,ncol=4,byrow=TRUE)
g<-graph.adjacency(P,weighted=TRUE)
plot(g)
```

See Figure 4.5.

Suppose that we want to estimate the probability of the return to state 1 having period 2. For this example, $P_{11} = P_{11}^{n} = .5$; thus, it is obvious that the period of state 1 is 2.

However, in practice, one does not know the actual values of the transition probabilities; thus, transition counts will be generated using the transition matrix (Equation 4.61), and based on those counts, the posterior distribution of the P_{11}^{n}, $n = 2,4,6,\ldots.$will be determined.

Based on the transition matrix P, I generated 100 observations from the appropriate multinomial distribution for each of the four conditional distributions resulting in the following transition counts:

$$Q = \begin{pmatrix} 0,48,0,52 \\ 48,0,52,0 \\ 0,52,0,48 \\ 47,0,53,0 \end{pmatrix}, \tag{4.79}$$

and from this, one can verify that the corresponding estimated transition matrix is

$$\hat{P} = \begin{pmatrix} 0,.48,0,.52 \\ .48,0,.52,0 \\ 0,.52,0,.48 \\ .47,0,.53,0 \end{pmatrix}. \tag{4.80}$$

Now consider the following powers of the estimated transition matrix:

$$\hat{P}^2 = \begin{pmatrix} .4748, 0, .5252, 0 \\ 0, .5008, 0, .4992 \\ .4752, 0, .5248, 0 \\ 0, .5012, 0, .4985 \end{pmatrix}, \tag{4.81}$$

$$\hat{P}^4 = \begin{pmatrix} .47501, 0, .5249899, 0 \\ 0, .50009978, 0, .4990003 \\ .4750090, .5249901, 0 \\ 0, .50000995, 0, .499005 \end{pmatrix}, \tag{4.82}$$

and it can be shown that higher even powers of $\hat{P}$ are essentially the same as $\hat{P}^2$; thus, since all the even powers of $\hat{P}$ are the same, it is sufficient to estimate only P_{11}^2.

Estimation will be done via the Bayesian approach, which is to estimate the probability that the chain returns to state 1 in two transitions.

Consider the first row of P_{11}^2, and then generate 100 observations to complete these four cell counts. Assuming the transition counts follow a multinomial distribution with mass function

$$f(n_{11}, n_{12}, n_{13}, n_{14} | P_{11}^2, P_{12}^2, P_{13}^2, P_{14}^2) = [100!/n_{11}!, n_{12}!, n_{13}!, n_{14}!] \prod_{j=1}^{j=4} P_{1j}^2, \tag{4.83}$$

where $\sum_{j=1}^{j=4} P_{1j}^2 = 100$ and $\sum_{j=1}^{j=4} n_{1j} = 100$, the 100 cell counts were generated using the R function rmultinom(1,100,prob) with prob=c(.53,0,.47,0), and the resulting realization is $n_{11} = 54$, $n_{12} = 0$, $n_{13} = 46$, $n_{14} = 0$. This implies that the marginal distribution of n_{11} is binomial $(100, P_{11}^2)$ with mass function

$$f(n_{11} | P_{11}^2) = \binom{100}{n_{11}} (P_{11}^2)^{n_{11}} (1 - P_{11}^2)^{100 - n_{11}}, \tag{4.84}$$

where $0 < P_{11}^2 < 1$.

How should P_{11}^2 be estimated? If one assumes a uniform prior for the unknown parameter, the posterior distribution of P_{11}^2 is beta with parameter vector $(n_{11} + 1, 100 - n_{11} + 1)$; thus, substituting $n_{11} = 54$, the posterior mean is $55/102 = .5392$. Remember that P_{11}^2 is the probability of a two-step transition to return to state 1, not the probability of the square of P_{11}!

Using WinBUGS to execute the Bayesian analysis, I generated 35,000 observations for the MCMC simulation, with a burn-in of 5,000 observations and a refresh of 100 with the results shown in Table 4.12.

Thus, the posterior mean is .5351, and the 95% credible interval is (.5254, .5449), which implies that state 1 has period 2. That is to say, the credible interval indicates that $P_{11}^2 > 0$. Because higher even powers of the matrix P^2 do not change, one would expect the Bayesian analysis for P_{11}^n, $n = 4, 6, \ldots$ to be the same as that portrayed in Table 4.12; thus, one is confident that state 1 has period 2. Also, shown in the table are the posterior mean and median which implies a symmetric posterior density for P_{11}^2.

WinBUGS Code 4.7 generates the Bayesian analysis:

WinBUGS Code 4.7

```
  model;
{
 for ( i in 1:100){
p[i]~dbin(p11,100)}
p11~dbeta(1,1)
}
list(
p = c(
56.0,53.0,48.0,55.0,50.0,
50.0,60.0,47.0,61.0,57.0,
56.0,49.0,48.0,55.0,54.0,
56.0,58.0,59.0,61.0,51.0,
50.0,48.0,43.0,52.0,51.0,
57.0,55.0,57.0,47.0,56.0,
56.0,54.0,60.0,50.0,53.0,
59.0,53.0,50.0,53.0,55.0,
50.0,58.0,47.0,61.0,56.0,
54.0,60.0,43.0,50.0,52.0,
57.0,57.0,60.0,53.0,63.0,
49.0,46.0,52.0,56.0,59.0,
58.0,54.0,46.0,51.0,49.0,
61.0,48.0,50.0,57.0,55.0,
57.0,59.0,51.0,43.0,58.0,
58.0,55.0,53.0,56.0,58.0,
46.0,56.0,54.0,52.0,53.0,
46.0,44.0,50.0,56.0,59.0,
49.0,50.0,57.0,57.0,53.0,
60.0,44.0,56.0,53.0,53.0))
```

TABLE 4.12

Posterior Analysis for the Period of State 1

Parameter	Mean	SD	Error	2 1/2	Median	97 1/2
P_{11}^2	.5351	.00496	.0000294	.5254	.5351	.5449

The list statement contains the data of 100 values generated from a binomial distribution with parameters (.54, 100), and the code shows a beta uniform prior placed on the unknown parameter P_{11}^2.

4.8 Ergodic Chains and Time Reversibility

Time reversibility is an interesting property of some Markov chains. The type of Markov chain we are interested in is the so-called ergodic chains, which are irreducible and aperiodic (all states have period 1) and have a finite average return time. All finite chains are ergodic if they are aperiodic and irreducible. Such chains are of interest because they possess a unique positive stationary distribution π. Such processes can exhibit the time reversibility property

$$\pi_i P_{ij} = \pi_j P_{ji}, \quad i, j = 1, 2, .., k, \tag{4.85}$$

where $(\pi_1, \pi_2, ..., \pi_k)$ is the stationary distribution, and P_{ij} is the one-step transition probability. Such processes have no directional bias, such as a random walk where the chain moves one unit to the right with probability q and one unit to the left with probability $1 - q$ with $q = 1/2$. If $q > 1/2$, the chain exhibits a bias that propels the process to the right, and the process has a directional bias. Time reversibility, as defined by Equation 4.85, implies a process such that its behavior in the future is the same as the process moving backwards; one cannot tell the difference.

Such processes invite many inferential challenges. For example, given the data from a finite, irreducible, and aperiodic Markov chain, is the process time reversible? In practice, the data would consist of transition counts of the chain, from which one can estimate the one-step transition probabilities and stationary distribution. For the Bayesian, our main interest will be in determining if the chain is time reversible, where inferences will be either the posterior estimation of the parameters

$$\zeta_{ij} = \pi_i P_{ij} - \pi_j P_{ji}, \quad i, j = 1, 2, .., k$$

or a test of the hypotheses

$$H_0: \zeta_{ij} = 0 \quad \text{versus} \quad H_1: \zeta_{ij} \neq 0$$

Two examples are described, where the first involves a chain which is known to be not time reversible, while the other will be a chain which is time reversible. Consider the example of social mobility described in Section 4.4 with the transition matrix

$$P = \begin{pmatrix} .45, .48, .07 \\ .05, .70, .25 \\ .01, .50, .49 \end{pmatrix}. \tag{4.61}$$

In this example, the posterior distribution of the stationary transition was determined using the data from Table 4.5, WinBUGS Code 4.6, and the results appearing in Table 4.6.

TABLE 4.13

Bayesian Analysis for Time Reversibility of Social Mobility

Parameter	Mean	SD	Error	2 1/2	Median	97 1/2
ζ_{12}	−.00253	.005433	.0000322	−.01347	−.00249	.00817
ζ_{13}	.02171	.00522	.0000324	.01181	.02161	.03218
ζ_{23}	−.00173	.0282	.000167	−.05476	−.00261	.05587

The data consist of realizations generated from the multinomial distribution with 200 observations for each row of Equation 4.61. An improper prior distribution (Equation 4.64) induces a Dirichlet posterior distribution for each row and a beta posterior for each transition probability. WinBUGS Code 4.6 contains Equations 4.66 and 4.68 for determining the stationary distribution of the chain. In order to specify the posterior distribution of the relevant parameters ζ_{ij}, a slight modification of the WinBUGS Code is necessary. The modified code appears as WinBUGS Code 4.8. I used 35,000 observations for the simulation, with a burn-in of 5,000. The relevant parameters are denoted by d11,d12, and d13, and the posterior analysis is reported in Table 4.13.

WinBUGS Code 4.8

```
model;
{
p11~dbeta(98,102)
p12~dbeta(84,116)
p13~dbeta(18,182)
p21~dbeta(14,186)
p22~dbeta(139,61)
p23~dbeta(47,153)
p31~dbeta(4,196)
p32~dbeta(99,101)
p33~dbeta(97,103)
x2<-(p32*(1-p11)+p12*p31)/(p32*p21-p31*(p22-1))
x3<-(1-p11-p21*x2)/p31
tot<-1+x2+x3
pi1<-(1/tot)
pi2<-x2*(1/tot)
pi3<-x3*(1/tot)
d12<-pi1*p12-pi2*p21
d13<-pi1*pi3-pi3*p31
d23<-pi2*p23-pi3*p32
}
```

For the most part, the posterior distributions appear symmetric about the posterior mean. It is interesting to note that the 95% credible interval for ζ_{13} excludes zero. Is this chain time reversible?

TABLE 4.14

Posterior Analysis for a Time-Reversible Chain

Parameter	Mean	SD	Error	2 1/2	Median	97 1/2
ζ_{12}	−.00369	.01212	.000067	−.02767	−.00360	.01987
ζ_{13}	−.04937	.00967	.0000578	−.06842	−.04941	−.03036
ζ_{23}	−.00362	.01228	.0000649	−.02744	−.00362	.00694
P_{12}	.4602	.03516	.0002931	.3921	.4599	.5285
P_{13}	.5403	.03513	.0000235	.4707	.5402	.6088
P_{21}	.5747	.03479	.000185	.5066	.5748	.6429
P_{22}	.2046	.02857	.000164	.1515	.204	.2631
P_{23}	.2198	.02933	.000157	.1647	.219	.2798
P_{31}	.4752	.03545	.0002102	.4053	.4751	.5448
P_{32}	.175	.02685	.0001583	.1255	.1738	.2309
P_{33}	.3496	.03368	.0001988	.2856	.3488	.4174
π_1	.3432	.01183	.000063	.3195	.3434	.3658
π_2	.2812	.01903	.0001176	.2443	.2809	.3185
π_3	.3757	.02516	.0001478	.3265	.3758	.4249

The second example involves a transition matrix of a chain which is time reversible. See example 3.22 on page 115 of Dobrow,[2] where the transition matrix is

$$P = \begin{pmatrix} 0, 2/5, 3/5 \\ 1/2, 1/4, 1/4 \\ 1/2, 1/6, 1/3 \end{pmatrix}. \tag{4.86}$$

It can be verified the chain is time reversible. In order to perform the Bayesian analysis, multinomial observations will be generated for each row of Equation 4.86, then based on these realizations, the estimated transition probabilities and estimated stationary distribution will be determined using WinBUGS Code 4.8. Using the improper prior the posterior distribution of the transition probabilities, stationary distribution, and the time-reversible parameters ζ_{ij} will be available and reported in a way similar to Table 4.13.

Two hundred multinomial observations are generated for each row of Equation 4.87 to give the transition count matrix Q, where

$$Q = \begin{pmatrix} 0, 92, 108 \\ 115, 41, 44 \\ 95, 35, 70 \end{pmatrix}. \tag{4.87}$$

The Bayesian analysis is executed with WinBUGS Code 4.9 with 5,000 observations for the simulation, a burn-in of 5,000, and a refresh of 100. The posterior analysis is reported in Table 4.14.

WinBUGS Code 4.9

```
   model;
{
p11~dbeta(1,99)
```

```
p12~dbeta(92,108)
p13~dbeta(108,92)
p21~dbeta(115,85)
p22~dbeta(41,159)
p23~dbeta(44,156)
p31~dbeta(95,105)
p32~dbeta(35,165)
p33~dbeta(70,130)
x2<-(p32*(1-p11)+p12*p31)/(p32*p21-p31*(p22-1))
x3<-(1-p11-p21*x2)/p31
tot<-1+x2+x3
pi1<-(1/tot)
pi2<-x2*(1/tot)
pi3<-x3*(1/tot)
d12<-pi1*p12-pi2*p21
d13<-pi1*pi3-pi3*p31
d23<-pi2*p23-pi3*p32
```

The stationary distribution estimated by the posterior mean is

$$\hat{\pi} = (.3432, .2812, .3757), \tag{4.88}$$

and the main parameters of interest, the ζ_{ij}, $i < j$, are estimated with the posterior median as −.00360, −.04941, and −.00362, respectively. Also, the posterior distribution of these parameters is evidently symmetric about the posterior mean, and their 95% credible intervals imply that the chain is time reversible. In order to show the uncertainty when using a smaller sample size for the multinomial realization, the example is repeated with a much smaller sample size of 50 (smaller than 200) for each row of Equation 4.86. The multinomial realizations for the three rows of the transition matrix (Equation 4.86) is portrayed in matrix

$$R = \begin{pmatrix} 0, 21, 29 \\ 28, 10, 12 \\ 23, 11, 16 \end{pmatrix}. \tag{4.89}$$

TABLE 4.15

Posterior Analysis for a Time-Reversed Chain: Size 50 Multinomial Realizations

Parameter	Mean	SD	Error	2 1/2	Median	97 1/2
ζ_{12}	−.01573	.0244	.000116	−.01538	.03106	.03106
ζ_{13}	−.0456	.0192	.0000892	−.08319	−.04577	−.00811
ζ_{23}	−.0145	.02565	.000119	−.06306	−.01498	.03805
π_1	.3382	.02407	.000106	.2886	.3391	.383
π_2	.2818	.03643	.000182	.2125	.2812	.3548
π_3	.3801	.04833	.000232	.2854	.3798	.4742

The Bayesian analysis is reported in Table 4.15 and should be compared to the results in Table 4.14. The latter analysis is based on realizations of size 50, whereas the results of the former posterior analysis is based on realizations of size 200.

Comparing Tables 4.14 and 4.15, one sees that the posterior means are approximately the same, but that the posterior standard deviations are much smaller for Table 4.14, and as a consequence, the 95% credible intervals are wider for Table 4.15. Overall, the conclusion about time reversibility based on Table 4.15 would be about the same to those based on Table 4.14. The student will be asked as an exercise to repeat the analysis with multinomial simulations of size 20 and compare those results to those in Table 4.15. Will the Bayesian analysis imply time reversibility?

Our next phase of inference for time reversibility is to test the null hypothesis H_0: $\zeta_{ij} = 0$ versus H_1: $\zeta_{ij} \neq 0$, where $\zeta_{ij} = \pi_i P_{ij} - \pi_j P_{ji}, i, j = 1, 2, \ldots, k$.

First, consider $\zeta_{12} = \pi_1 P_{12} - \pi_2 P_{21}$ and a test of

$$H_0: \zeta_{12} = 0 \quad \text{versus} \quad H_1: \zeta_{12} \neq 0, \tag{4.90}$$

where the test is based on the data generated with 200 multinomial observations for each row of the transition matrix (Equation 4.87), and the Bayesian analysis is recorded in Table 4.13. Note that the posterior mean and standard deviation are

$$E(\zeta_{12}|\text{data}) = -.00253$$

and

$$\sigma(\zeta_{12}|\text{data}) = .005433, \text{ respectively}$$

Therefore, I will assume that $\zeta_{12} \sim \text{normal}(\mu, \sigma^2)$, where $\mu = -.00253$ and $\sigma^2 = .000029517$. Using WinBUGS Code 4.8, one can show that the density of ζ_{12} appears to be in the shape of a normal density; thus, for the purpose of testing H_0 versus H_1, it is assumed that ζ_{12} is indeed normally distributed.

For reviewing the Bayesian approach to testing hypotheses, the reader is referred to Section 2.5.3. For the Bayesian approach, it is assumed that $\zeta \sim \text{normal}(\mu, \sigma^2)$, where $\mu \sim \text{normal}(0, v^2)$, and v^2 is known; thus, the test will be described as

$$H_0: \mu = 0 \quad \text{versus} \quad H_0: \mu \neq 0 \tag{4.91}$$

The Bayesian test is implemented by computing the posterior probability of the null and alternative hypotheses. The test is based on the predictive density of $\zeta = \pi_1 P_{12} - \pi_2 P_{21}$, where $\zeta \sim \text{normal}(\mu, \sigma^2)$ and $\mu \sim \text{normal}(0, v^2)$; therefore, the predictive density is

$$f(\zeta) = \rho_0 f(\zeta|\mu = 0) + \rho_1 \int f(\mu) f(\zeta|\mu)\, d\mu, \tag{4.92}$$

where ρ_0 is the prior probability of the null hypothesis, ρ_1 is the prior probability of the alternative hypothesis, and $\rho_0 + \rho_1 = 1$. In addition, the prior density of μ under the alternative is

$$f(\mu) = \left[1/\sqrt{2\pi v}\right] \exp -(1/2v^2)\mu^2, \tag{4.93}$$

and

$$f(\zeta|\mu) = \left[1/\sqrt{2\pi\sigma}\right] \exp(1/2\sigma^2)(\zeta - \mu)^2, \tag{4.94}$$

TABLE 4.16

Posterior Probability of Time Reversibility

ν	ω_0
1	.9991
3	.99801
5	?
10	?
100	.9979
.0024	.9931

is the conditional density of γ given μ. Thus, it can be shown that the predictive density is

$$f(\zeta) = \rho_0\left[1/\sqrt{2\pi}\sigma\right]\exp\left(-1/2\sigma^2\right)\zeta^2 + \rho_1\left[\nu\sigma/\sqrt{2\pi(\sigma^2+\nu^2)}\exp\left(-1/\left[2(\sigma^2+\nu^2)\right]\right)\right]\zeta^2 \quad (4.95)$$

and that the posterior probability of the null hypothesis is

$$\omega_0 = \rho_0\left[1/\sqrt{2\pi}\sigma\right]\exp\left(1/2\sigma^2\right)\zeta^2 \div f(\zeta). \quad (4.96)$$

Using the information from Table 4.15, $\zeta = -.01573$ and letting $\rho_0 = \rho_1 = .5$, the posterior probability ω_0 of the null hypothesis is computed for various values of ν, the standard deviation of the prior distribution for μ, and is reported in Table 4.16. I am assuming that σ and ν are known. For example, when $\nu = 3$, the probability of null hypothesis (time reversibility) is .99801. This is not surprising since the observed value of the parameter that measures time reversibility is $-.01573$ with a standard deviation of 0.0244. It is seen that the evidence is very strong for the conclusion that the chain is time reversible, regardless of what is used for the prior distribution of μ (Equation 4.93) under the alternative. The reader is invited to verify the values of ω_0 in Table 4.16 and to complete the entries designated by "?" I used Equation 4.96.

The effect of ν on ω_0 is negligible!

4.9 Absorbing States of a Chain

The last topic to be presented is that of estimating the probability of absorption using the Bayesian approach. The reader can review the topic of absorbing starts of a Markov chain by referring to Section 3.6.4, where the concept is introduced with the gambler's ruin problem with transition matrix

$$P = \begin{pmatrix} 1,0,0,0,....0,0,0 \\ q,0,p,0,...0,0,0 \\ 0,q,0,p,0,...,0 \\ . \\ 0,0,0,0,.,0,p,0 \\ 0,0,0,0,.,q,0,p \\ 0,0,0,0,....,0,1 \end{pmatrix} \tag{3.69}$$

where the gambler begins with capital k dollars (the process is at state k) and wins \$1 with probability p and loses \$1 with probability $q = 1 - p$. When the gambler's stake is 0 (the process is at state 0), the gambler has lost all, or the gambler can win the total pot of N dollars (the process is in state N). There are three communicating classes $\{0\}, \{1, 2, ..., N - 1\}$ and $\{N\}$; that is, 0 communicates only with 0, N communicates only with N, and the remaining $N - 1$ states communicate only with each other, but not with 0 or N. The states 0 and N are called absorbing states.

There are many interesting aspects of the dynamics of the gambler's ruin problem, including (1) the probability the gambler either wins total capital of N or loses all capital and (2) the average length of time in order to win or lose all capital. For the Bayesian, the objective is to find the posterior distribution of the probability that the gambler wins the total pot of N or the posterior distribution that the gambler loses all. Also of interest is estimating the average time to win the total pot of N dollars.

For the first objective, it will be assumed that p has a particular value, then using the multinomial distribution, the transition counts will be generated for the second row of the transition matrix; thus, the cell count for cell (1,2) will have a binomial distribution with parameters $n = 200$ and p. Assuming an improper prior for p induces a beta posterior 3 the gambler is ruined is given by

$$a_k = \left[(q/p)^N - (q/p)^k\right] / \left[(q/p)^N - 1\right], \quad p \neq q. \tag{3.86}$$

Thus, our objective is to determine the posterior distribution of a_k, which is the probability that the gambler is ruined, given the gambler began with a stake of k dollars. Since this probability is a function of p, N, and k, the posterior distribution of p induces the posterior distribution of a_k.

Assuming $N = 5$, $k = 2$, and $p = .6$, 200 multinomial observations were generated from the second row of Equation 3.69 with the following vector of cell counts (94, 0, 106, 0, 0, 0); thus, assuming an improper prior for p, the posterior distribution of p is beta(106, 94). Using WinBUGS Code 4.10, the posterior analysis is executed with 35,000 observations and a burn-in of 5,000, and the results are reported in Table 4.17.

TABLE 4.17
Bayesian Analysis for Gambler's Ruin

Parameter	Mean	SD	Error	2 1/2	Median	97 1/2
a_k	.527	.0843	.000434	.3653	.5268	.6906
P	.5298	.0350	.000129	.4611	.53	.5979

WinBUGS Code 4.10

```
   model;
{
p~dbeta(106,94)
q<-1-p
N<-5
k<-2
ak1<-pow(q/p,N)-pow(q/p,k)
ak2<-pow(q/p,N) -1
ak<-ak1/ak2
dk<-k/(q-p)-(N/(q-p))*(1-pow(q/p,k))/(1-pow(q/p,N))
}
```

Thus, the probability of the gambler's ruin is estimated as .527 with the posterior mean, a posterior standard deviation of 0.0843, and a 95% credible interval of (.3653, .6906). The multinomial observation vector (94, 0, 106, 0, 0, 0) was generated assuming $p = .6$; thus, the posterior mean of .5298 somewhat underestimates the value of p used to generate the cell counts. The right end point of .5979 of the 95% credible interval is barely less than $p = .6$. Of course, with another realization, one would expect somewhat different results for the posterior analysis.

Consider the five realizations for the second row of the transition matrix using 200 observations and $p = .6$ (Table 4.18).

These five realizations would result in different posterior analyses compared to the one reported in Table 4.17, and the student will be asked to execute those analyses as exercises at the end of the chapter.

One remaining topic to be presented is to estimate the average duration of play for the gambler. Pages 24–30 of Bailey[7] present important information about the gambler's ruin problem which includes the derivation of the following formula for expected duration of the gambler's play. Suppose the gambler begins with capital k and that N is the total pot, then the average duration of play (until a total loss of capital or winning the total pot N) is given by

$$d_k = k/(q-p) - N/(q-p)\left[1-(q/p)^k\right]/\left[1-(q/p)^N\right]. \tag{4.97}$$

The Bayesian analysis is based on the realization (94, 0, 106, 0, 0, 0) of the second row of the transition matrix and executed with WinBUGS Code 4.10 (refer to the line in the code starting with dk) assuming $N = 5$ and $k = 2$. The posterior analysis generated 35,000

TABLE 4.18

Five Multinomial Realizations for Gambler's Ruin

	1	2	3	4	5
P_{10}	76	78	72	78	84
P_{11}	0	0	0	0	0
P_{12}	124	122	128	122	116
P_{13}	0	0	0	0	0
P_{14}	0	0	0	0	0
P_{15}	0	0	0	0	0

TABLE 4.19

Bayesian Analysis for Average Duration of Gambler's Ruin

Parameter	Mean	SD	Error	2 1/2	Median	97 1/2
a_k	.5274	.0838	.000405	.3647	.5275	.6897
d_k	6.033	0.0875	0.000452	5.784	6.067	6.098
P	.5297	.0350	.000169	.4613	.5298	.598

TABLE 4.20

Posterior Distribution of the Gambler's Ruin

Parameter	Mean	SD	Error	2 1/2	Median	97 1/2
P	.6099	.03427	.000189	.5415	.6104	.676
q_{24}	.00729	.001492	.0000082	.004593	.007221	.01044

observations for the simulation with a burn-in of 5,000, and the results are reported in Table 4.19.

Therefore, the average duration of play for the gambler is 6.033 moves, with a posterior standard deviation of 0.0875, and a 95% credible interval of (.5784, 6.098). It appears that the posterior distribution of d_k is symmetric about the posterior mean. Exercises at the end of the chapter will ask the student to repeat the Bayesian analysis with several realizations from Table 4.18 and various values of N and k.

One last topic to be presented is that of estimating the probability distribution of the gambler's ruin at trial n. Let q_{kn} be the probability that the gambler will lose their stake at trial n beginning with capital k at the start of the gamble, then it can be shown that

$$q_{kn} = (k/n)\binom{n}{(n-k)/2} p^{(n-k)/2} q^{(n+k)/2}, \quad 1 < k < N-1, \quad n > 1, \tag{4.98}$$

where $q_{0n} = q_{Nn} = 0$ for $n \geq 1$, while for $n = 0$, $q_{00} = 1$ and $q_{k0} = 0$. Also, $(n-k)/2$ must be an integer in the interval [0,n]. Equation 4.98 was derived by Bailey (pages 31–34).[7]

I executed the Bayesian analysis with 35,000 observations for the simulation and a burn-in of 5,000. Assuming $n = 4$ and $k = 2$, the posterior distribution for p and q_{24} is reported in Table 4.20. The prior distribution for p is assumed to be improper, and I used the second realization from Table 4.18, implying that the posterior distribution of p is beta(122, 78).

Thus, when the gambler is at the fourth bet (has a $4 stake), the probability of the gambler's ruin is estimated as .007221 with the posterior median of .0729, a posterior standard deviation of 0.001492, and a 95% credible interval of (.004593, .01044).

4.10 Comments and Conclusions

A brief review of Chapter 4 is now presented and begins with an example from Diaconis[3] about biased coin tossing. The transition probability matrix is given by Equation 4.4. The next example is based on information about rainy days; that is, if today it rains, what are the chances that it will rain tomorrow? And if it did not rain today, what is the probability that it will rain tomorrow, etc.? The transition matrix (Equation 4.7) is used to generate future realizations using the R function Markov. Bayesian inferences consist of testing the hypothesis $P_{11} = .7$, which is the actual value used to generate samples from the chain. Assuming an improper prior for P_{11}, the posterior distribution of P_{11} is beta, and it is shown that the posterior probability of the null hypothesis is $P_0 = .98$. The Bayesian predictive mass function was derived and used to forecast future observations for the transition counts of the chain. WinBUGS Code 4.3 was used to generate the future observations.

Next to be considered is how to compute the n-step transition probabilities from the one-step transition matrix. This computation was illustrated with the random walk and implemented with the R function matrixpower, which computes the nth power of the one-step transition matrix P. Multinomial simulations are used to generate transition counts for the one-step transition matrix, which in turn provides estimates of the one-step transition matrix of the random walk. The function matrix power then computes the sixth step transition matrix with entries that estimate the sixth step transition matrix. Assuming an improper prior for the one-step transition matrix, and assuming multinomial distributions for the cell counts of each row of P, it is known that the distribution of each row of P is Dirichlet. The challenge for the Bayesian is to determine the posterior distribution of the entries of the sixth power of P. Section 4.3 is concluded with a specification of the Bayesian estimate of a joint probability of several random variables of the chain involving times 5, 6, 9, and 17 of the process.

Section 4.4 emphasizes the limiting distribution of a chain and is illustrated with the forest fire index example. The R function matrix power computes the limiting distribution of the chain by computing higher and higher powers of the one-step transition matrix P. If the limiting distribution exists, the entries of higher powers of P will stabilize. Bayesian inferences consist of testing hypotheses about the fire index transition matrix and predicting future observations with the Bayesian predictive density.

If a chain has a stationary distribution, it is of interest to be able to estimate it from sample data. If one knows the one-step transition matrix P, and if a stationary solution exists, it is known to be the solution of a system of linear equations. The primary goal of this section was to provide estimates of the stationary distribution, and this was illustrated with the social mobility example. For information from the data, multinomial realizations were generated for each row of the chain, then assuming an improper prior, the marginal posterior distribution of each transition probability is determined to be a beta. The system of equations in Equation 4.61 determines the stationary distribution as a function of the nine cell entries of P. WinBUGS Code implements the Bayesian estimation of the stationary distribution, and the results are reported in Table 4.6. Also, an R program was used to compute the stationary distribution of the chain and compared to the Bayesian results.

Section 4.6 introduces the idea of a transfer graph which is a representation of a Markov chain. The R package igraph is employed to implement such graphs and provides one with additional information about the behavior of the process. The section includes a discussion of Bayesian inferences for irreducible chains; that is, those where all the states communicate, that is, there is only one class. Bayesian inferences are demonstrated with the 3×3 irreducible chain with transition matrix (Equation 4.70). Specifically, the entries of the third row of the chain are estimated with the Bayesian paradigm in the usual way with multinomial observations generated for the third row, and the results are reported in Table 4.8.

It is well known that there are two types of states in a Markov chain, transient and recurrent; such states are defined in Section 4.6.3. Example of a chain with both states is the one with three states and transition matrix (Equation 4.71). It is easy to show that 1 and 2 are recurrent, but that 3 is transient. Realizations generated via the R function Markov illustrate that 3 is transient and that 1 is recurrent. For the Bayesian analysis, tests of hypotheses about the third row of the transition matrix are carried out. The social class example (Equation 4.61) with three states is irreducible and employed to illustrate the average return time to a particular state. The goal for the Bayesian is to estimate those average return times, and WinBUGS Code 4.6 is executed for the posterior analysis with the results reported in Table 4.11.

Section 4.7 explains the Bayesian approach to making inferences about the period of state of a Markov chain. The example has transition matrix (Equation 4.7) where it is known that each state has period 2. The problem for the investigator is to test the hypothesis that the period of, say, state 1 is period 2. This is a very interesting example of Bayesian inference! The posterior analysis is reported in Table 4.12.

Time reversibility is the subject of Section 4.8 and is an interesting property of those chains that exhibit it. Such chains demonstrate the same behavior whether looking into the future or going backward to the past. The concept is defined, and two examples are explored: one where it is known that there is no time reversibility and the other where there is time reversibility. When time reversibility is present for a three-state chain, there are three conditions that must be satisfied. Therefore, the Bayesian approach is to test the hypothesis that the chain is time reversible. One assumes that the chain is irreducible and uses the Bayesian approach to testing hypotheses, see Section 2.5.3. Using the stationary distribution of the chain is essential, and WinBUGS Code 4.9 implements the analysis, and the posterior analysis is reported in Table 4.16.

Lastly, the chapter concludes with the Bayesian approach to estimating the absorbing states of a Markov chain. The basic idea of absorbing states is well illustrated with the gambler's ruin problem, where there are three communicating classes: {1}, the gambler has lost all; $\{N\}$, the gambler has won the pot; and the rest is $\{2, 3, \ldots, N-1\}$. The two absorbing states are 1 and N, where $N = 5$ and $k = 2$ (the initial capital), and the Bayesian analysis consists of estimating the probability that the gambler will lose and estimate the average duration of the game. See Tables 4.19 and 4.20 for reports of the posterior analysis.

Bayesian inference for Markov chains is an active area of interest, and what follows are several references that are applicable to the presentation given here. When using Jeffrey's prior for the inference of a Markov chain, Assodou and Essebbar[8] provide an interesting perspective. With regard to time-reversible chains, see Diaconis and Rolles,[9] and if you are interested in a general account of Bayesian inference for Markov chains, refer to Eichelsbacher and Ganesh[10] as well as the book by Insua, Ruggeri, and Wiper.[11] Also of interest is the relation between Markov chains and associated decision problems, and Martin[12] details the relationship. The empirical Bayes approach to inference is not taken in this book; however, it is adopted by Meshkani and Billard,[13] who explain the subtleties of that method. Lastly, Welton and Ades[14] deal with the important topic of using partially observed data to estimate the transition probabilities of a Markov chain. For a more complete list of Bayesian inference for Markov chains, see pages 78–81 of Insua, Ruggeri, and Wiper.[11]

4.11 Exercises

1. Write a two-page essay summarizing Chapter 4.
2. Refer to the one-step transition matrix in Equation 4.4, the Diaconis example of a biased coin.
   a. To what extent is the coin biased?
   b. Using the Markov function of R, generate 200 observations with initial value heads.
   c. Using the transition counts generated in b, give the usual estimates of P_{11} and P_{12}.
3. Consider the rainy day example with the transition matrix in Equation 4.7, and using the Markov function of R, generate 250 observations from this chain with initial value 2.
   a. How many times is state 1 visited?
   b. How many times is state 2 visited?
   c. Using WinBUGS Code 4.2, validate the posterior analysis for P_{11} reported in Table 4.2. Using 35,000 observations with a burn-in of 5,000, execute the Bayesian analysis.
   d. How do the results of Table 4.1 compare with those of Table 4.2.
4. Derive the predictive mass function (Equation 4.17).
5. Refer to the transition matrix in Equation 4.21, the transition matrix for the random walk on a cycle. Using the matrixpower function of R, confirm the entries of the matrix P^6, the six-state transition matrix corresponding to Equation 4.21.
6. Use the R function rmultinom to generate cell counts for the first row of the transition matrix P in Equation 4.21 with $n = 50$. Using this as the sample information, the cell counts in the first row will follow a multinomial distribution, and assuming an improper prior distribution for the first row entries of P, what is the Bayesian posterior distribution of P_{01}? Execute the Bayesian analysis with 35,000 observations for the simulation with a burn-in of 5,000. Your results should be similar to those reported in Table 4.2.
7. Refer to the transition probability matrix in Equation 4.33.
   a. Use the R Code 4.2 and verify the entries of $P^{17}$and P^{18}.
   b. What is the long-term risk of the forest fire index with 5 states?

8. Duplicate the five multinomial realizations appearing in Table 4.4 of the first row of the transition matrix in Equation 4.33 using a total count of 100.
9. For the forest fire index example, consider the test of the hypothesis H_0: $P_{11} = .755$, $P_{12} = .118$, $P_{13} = .172$, $P_{14} = .109$, $P_{15} = .026$; based on Equations 4.49 through 4.58, show that the posterior probability of the null hypothesis is $P_0 = .1845$.
10. Refer to the probability transition matrix in Equation 4.61, the social mobility example, and using R Code 4.3, verify the stationary distribution as $\pi = (.0623, .6234, .3141)$. In the long term, what is the probability that a person will be in the middle class?
11. a. Using the first realization for the social mobility example (Table 4.5) and assuming the improper prior distribution in Equation 4.63 for the first row of the transition probability matrix P in Equation 4.61, show that the posterior distribution of the first row is Dirichlet(98, 84, 18).
    b. Based on the third realization reported in Table 4.5, show that the posterior distribution of the first row is Dirichlet(86, 106, 8).
    c. What are the implications for having two different Dirichlet distributions for the posterior distribution of the first row of P?
12. Refer to the social mobility example with transition matrix P in Equation 4.61. Based on WinBUGS Code 4.6 and generating 40,000 observations for the simulation with a burn-in of 5,000, execute a Bayesian analysis for the stationary distribution of the chain. Refer to Table 4.6. Your results should be similar.
13. Using R Code 4.4, verify the transition graph in Figure 4.3. You will need to download the package "igraph" to the R platform.
14. Based on the transition graph in Figure 4.4, explain why the chain in Equation 4.69 is irreducible.
15. Based on the first multinomial realization (Table 4.7) of the third row of the chain in Equation 4.69 and assuming an improper prior distribution for the third row of Equation 4.70, do the following:
    a. Show that the posterior distribution probabilities of the third row are as follows:

$$P_{31} \sim \text{beta}(1,102),\ P_{32} \sim \text{beta}(31,72),\ P_{33} \sim \text{beta}(71,32).$$

    b. Verify the posterior analysis of Table 4.7.
    c. What is the 95% credible interval for P_{31}?
    d. Plot the posterior density of P_{31}.
    e. Based on the density of P_{31}, is the distribution symmetric?

16. Refer to the Markov chain with transition probability matrix in Equation 4.72.
    a. Show that states 1 and 2 are recurrent.
    b. Show that state 3 is transient.
    c. Using the R function Markov, generate 200 observations from the chain with matrix in Equation 4.73 and initial value of 1.
    d. How does the simulation of item c compare to that appearing in Table 4.9?
17. Verify Equation 4.74 for the average return time to a particular state.
18. Refer to the chain with transition matrix in Equation 4.61, the irreducible chain of the social mobility example.
    a. Execute the Bayesian analysis with WinBUGS Code 4.6 using the information reported in Table 5.4. Perform the Bayesian analysis with 35,000 observations and a burn-in of 5,000.
    b. Verify the results of Table 4.11, which reports the Bayesian analysis for the first return times for each of the three states (in lower, middle, and higher social classes).
    c. Are the results of Table 4.11 reasonable? Explain in detail and justify your answer.
    d. Explain why R Code 4.5 computes the average return time to the states of a Markov chain.
    e. Compare the Bayesian estimate of the average return time (reported in Table 4.11) to that computed by WinBUGS Code 4.8.
19. Refer to the transition probability matrix in Equation 4.77.
    a. Show that state 1 has period 2.
    b. Using the igraph package with R Code 4.6, generate the transition graph of the chain in Equation 4.77. Your results should look like Figure 4.5.
    c. Does the graph show that each state has period 2? Why?
20. Based on the entries of the estimated transition matrices of P in Equation 4.80, namely, $\hat{P}^2$(Equation 4.81) and $\hat{P}^4$(Equation 4.82), is the period of each state of P of period 2?
21. Verify the Bayesian analysis reported in Table 4.12 of the 1-1 element of the two-step transition matrix P^2. Do the results imply that each state of the chain in Equation 4.17 is of period 2? Explain your answer in detail.
22. The time reversibility of a Markov chain is defined by Equation 4.85. Does the Bayesian analysis reported in Table 4.13 support the conjecture that the chain is time reversible?
23. Explain how the posterior probability ω_0 of the null hypothesis in Equation 4.91 of time reversibility is computed. In your answer, refer to Equations 4.92 through 4.96.

24. Refer to the gambler's ruin transition matrix in Equation 3.68.
    a. How many communicating classes are there?
    b. What states are absorbing?
    c. Explain the derivation of Equation 3.85, the probability that the gambler will be ruined.
25. Using WinBUGS Code 4.10, execute the Bayesian analysis with 35,000 observations for the simulation with a burn-in of 5,000 and verify the posterior analysis of Table 4.17.
26. Explain the derivation of Equation 4.87, the average duration of play in the gambler's ruin problem. Assuming the total capital is $N = 5$, the starting capital is $k = 2$, and the probability of winning $1 is $p = .6$, use WinBUGS Code 4.10 to verify the posterior analysis for the average duration of play reported in Table 4.19.
27. Derive Equation 4.87, the probability that the gambler will be ruined at stage n of the game. Refer to pages 31–37 of Bailey.

References

1. Markov, A. A. 1906. Rasprostranenie zakona bol'shih chisel na velichiny, zavisyaschie drug ot druga. *Izvestiya Fiziko-matematicheskogo obschestva pri Kazanskom universitete, 2-ya seriya* 15:135–156.
2. Dobrow, R. P. 2016. *Introduction to Stochastic Processes with R*. New York: John Wiley & Sons.
3. Diaconis, P. 2007. Dynamical bias in coin tossing. *SIAM Review* 49(2):211–235.
4. Martell, D. L. 1999. A Markov chain model of day to day changes in the Canadian Forest Fire Weather Index. *International Journal of Wildland Fire* 9(4):265–273.
5. Ross, S. 1996. *Stochastic Processes*. New York: John Wiley & Sons.
6. DeGroot, M. H. 1970. *Optimal Statistical Decisions*. New York: McGraw-Hill.
7. Bailey, N. T. J. 1964. *The Elements of Stochastic Processes*. New York: John Wiley & Sons.
8. Assodou, S., and Essebbar, B. 2003. A Bayesian model for Markov Chains via Jeffrey's prior. *Communications in Statistics: Theory and Methods* 32:2163–2184.
9. Diaconis, P., and Rolles, S. 2006. Bayesian analysis for reversible Markov chains. *Annals of Statistics* 34:1270–1292.
10. Eichelsbacher, P., and Ganesh, A. 2002. Bayesian inference for Markov chains. *Journal of Applied Probability* 39:91–99.
11. Insua, D. R., Ruggeri, F., and Wiper, M. P. 2012. *Bayesian analysis of Stochastic Process Models*. New York: John Wiley & Sons.
12. Martin, J. J. 1967. *Bayesian Decision Problems and Markov Chain*. New York: John Wiley & Sons.
13. Meshkani, M. R. and Billard, L. 1992. Empirical Bayes estimators for a finite Markov chain. *Biometrika* 79:185–193.
14. Welton, N. J., and Ades, A. E. 2005. Estimation of Markov chain transition probabilities and rates from fully and partially observed data: Uncertainty propogation, evidence synthesis and model calibration. *Medical Decision Making* 25:633–645.

5

Examples of Markov Chains in Biology

5.1 Introduction

Bayesian inferential techniques will be employed to gain a deeper understanding of the mechanism of various biological phenomena. Several examples will illustrate the Bayesian approach to making inferences: (1) an example of inbreeding in genetics; (2) the general birth and death process; (3) the logistic growth process; (4) a simple model for an epidemic; (5) the chain binomial model and Greenwood and the Reed–Frost versions of the epidemic models; (6) several genetic models, including the Wright model; and (7) the Ehrenfest model for diffusion through a membrane.

Bayesian inferences will include determining the posterior distribution of the relevant parameters, testing hypotheses about those parameters, and determining the Bayesian predictive distribution of future observations. Such Bayesian procedures will closely follow those presented in Chapter 4 and will comprise generating simulations from the chain, displaying the associated transition graph for the chain of each example, and employing the appropriate R Code and WinBUGS Code for the analysis.

5.2 Inbreeding Problem in Genetics

Inheritance depends on the information contained in the chromosomes that are passed down to future generations. We have two sets of chromosomes, one from the mother and one from the father. Certain locations of the chromosome contain detailed information about the physical characteristics of the individual. Chemicals that make up the chromosome at specific locations (loci) are called genes, and at each locus, the genes are manifested in one of several forms called alleles. For more knowledge about the inbreeding example, refer to pages 76–78 of Allen,[1] who in turn references Hoppensteadt.[2]

Suppose there are two forms of alleles for a given gene symbolized by *a* and A; thus, humans could then have one of three types, namely *aa*, AA, or A*a*, and are called genotypes of the locus. In addition, the two genotypes AA and *aa* are denoted as homozygous, while A*a* is referred to as a heterozygous genotype. The first statisticians to investigate the inbreeding problem as a discrete-time Markov chain (DTMC) are Bailey[3] and Feller.[4] The following description of the inbreeding problem closely follows Allen,[1] where the appropriate chain has six states: (1) AA × AA, (2) AA × A*a*, (3) A*a* × A*a*, (4) A*a* × *aa*, (5) AA × *aa*, and (6) *aa* × *aa*. The laws of inheritance imply the following makeup of the next generation: (1) If the parents are both of type AA, the offspring will be AA individuals, so that the crossing of brother and sister will be of one type. (b) Now suppose the parents are type 2, namely, AA × A*a*, and the offspring will occur in the following proportions: 1/2 AA and 1/2 A*a*; therefore, the crossing of brother and sister will be 1/4 type {AA × AA}, 1/2 type {AA × A*a*}, and 1/4 type {Aa × A*a*}. (c) Lastly, if the parents are of type A*a* × A*a*, the offspring are in the proportion of 1/4 type AA, 1/2 type A*a*, and 1/4 type *aa*; thus, brother and sister mating will give 1/16 type {AA × AA}, 1/4 type {AA × A*a*}, 1/4 type {A*a* × A*a*}, 1/4 type { A*a* × *aa*}, 1/8 type {AA × *aa*}, and 1/16 type {*aa* × *aa*}. It can be shown that the transition matrix is

$$P = \begin{pmatrix} 1, 0.0, 0.0, 0.0, 0.0, 0.0 \\ 1/4, 1/2, 1/4, 0, 0, 0.0 \\ 1/16, 1/4, 1/4, 1/4, 1/8, 1/16 \\ 0.0, 0.0, 1/4, 1/2, 0.0, 1/4 \\ 0.0, 0.0, 1.0, 0.0, 0.0, 0.0 \\ 0.0, 0.0, 0.0, 0.0, 0.0, 1.0 \end{pmatrix}. \tag{5.1}$$

Recall that the states of the process are as follows:
1: AA × AA, 2: AA × A*a*, 3: A*a* × A*a*, 4: A*a* × *aa*, 5: AA × *aa*, and 6: *aa* × *aa*.

Thus, referring to the first row of P, state AA × AA is absorbing; that is, once the chain is in AA × AA, it stays there. The first row of P corresponds to the offspring of random brother–sister mating with parents of type 1 (AA × AA); that is, the offspring of inbreeding can only be of the same type as that of the parents.

One can show that the chain is irreducible and had three communicating classes: {1}, {6}, and {2,3,4,5}. Recall the definition of recurrent and transient; it can be shown that the first two classes are positive recurrent, but the third class is transient.

Recall R Code 4.1 for the simulation of a Markov chain with commands:

R Code 4.1

```
> markov <- function(init,mat,n,labels) {
+ if (missing(labels)) labels <- 1:length(init)
+ simlist <- numeric(n+1)
+ states <- 1:length(init)h
+ simlist[1] <- sample(states,1,prob=init)
+ for (i in 2:(n+1))
+ { simlist[i] <- sample(states,1,prob=mat[simlist[i-1],]) }
+ labels[simlist]
+ }
> P<- 1matrix (c(.51,.49,.49,.51),nrow=2,ncol=2,byrow=TRUE)
> init<=c(1,0)
```

For example, with a starting value of 1 (AA × AA) and n = 100, the Markov function markov(init,P,n) produces the realization

1 1 1 1 1 1 1 1 1 1 1 1 1 1 1 1 1 1 1 1 1 1 1 1 1 1 1 1 1 1 1 1 1 1 1 1 1 1 1 1 1 1 1 1 1 1 1 1
1 1 1 1 1 1 1 1 1 1 1 1 1 1 1 1 1 1 1 1 1 1 1 1 1 1 1 1 1 1 1 1 1 1 1 1 1 1 1 1 1 1 1 1 1 1 1 1
1 1 1 1 1,

and as it should, the process stays in state 1 for 100 times.

On the other hand, when the initial value is 2 (AA × A*a*), the following 99 transitions remain in state 1:

21111111111111111111111111111111111111111111111111
111111111111111111111111111111111111111111111111,11
1.

When studying the behavior of the chain in Equation 5.1, additional information is available by viewing the corresponding transition graph, which is implemented with R Code 5.1:

R Code 5.1

```
library(igraph)
P<-matrix(c(1,0,0,0,0,0,
       1/4,1/2,1/4,0,0,0,
       1/16,1/4,1/4,1/4,1/8,1/16,
       0,0,1/4,1/2,0,1/4,
       0,0,1,0,0,0,
       0,0,0,0,0,1), nrow=6, ncol=6, byrow=TRUE)
> g<-graph.adjacency(P,weighted=TRUE)
> plot(g)
```

Refer to Figure 5.1, which clearly shows that states 1 and 6 are absorbing and positive recurrent, but that the remaining are transient. Once the process is in one of the states 2, 3, 4, or 5, there is a positive probability that the process will be absorbed.

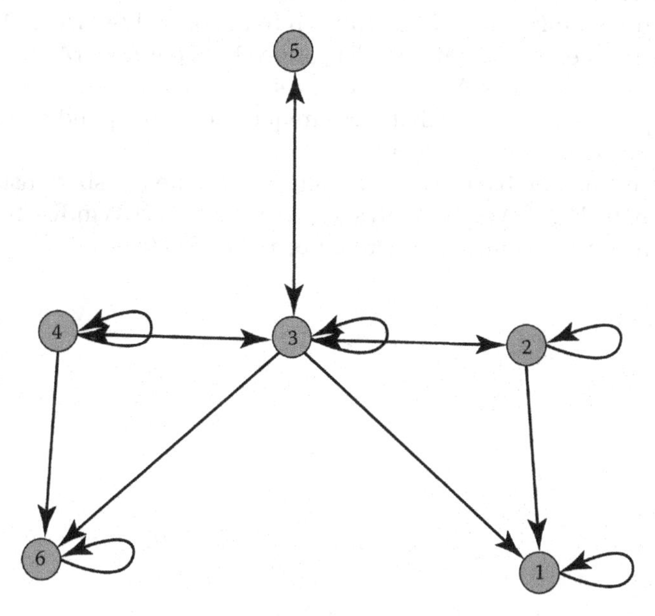

FIGURE 5.1
Markov process for inbreeding.

TABLE 5.1
Five Realizations for Inbreeding Example

	R1	R2	R3	R4	R5
n_{31}	2	1	0	0	0
n_{32}	1	2	1	3	2
n_{33}	3	3	5	3	4
n_{34}	2	1	2	1	2
n_{35}	0	2	0	2	2
n_{36}	2	1	2	1	0

For the Bayesian analysis, five multinomial realizations of size 10 of the third row of P are generated with the R command: multinom(5,10,prob), where prob<-c(1/16,1/4,1/4,1/4,1/8,1/16), and the results reported in Table 5.1.

The realizations of Table 5.1 will be used as the sample information for the Bayesian analysis of estimating the transition probabilities of the third row of P; thus, it will be necessary to assign a prior distribution to these unknown parameters. Consider the improper prior density

$$\xi(P_{31}, P_{32}, P_{33}, P_{34}, P_{35}, P_{36}) \propto \prod_{j=1}^{j=6} P_{3j}^{-1}, \tag{5.2}$$

where $\sum_{j=1}^{j=6} P_{3j} = 1$ and $0 < P_{3j} < 1$, $j = 1, 2, 3, 4, 5, 6$.

Using the first realization, it can be shown that the posterior distribution of the P_{3j}, $j = 1, 2, 3, 4, 5, 6$ is Dirichlet with parameter (2,1,3,2,0,2).

Recall the interpretation of the states of this chain: For example, consider the parents with type Aa × Aa, then their offspring will occur with frequencies 1/4 AA, 1/2 *Aa*, and 1/4 *aa*. Consequently, the brother–sister mating will produce 1/16 of type 1 (AA × AA), 1/4 of type 2 (AA × A*a*), 1/4 of type 3 (A*a* × A*a*), 1/4 of type 4 (A*a* × *aa*), 1/8 of type 5 (AA × *aa*), and, finally, 1/16 of type 6 (*aa* × *aa*). Note that these frequencies correspond to the third row of P and are dictated by the laws of inheritance.

Our main objective for the Bayesian analysis is to estimate the six transition parameters P_{3j}, $j = 1, 2, 3, 4, 5, 6$. The Bayesian analysis is executed with WinBUGS Code 5.1 using 35,000 observations for the simulation with a burn-in of 5,000:

WinBUGS Code 5.1

```
model;
{
p31~dbeta(2,8)
p32~dbeta(1,9)
p33~dbeta(3,7)
p34~dbeta(2,8)
p35~dbeta(0,10)
p36~dbeta(2,8)
psum<-p31+p32+p33+p34+p35+p36
}
```

TABLE 5.2

Posterior Analysis of Inbreeding Example

Parameter	Mean	SD	Error	2 1/2	Median	97 1/2
P_{31}	.1998	.1215	.000713	.02789	.1782	.489
P_{32}	.09442	.08948	.000489	.00281	.07413	.3354
P_{33}	.2998	.1378	.000788	.0756	.2859	.5975
P_{34}	.2006	.1209	.000726	.02912	.1799	.4863
P_{35}	0	0	0	0	0	0
P_{36}	.1991	.12	.000700	.0279	.1786	.482
$P_{3.}$	.999	.266	.001358	.5284	.9823	1.559

The Bayesian analysis is reported in Table 5.2.

The last row of Table 5.2 is the sum of the transition probabilities and serves as a check on the analysis! What is the interpretation of the posterior analysis? For example, for the transition probability P_{33} (the probability that the offspring of brother–sister mating with parents of type 3 genetic disposition A*a* × A*a* will result in an offspring with type 3 genotypes) has a posterior mean of .2998 with a 95% credible interval (.0756,.5975). The posterior distributions appear to be symmetric and the Monte Carlo Markov chain error is sufficiently small to have confidence in the simulation. What does the posterior analysis imply about inbreeding?

Of course, a different posterior analysis will result if one uses the second column of Table 5.1, which is a different multinomial simulation of the third row of *P*. Exercises at the end of the chapter will ask the student to execute their own Bayesian analysis based on realizations 2 through 5 appearing in Table 5.1.

Of primary interest to the investigator is testing the hypothesis that the laws of inheritance are indeed those used to generate the multinomial realizations of Table 5.1. The laws of inheritance are those specified by the transition matrix *P* in Equation 5.1. Thus, consider a test of the null hypothesis

$$H_0: P_{31} = 1/16,\ P_{32} = 1/4,\ P_{33} = 1/4,\ P_{34} = 1/4,\ P_{35} = 1/8,\ P_{36} = 1/16$$

versus the alternative that H_0 is not true. H_0 is the hypothesis implied by the law of inheritance for the offspring of brother–sister mating whose parents are of type 3, namely, A*a* × A*a*. Therefore, one needs to compute the posterior probability of the null hypothesis.

I will use the first multinomial realization of Table 5.1 for the sample information; thus, let

$$n = (n_{31}, n_{32}, n_{33}, n_{34}, n_{35}, n_{36})$$

and the third row of *P*

$$P_3 = (P_{31}, P_{32}, P_{33}, P_{34}, P_{35}, P_{36}).$$

The conditional mass function of the observations *n* given the unknown parameters P_3 is the multinomial mass function

$$f(n|P_3) = \left[n_{3.}!/\prod_{j=1}^{j=6} n_{3j}! \right] \prod_{j=1}^{j=6} P_{3j}^{n_{3j}}, \tag{5.3}$$

where P_{3j} are probabilities and $\sum_{j=1}^{j=6} P_{3j} = 1$. Note that for the first realization of Table 5.1, $n_{31} = 2$, $n_{32} = 1$, $n_{33} = 3$, $n_{34} = 2$, $n_{35} = 0$, $n_{36} = 2$, and the transition count total is $n_{3.} = 10$.

Recall that the posterior probability of the null hypothesis is

$$p_0 = \pi_0 f(n|P_{3.} = (1/16, 1/4, 1/4, 1/4, 1/8, 1/16))/\zeta(n), \tag{5.4}$$

where

$$\zeta(\mathrm{n}) = \pi_0 f(n|P_3 = (1/16, 1/4, 1/4, 1/4, 1/8, 1/16)) + g(n)$$

and

$$g(n) = \pi_1 \int \xi(P_3) f(n|P_3) dP_3. \tag{5.5}$$

Also, π_0 is the prior probability of the null hypothesis and $\pi_1 = 1 - \pi_0$, the prior probability of the alternative. Also $\xi(P_3)$ is the prior density of P_3 under the alternative hypothesis. How should the prior density under the alternative be selected? I assume that the prior density under the null hypothesis is Dirichlet with parameter $(\alpha_{1.}, \alpha_2, \alpha_3, \alpha_4, \alpha_5, \alpha_6) = (1.25, 5, 5, 5, 2.5, 1.25)$, which gives the prior mean under the alternative as (.0625,.25,.25,.25,.125,.0625).

One can show that

$$\int \xi(P_3) f(n|P_3) \mathrm{d}P_3$$

$$= \left\{ \left[n! / \prod_{j=1}^{j=6} n_{3j}! \right] \Gamma\left(\sum_{j=1}^{j=6} \alpha_j \right) \sum_{j=1}^{j=6} \Gamma\left(n_{3j} + \alpha_j + 1\right) \right\} \Big/ \left\{ \prod_{j=1}^{j=6} n_{3j}! \prod_{j=1}^{j=6} \Gamma\left(\alpha_j\right) \Gamma\left(\sum_{j=1}^{j=6} \left(n_{3j} + \alpha_j + 1\right) \right) \right\} \tag{5.6}$$

In addition, it can be shown that $\zeta(n) = .000280001$ and, consequently, that $p_0 = .9999$; thus, the data support the inbreeding probabilities implied by the laws of inheritance.

5.3 Wright Model for Genetics

The following idealized genetics model was introduced by Wright[5] in order to investigate the variation in gene frequency under the influence of mutation and selection. This initial version of the model is the less complex in that it describes the simple haploid process with random reproduction and does not account for mutation and selection forces. The population size is fixed at $2N$ genes made up of type a and type A individuals. If the present generation contains j-type a genes and $2N - j$-type A genes, the next generation results in a type a gene or a type A with probabilities

$$p_j = j/2N,\ q_j = 1 - j/2N, \tag{5.7}$$

respectively. The genes of successive generations are selected with replacement inducing a Markov chain $\{X(n) = 0, 1, 2, ..., 2N\}$, where $X(n)$ is the number of type a genes. The transition matrix is given by

$$\Pr[X(n+1) = k|X(n) = j] = \binom{2N}{k} p_j^k q_j^{2N-k}, j,k = 0,1,2,...,2N. \tag{5.8}$$

Thus, the conditional distribution of $X(n + 1)|X(n)$ is binomial with parameters $2N$ and p_j. For additional information about the biology of this process, see Fisher.[6] It is obvious that 0 and $2N$ are absorbing states; that is, if $X(n) = 0$, there are no type a genes, and when $X(n) = 2N$, there are $2N$ type a genes. An interesting question is that of fixation; that is, assuming an initial state of i type a genes, what is the probability that the process will be all type a genes or all type A genes? The rate of approach to fixation is also something to be explored.

A more realistic model takes into account the possibility of mutations; therefore, prior to the formation of a new generation, each gene has a chance to mutate. The model incorporating mutation is as follows: The probability of mutation $a \to$ A is denoted by α_1 and the probability of mutation A $\to a$ is α_2. As in the previous model, assume that the composition of the next generation is determined by $2N$ independent trials according to the binomial model in Equation 5.8. Assume that the parent population has j-type a genes, where the probability of a type a gene is

$$p_j = (j/2N)(1 - \alpha_1) + (1 - j/2N)\alpha_2, \tag{5.9}$$

and the probability of a type A gene is

$$q_j = (j/2N)\alpha_1 + (1 - j/2N)(1 - \alpha_2). \tag{5.10}$$

Thus, the transition matrix for the model that takes into account mutation is given by Equation 5.8, where the p_j and q_j, $j = 0, 1, 2, ..., 2N$, are defined by Equations 5.9 and 5.10, respectively. It is assumed that the mutation forces act first, after which a new gene is specified by selecting at random from the population. It should be observed that the chance of selecting a type a gene is $1/2N$ times the number of type a genes; therefore, the average probability of selecting an a-type gene is $1/2N$ times the average number of a genes which is $j(1 - \alpha_1) + (2N - j)\alpha_2$, which in turn implies Equation 5.9.

Note that the probabilities in Equations 5.9 and 5.10 of the transition matrix in Equation 5.8 depend on the unknown mutation rates α_1 and α_2; thus, the goal of the Bayesian analysis will be to estimate these mutation rates and to test hypotheses about these unknown parameters. One must generate an observation for the process with the 1×7 first row of the transition matrix:

$$(1-\alpha_2)^6, 6\alpha_2(1-\alpha_2)^5, 15\alpha_2^2(1-\alpha_2)^4, 20\alpha_2^3(1-\alpha_2)^3, 15\alpha_2^4(1-\alpha_2)^2, 6_2^5(1-\alpha_2), \alpha_2^6$$

To generate the observations, a value for α_2 must be assigned. Note that the first row of the transition matrix in Equation 5.8 depends only on α_2; thus, our approach is to generate multinomial realizations for the first row of the transition matrix. Assuming that the mutation rate for $A \to a$ is $\alpha_2 = .07$, the first row probabilities are

$$.647,\ .2992,\ .05498,\ .005518,\ .000311,\ .000009,\ .0000001, \tag{5.11}$$

Then, in a similar way, generate realizations for the seventh row of the transition matrix. The components of the 1×7 seventh row will only depend on α_1, namely,

$$\begin{aligned}&\alpha_1^6, 6(1-\alpha_1)\alpha_1^5, 15(1-\alpha_1)^2\alpha_1^4, 20(1-\alpha_1)^3\alpha_1^3, 15(1-\alpha_1)^4\alpha_1^2, 6(1-\alpha_1)^5\alpha_1, (1-\alpha_1)^6\\ &= .000001, .000054, .001215, .01458, .098415, .354294, .531441,\end{aligned} \tag{5.12}$$

when $\alpha_1 = .1$.

Our goal is to estimate the mutation rates, based on multinomial realizations for the first and seventh rows of the transition matrix in Equation 5.8.

Consider the first row of the transition rule; then the following R Code is used to generate multinomial realizations of size 100 for the mutation example:

```
> p<-c(.647,.299,.05498,.005518,.000311,.000009,.0000001)
> rmultinom(5,100,p)
```

The usual way to estimate α_2 is to let $68/100 = .68 = (1-\alpha_2)^6$, which implies the estimate is $\alpha_2 = .06223$, which is quite close to the value of .07 used to generate the realizations of Table 5.3. What is the Bayesian approach to estimating α_2? It will be indirect as follows. Let P_{0j}, $j = 0,1,2,\ldots,6$, be the transition probabilities for the first row of the transition matrix in Equation 5.8, where

$$P_{00} = (1-\alpha_2)^6. \tag{5.13}$$

Suppose that the prior distribution for the first row is improper.

Then, the posterior distribution for the first row is Dirichlet with parameter (68,28,3,1,0,0,0) and the marginal posterior of P_{00} is beta (68, 72). I executed the Bayesian analysis with WinBUGS Code 5.2 using 35,000 observations for the simulation with a burn-in of 5,000, and the posterior analysis is reported in Table 5.4.

TABLE 5.3

Multinomial Realizations for Mutation Example

R1	R2	R3	R4	R5
68	67	65	62	63
28	27	27	34	35
3	4	7	3	2
1	2	1	1	0
0	0	0	0	0
0	0	0	0	0
0	0	0	0	0

TABLE 5.4

Posterior Analysis for the Mutation Rate α_2

Parameter	Mean	SD	Error	2 1/2	Median	97 1/2
α_2	.06251	.01069	.0000687	.04309	.06192	.08524
p_{00}	.6802	.04613	.000296	.5859	.6814	.7677

WinBUGS Code 5.2

```
model;
{
p00~dbeta(68,32)
alpha2<-1-pow(p01,.16666)
}
```

The posterior mean for mutation A $\rightarrow$ *a* with rate α_2 is .06251 with a 95% credible interval of (.043,.085), and the posterior density appears to be symmetric about the mean, as seen from Figure 5.2. Also, the distribution appears to be normally distributed in appearance, which is an attribute that should be taken into account for testing hypotheses about α_2. The student will be asked to perform a similar posterior analysis for estimating the rate α_1 for the mutation *a* $\rightarrow$ A. It is interesting to note that the credible interval for α_2 does contain .07, the value used to generate the data, the first realization in Table 5.3.

One can show that if $\alpha_1\alpha_2 > 0$, then the fixation will not occur in any state, and the reader will be asked to investigate the fixation question by performing a Bayesian analysis.

Selection is another genetic force that should be taken into account, and the topic is presented on page 57 of Karlin and Taylor.[7] Suppose the selection forces are in favor of, say, the *a*-type gene; that is, suppose we want to impose a selection advantage of *a*-type genes over A-type genes, in such a way that the selected number of offspring have an expectation proportional to 1 + *s* and 1, respectively, where *s* is small and positive. Thus, in the binomial transition matrix (Equation 5.8), p_j and q_j given by Equations 5.9 and 5.10, respectively, are replaced by

$$p_j = (1 + s)j/(2N + sj) \tag{5.14}$$

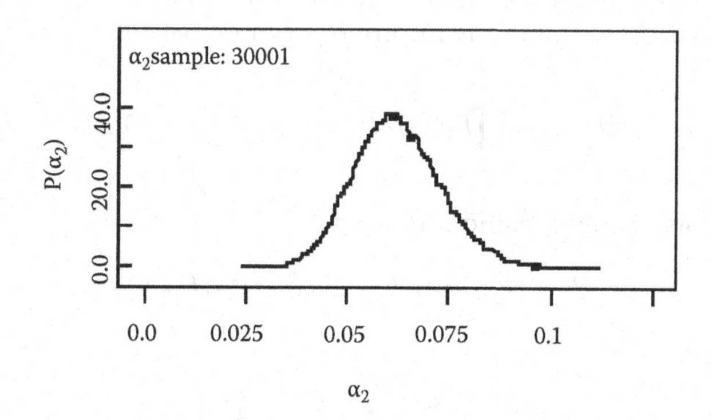

FIGURE 5.2
Posterior density of α_2.

and

$$q_j = 1 - p_j = (2N - j)/(2N + sj), \tag{5.15}$$

respectively.

It can be shown that if the parent population consists of j type a genes and $2N - j$ type A genes, the next generation will on the average have $2Np_j$ type a genes and $2Nq_j$ type A genes. Also, at the $(n + 1)$ generation, the ratio of the expected size of the population of type a genes to type A genes is

$$(1 + s)j/(2N - j) = [(1 + s)/s]r,$$

where r = [number of a-type genes in the nth generation ÷ number of type A genes in the nth generation].

For the Bayesian, the main goal is to estimate the selection force s involved in the binomial probabilities in Equations 5.14 and 5.15 of the binomial transition matrix (Equation 5.8).

It can be shown that the second row of the transition matrix is

$$\begin{aligned} P_{10} &= [5/(6+s)]^6, \\ P_{11} &= 6[(1+s)/(6+s)][5/(6+s)]]^5, \\ P_{12} &= 15[(1+s)/(6+s)]^2[5/(6+s)]^4, \\ P_{13} &= 20[(1+s)/(6+s)]^3[5/(6+s)]^3, \\ P_{14} &= 15[(1+s)/(6+s)]^4[5/(6+s)]^2, \\ P_{15} &= 6[(1+s)/(6+s)]^5[5/(6+s)], \\ P_{16} &= [(1+s)/(6+s)]^6. \end{aligned} \tag{5.16}$$

For the Bayesian analysis, one must assign a value to the selection force s; thus let $s = 1$, then the second row of the transition matrix is as follows:

$$P_{10} = .1328,\ P_{11} = .3188,\ P_{12} = .3187,\ P_{13} = .17,\ P_{14} = .05099,\ P_{15} = .008158,\ P_{16} = .000544 \tag{5.17}$$

The Bayesian approach will generate a multinomial realization for the second row of the transition matrix; then using those as data to form the likelihood function, the selection force s will be estimated assuming the improper prior density

$$\xi(P_1) \propto \prod_{j=0}^{j=6} P_{1j}^{-1},\ 0 < P_{1j} < 1,\ \sum_{j=0}^{j=6} P_{1j} = 1, \tag{5.18}$$

where the second row of the transition matrix is

$$P_1 = (P_{10}, P_{11}, P_{12}, P_{13}, P_{14}, P_{15}, P_{16}). \tag{5.19}$$

Consider

$$P_{10} = [5/(6+s)]^6 = .1328 \tag{5.20}$$

TABLE 5.5
Five Multinomial Realizations of Selection Force $s = 1$

	R1	R2	R3	R4	R5
n_{10}	15	15	13	16	7
n_{11}	26	33	29	37	31
n_{12}	35	27	37	26	44
n_{13}	17	18	14	15	13
n_{14}	6	2	6	6	4
n_{15}	1	5	1	0	1
n_{16}	0	0	0	0	0

Now solving Equation 5.20 for s gives $s = .99999634$. What is the Bayesian approach to estimating s? The first five multinomial realizations for the second row of the transition matrix will be performed with a total of $n = 100$ counts as revealed in Table 5.5. The following R Code was employed to generate the transition counts of the table:

```
p<-c(.1328,.3188,.3187,.17,.05099,.008158,.000544)
rmultinom(5,100,p),
```

where p is the vector of transition probabilities of the second row of the transition matrix. With the improper prior density for these transition probabilities and based on the first realization of Table 5.5, it is well known that the posterior distribution of P_1 of Equation 5.18 is Dirichlet (15,26,35,17,6,1,0). See pages 48–50 of Degroot.[8]

The Bayesian analysis is executed with WinBUGS Code 5.3 using 70,000 observations for the simulation and a burn-in of 5,000, and the posterior distribution for P_{10} and s is reported in Table 5.6.

WinBUGS Code 5.3

```
model;
{
p10~dbeta(15,85)
s<-5/(pow(p10,.16667))-6
}
```

The posterior mean of the selection force is 0.8975 with a 95% credible interval of (0.4046,1.503), and it appears that the distribution is symmetric about the posterior mean. The value of 0.897 for the posterior mean is fairly close to the value of $s = 1$ used to generate the transition counts for the second row, but this is not surprising since one would expect a

TABLE 5.6
Posterior Analysis for Selection

Parameter	Mean	SD	Error	2 1/2	Median	97 1/2
P_{10}	.15	.03546	.000143	.0875	.1476	.2264
s	0.8975	0.2799	0.001134	0.4046	0.878	1.503

sample variation in the various realizations as portrayed in Table 5.5. In order to test the hypothesis $s = 1$ versus $s \neq 1$, one could test the hypothesis that $P_{01} = .1328$. See Equation 5.16, which expresses the transition probabilities in terms of the selection force s.

5.4 Birth and Death Process

A general birth and death process is formulated as a DTMC. We consider a finite population of maximum size N and a chain $\{X(n), n = 0, 1, 2, \ldots\}$ with state space $\{0, 1, 2, \ldots, N\}$, where $X(n)$ is the size of the population at time n. The birth and death probabilities b_i and d_i depend on the size of population where

$$\begin{aligned} P_{ij} &= \Pr\{X(n+1) = j \mid X(n) = i\} \\ &= b_i,\ j = i+1, \\ &= d_i,\ j = i-1, \\ &= 1 - (b_i + d_i), \\ &= 0, \text{ otherwise,} \end{aligned} \tag{5.21}$$

and $i = 1, 2, \ldots$; $P_{00} = 1$; and $P_{0j} = 0, j \neq 0$. Also, note that $P_{N,N+1} = b_N = 0$; therefore the $(N + 1) \times (N + 1)$ transition matrix is given by

$$P = \begin{pmatrix} 1,0,0,0,0,\ldots,,,,,,,,,,,,,,,,,,,,,,,,,,,,,0 \\ d_1, 1-(d_1+b_1), b_1, 0,0,0,0,0,0,0,\ldots,,0 \\ 0, d_2, 1-(d_2+b_2), b_2, 0,0,0,0,0,0,\ldots,0 \\ . \\ . \\ 0,0,0,\ldots\ldots\ldots, 0, d_{N-1}, 1-(d_{N-1}+b_{N-1}), b_{N-1} \\ 0,0,\ldots\ldots\ldots\ldots\ldots\ldots\ldots\ldots\ldots, 0, d_N, 1-d_N \end{pmatrix}. \tag{5.22}$$

Thus, the population increases by one, decreases by one, or remains the same. There are two communicating classes where $\{0\}$ and $\{1, 2, \ldots, N\}$, where 0 is an absorbing state and the remaining are transient. In addition, it can be shown that there is a unique stationary distribution.

As an example of a birth and death process, let $N = 10$ and $b_i = bi$, $i = 1,2,\ldots,9$, $d_i = di$, $i = 1,2,\ldots,10$, where b and d are constants.

Three cases are considered: (1) $b = .02 < .03 = d$, (2) $b = .025 = d$, and (3) $b = .03 > .02 = d$.

For the first case, the transition matrix is given by

$$P = \begin{pmatrix} 1,0,0,0,0,0,0,0,0,0,0 \\ .03,.95,.02,0,0,0,0,0,0,0,0 \\ 0,.06,.9,.04,0,0,0,0,0,0,0 \\ 0,0,.09,.85,.06,0,0,0,0,0,0 \\ 0,0,0,.12,.8,.08,0,0,0,0,0 \\ 0,0,0,0,.15,.75,.10,0,0,0,0 \\ 0,0,0,0,0,.18,.70,.12,0,0,0 \\ 0,0,0,0,0,0,.21,.65,.14,0,0 \\ 0,0,0,0,0,0,0,.24,.60,.16,0 \\ 0,0,0,0,0,0,0,0,.27,.55,.18 \\ 0,0,0,0,0,0,0,0,0,.30,.70 \end{pmatrix}. \tag{5.23}$$

Recall in R Code 5.2 the Markov function that generates observations from the chain with transition matrix P:

R Code 5.2

```
markov <- function(init,mat,n,labels) {
+ if (missing(labels)) labels <- 1:length(init)
+ simlist <- numeric(n+1)
+ states <- 1:length(init)
+ simlist[1] <- sample(states,1,prob=init)
+ for (i in 2:(n+1))
+ { simlist[i] <- sample(states,1,prob=mat[simlist[i-1],]) }
+ labels[simlist]
+ }
```

The first simulation has initial value 2 and the process is absorbed by the state 1.

```
> init<-c(0,1,0,0,0,0,0,0,0,0,0)
> markov(init,p,100)
2 2 2 2 2 2 2 2 2 2 2 2 3 3 3 3 3 3 3 3 3 3 3 3 3 2 2 2 2 2 2 2 2 2 2 2 2
2 2 2 2 2 2 2 2 2 2 2 2 2 2 2 2 2 1 1 1 1 1 1 1 1 1 1 1 1 1 1 1 1 1 1 1 1
1 1 1 1 1 1 1 1 1 1 1 1 1 1 1 1 1 1 1 1 1 1 1 1 1 1 1
```

The second simulation begins with the state 3 and stays in state 2 for the remaining observations:

```
> init<-c(0,0,1,0,0,0,0,0,0,0,0)
>init<-c(0,0,1,0,0,0,0,0,0,0,0)
>markov(init,p,100)

3 3 3 3 4 4 3 3 3 3 3 3 3 3 3 3 3 3 3 3 3 3 3 3 3 3 3 3 3 3 3 3 3 3 3 3 3 3 3 3 3
3 3 3 3 3 3 3 3 3 3 3 3 2 2 2 2 2 2 2 2 2 2 2 2 2 2 2 2 2 2 2 2
2 2 2 2 2 2 2 2 2 2 2 2 2 2 2 2 2 2 2 2 2 2 2 2 2 2 2.
```

The student should experiment with the Markov R function with additional simulations using different initial values. Also of interest is the R function stationary, which when applied to the transition P gives the stationary distribution π reported in Table 5.7:

```
stationary <- function(P) {
+ x = eigen(t(mat))$vectors[,1]
+ as.double(x/sum(x))
+ }
>
> stationary(P)
```

It will be left to the student to find the Bayesian approach to determining the stationary distribution as an exercise at the end of the chapter. The reader is referred to Equations 4.66 through 4.68 of Chapter 4 and to WinBUGS Code 4.6.

Our last goal for this example of a birth and death process is to estimate the birth and death rates using a Bayesian approach, whereby multinomial realizations will be generated from the first three components of the second row of P, namely, (.03,.95,.02), which are the values assigned to the vector $(d_1, 1-(b_1+d_1), b_1)$. The R Code

```
p<-c(.03,.95,.02)
> rmultinom(5,100,p)
```

generates the following five realizations of size 100 depicted in Table 5.8.

TABLE 5.7

Stationary Distribution: Birth and Death Process

π_0	.001782713
π_1	.009306754
π_2	.032048848
π_3	.032048848
π_4	.217523222
π_5	.481424409
π_6	.120759090
π_7	.033324661
π_8	.009965581
π_9	.003265024
π_{10}	.001287085

TABLE 5.8
Five Realizations for Birth and Death Process

	R1	R2	R3	R4	R5
d_1	5	1	1	1	2
$1-(d_1+b_1)$	93	96	97	96	91
b_1	2	3	2	3	7

Based on the fifth realization, the usual estimates of the death and birth rates d_1 and b_1 are .02 and .07, respectively, compared to the values of .03 and .02, respectively, used to generate those realizations. Also apparent is the variation demonstrated between the five realizations. Recall that the Bayesian analysis is based on assuming that the realizations have multinomial distributions and an improper prior is assigned to the unknown parameters d_1 and b_1, that the prior density is

$$g(b_1, d_1) \propto 1/b_1 d_1, 0 < b_1 < 1, 0 < d_1 < 1 \tag{5.24}$$

Based on the fifth realization and the prior density (Equation 5.24), the posterior density of b_1 is beta (7,93) and that of d_1 is beta (2,98), and the complete Bayesian analysis is executed with WinBUGS Code 5.4:

WinBUGS Code 5.4

```
model;
{
d1~dbeta(7,93)
b1~dbeta(2,98)
}
```

I executed the analysis with 70,000 observations for the simulation and a burn-in of 5,000. The actual value of the birth rate is .02, and the posterior mean is .02006 with a 95% credible interval of (.0023,.0552), but the actual value of the death rate is .03 and is estimated with the posterior median of .0672 with a 95% credible interval (.0287,.1269). Note also that the credible interval for d_1 does include .03! Both posterior distributions appear to be skewed to the right. See the posterior density of b_1 shown in Figure 5.3.

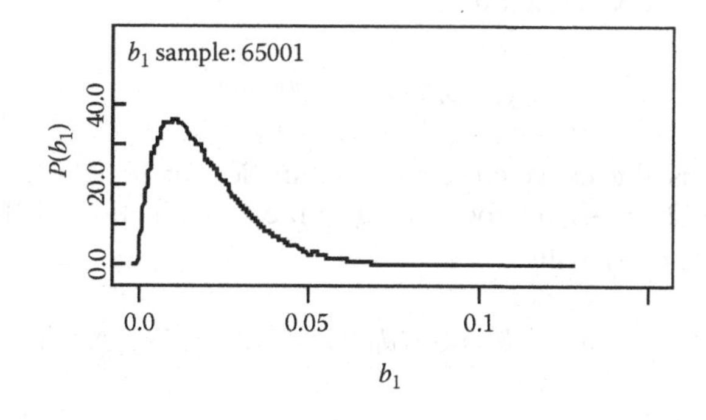

FIGURE 5.3
Posterior density of b_1.

The last aspect of Bayesian inference for the birth and death process is to derive the predictive distribution of the multinomial distribution used to generate realizations (Table 5.8) for the counts of the birth and death process (Equation 5.21) with transition matrix (Equation 5.23).

In general, the predictive density of the future transition counts (m_1, m_2, m_3) of the first three components of the first row of P (Equation 5.23):

$$\int\int\int [m!/m_1!m_2!m_3!]\left[\Gamma\left(\sum_{j=1}^{j=3} m_j + \alpha_j\right)/\prod_{j=1}^{j=3}\Gamma\left(m_j + \alpha_j\right)\right]\prod_{j=1}^{j=3} P_j^{m_j+n_j+\alpha_j-1}\, dP_{11}, dP_{12}, dP_{13}$$

$$= \left[m/\prod_{j=1}^{j=3} m_j!\right]\left[\Gamma\left(\sum_{j=1}^{j=3}\left(n_j + \alpha_j\right)\right)\right]/\left[\prod_{j=1}^{j=3}\Gamma(n_j + \alpha_j)\right] \tag{5.25}$$

$$\left[\prod_{j=1}^{j=3}\Gamma\left(n_j + m_j + \alpha_j\right)/\Gamma\left(\sum_{j=1}^{j=3}\left(n_j + m_j + \alpha_j\right)\right)\right],$$

where the posterior density of (P_{11}, P_{12}, P_{13}) is Dirichlet $(n_1 + \alpha_1, n_2 + \alpha_2, n_3 + \alpha_3)$, and the corresponding prior density is Dirichlet $(\alpha_1, \alpha_2, \alpha_3)$. Note that by letting the alpha hyper parameters be zero, one is in effect assuming an improper prior for the three transition parameters, which simplifies the predictive density to

$$g(m_1, m_2, m_3) = A/B, \tag{5.26}$$

where

$$A = [m!/m_1!m_2!m_3!]\Gamma(100)\Gamma(2 + m_1)\Gamma(91 + m_2)\Gamma(7 + m_3)$$

and

$$B = \Gamma(2)\Gamma(91)\Gamma(7)\Gamma(100 + m_1 + m_2 + m_3).$$

Recall that the cell count m_1 corresponds to the death rate d_1; m_2, to $1 - (d_1 + b_1)$; and, m_3, to b_1. For example, it can be shown that

$$g(2,3,5) = .000008392 \tag{5.27}$$

Of special interest is the expected time to extinction of the birth and death process. According to page 122 of Allen,[1] the average time to extinction of the birth and death process $\{X(n), n = 0, 1, 2, \ldots\}$ with

$$X(0) = m \geq 1,\ b_0 = 0 = d_0,\ b_i > 0,\ i = 1, 2, \ldots, N - 1,$$

and $d_i > 0$, $i = 1, 2, \ldots, N$ is given by

$$\mu_m = 1/d_1 + \sum_{i=2}^{i=N}(b_1...b_{i-1}/d_1...d_i), \quad m = 1$$
$$= \mu_1 + \sum_{s=1}^{s=m-1}[d_1...d_s/b_1..b_s]\left[\sum_{i=s+1}^{i=N}[b_1...b_{i-1}/d_1...d_i]\right], \quad m = 2,...,N. \tag{5.28}$$

Using Equation 5.28, it can be shown that the average time to extinction is $\mu_1 = 54.7933$ time points. Note that this estimate does not have a standard error attached, and this problem can be avoided using Bayesian inferential techniques. Is this value reasonable? Remember that if the population size is 1, one would need a death for extinction, but the probability of a death is .03, a small chance that extinction is imminent. Also, recall that the probability that the population size remains at 1 with probability .95 at each time point.

The Bayesian approach will generate multinomial data using the transition matrix P of Equation 5.23. I generated a multinomial realization of size 100 for the transition counts corresponding to the 10 population sizes which correspond to the various birth and death rates listed in columns 2 and 4. For example, corresponding to the population of size 5 with death rate of .15 and birth rate of .10, the multinomial realization of size 100 is (11,76,13). Thus, the usual estimate of the death rate is .11 compared to the actual value of .15. Of course, we would expect the estimated rates to differ from the actual.

Assuming an improper prior density of the birth and death rates, the posterior distribution of the transition probabilities (the birth and death rates) will each be beta. For example, the posterior distribution of d_2 is beta (8,92), which has a posterior mean of .08. I executed the posterior analysis with WinBUGS Code 5.5 with 70,000 observations for the simulation and a burn-in of 5,000.

WinBUGS Code 5.5

```
model;
{
d1~dbeta(2,98)
b1~dbeta(1,99)
d2~dbeta(8,92)
b2~dbeta(6,94)
d3~dbeta(7,93)
b3~dbeta(2,98)
d4~dbeta(15,85)
b4~dbeta(4,96)
d5~dbeta(11,89)
b5~dbeta(13,87)
d6~dbeta(17,83)
b6~dbeta(17,83)
d7~dbeta(22,78)
b7~dbeta(14,86)
d8~dbeta(23,77)
b8~dbeta(18,82)
d9~dbeta(27,73)
b9~dbeta(16,84)

d10~dbeta(28,72)
```

TABLE 5.9

Birth and Death Process

Parameter	Mean	SD	Error	2 1/2	Median	97 1/2
μ_1	134.4	465	1.702	21.53	76.55	574.4

```
b12<-b1*b2
b13<-b12*b3
b14<-b13*b4
b15<-b14*b5
b16<-b15*b6
b17<-b16*b7
b18<-b17*b8
b19<-b18*b9

d12<-d1*d2
d13<-d12*d3
d14<-d13*d4
d15<-d14*d5
d16<-d15*d6
d17<-d16*d7
d18<-d17*d8
d19<-d18*d9
d110<-d19*d10
mu1<-(1/d1)+(b1/d12)+(b12/d13)+(b13/d14)+(b14/d15)+(b15/d16)+(b16/d17)
+(b17/d18)+(b18/d19)+(b19/d110)
}
```

The Bayesian analysis is reported in Table 5.9.

The posterior distribution of μ_1 is extremely skewed to the right. Refer to Equation 5.28 for μ_1, which shows that the numerator and denominator of the terms are extremely small sometimes in very large values for the simulation. Because of the asymmetry of the posterior distributions, I recommend the median of 76.55 time points as the estimate of the time to extinction. This implies that on the average, it will take 77 time points to reach a population size of 0, starting with a population of size 1! This is because the probability of a death is only .03. Note that one is assuming the time between events so that in small time intervals, at most 1 event will occur.

5.5 Logistic Growth Process

The logistic growth process is a variation of the birth and death process that satisfies

$$b_i - d_i = ri(1 - i/K), \quad i = 0,1,2,\ldots,N, N > K, \tag{5.29}$$

where r is the intrinsic growth rate, K is the carrying capacity, b_i is the birth rate when the population is of size i, and d_i the corresponding death rate. The process is observed at various time points, such that the time between time points is sufficiently small so that $\{d_i + b_i\} \le 1$. It can be seen from Equation 5.29 that the behavior of the process varies according to the size of the population. Note that when $i = 0$ or $i = K$ (the carrying capacity), the birth and death rates are the same. Two cases are considered:

1. $b_i = r(i - i^2/2K), \quad d_i = ri^2/2K, \quad i = 0,1,2,\ldots,2K$
2. $b_i = ri, \quad i = 0,1,2,\ldots,N-1; \quad d_i = ri^2/K, \quad i = 0,1,2,\ldots.,N.$ (5.30)

For the first case, the maximum population size is $N = 2K$, and the birth probability increases when the population size is $<K$ and decreases for population sizes $>K$, but the death rate is always increasing. On the other hand, for the second case, both the birth and death probabilities increase with the population size. Refer to pages 123–127 of Allen[1] for additional details about the logistic growth process.

For the Bayesian analysis, interest is restricted to the second case with $N = 20$, $K = 10$, $r = .004$, and $b_i = ri$, $d_i = ri^2/K$. The one-step 21×21 transition matrix for the second case is given in the following by P, see Equation (5.31). The analysis consists of generating multinomial realizations of size 1000 for the nonzero entries of each row of P. In this way, the sample information is available, then this is combined with the prior information for the unknown parameters (the transition parameters of the transition matrix of which there are three for each row of P). As before, an improper prior distribution is assigned, which, when combined with the realizations of Table 5.10, results in Dirichlet distributions for the transition rows of P:

TABLE 5.10

Multinomial Realizations for the Logistic Growth Model

Population Size	d	$1-b-d$	b	n_1	n_2	n_3
1	.0004	.9956	.004	1	992	7
2	.0016	.9904	.008	1	996	3
3	.0036	.9844	.012	5	988	7
4	.0064	.9776	.016	4	977	19
5	.01	.9700	.020	9	964	27
6	.0144	.9616	.024	15	961	24
7	.0196	.9524	.028	18	952	30
8	.0256	.9242	.032	26	943	31
9	.0324	.9316	.036	29	935	36
10	.04	.92	.04	48	912	40
11	.0484	.9076	.044	41	914	41
12	.0576	.8944	.048	62	888	50
13	.0676	.8804	.052	74	880	46
14	.0784	.8656	.056	63	882	55
15	.09	.85	.06	84	859	57
16	.1024	.8336	.064	93	841	66
17	.1156	.8164	.068	109	824	67
18	.1296	.7984	.072	132	801	67
19	.1444	.7790	.076	139	781	80
20	.16	.84		154	846	

$$P = \begin{pmatrix}
1.0, 0.0, 0.0, 0.0, 0.0, 0.0, 0, 0, 0, 0, 0, 0, 0, 0, 0, 0, 0, 0, 0, 0, 0 \\
.0004, , 9966, .004, 0, 0, 0, 0, 0, 0, 0, 0, 0, 0, 0, 0, 0, 0, 0, 0, 0, 0 \\
0, .0016, .9904, .008, 0, 0, 0, 0, 0, 0, 0, 0, 0, 0, 0, 0, 0, 0, 0, 0, 0 \\
0, 0, .0036, .9844, .012, 0, 0, 0, 0, 0, 0, 0, 0, 0, 0, 0, 0, 0, 0, 0, 0 \\
0, 0, 0, .0064, .9776, .016, 0, 0, 0, 0, 0, 0, 0, 0, 0, 0, 0, 0, 0, 0, 0, \\
0, 0, 0, 0, .01, .97, .02, 0, 0, 0, 0, 0, 0, 0, 0, 0, 0, 0, 0, 0, 0 \\
0, 0, 0, 0, 0, .0144, .9616, .024, 0, 0, 0, 0, 0, 0, 0, 0, 0, 0, 0, 0, 0 \\
0, 0, 0, 0, 0, 0, .0196, .9524, .028, 0, 0, 0, 0, 0, 0, 0, 0, 0, 0, 0, 0 \\
0, 0, 0, 0, 0, 0, 0, .0256, .9424, .032, 0, 0, 0, 0, 0, 0, 0, 0, 0, 0, 0 \\
0, 0, 0, 0, 0, 0, 0, 0, .0324, .9316, .036, 0, 0, 0, 0, 0, 0, 0, 0, 0, 0 \\
0, 0, 0, 0, 0, 0, 0, 0, 0, .04, .92, .04, 0, 0, 0, 0, 0, 0, 0, 0, 0 \\
0, 0, 0, 0, 0, 0, 0, 0, 0, 0, .0484, .9076, .044, 0, 0, 0, 0, 0, 0, 0, 0 \\
0, 0, 0, 0, 0, 0, 0, 0, 0, 0, 0, .0576, .8944, .048, 0, 0, 0, 0, 0, 0, 0 \\
0, 0, 0, 0, 0, 0, 0, 0, 0, 0, 0, 0, .0676, .8804, .052, 0, 0, 0, 0, 0, 0 \\
0, 0, 0, 0, 0, 0, 0, 0, 0, 0, 0, 0, 0, .0784, .8656, .056, 0, 0, 0, 0, 0 \\
0, 0, 0, 0, 0, 0, 0, 0, 0, 0, 0, 0, 0, 0, .09, .85, .06, 0, 0, 0, 0 \\
0, 0, 0, 0, 0, 0, 0, 0, 0, 0, 0, 0, 0, 0, 0, .1024, .8336, .064, 0, 0, 0 \\
0, 0, 0, 0, 0, 0, 0, 0, 0, 0, 0, 0, 0, 0, 0, 0, .1156, .8164, .068, 0, 0 \\
0, 0, 0, 0, 0, 0, 0, 0, 0, 0, 0, 0, 0, 0, 0, 0, 0, .1296, .7984, .072, 0 \\
0, 0, 0, 0, 0, 0, 0, 0, 0, 0, 0, 0, 0, 0, 0, 0, 0, 0, .1444, .7796, .076 \\
0, 0, 0, 0, 0, 0, 0, 0, 0, 0, 0, 0, 0, 0, 0, 0, 0, 0, 0, .16, .84
\end{pmatrix}. \tag{5.31}$$

The first column of Table 5.10 designates the population size; the second, third, and fourth rows are the transition probabilities (the death and birth rates) of the transition matrix of the logistic growth model (Equation 5.30), while the last three columns of Table 5.10 are the corresponding multinomial realizations. Each realization is based on a sample of size 1000, and the realizations provide sample information for the Bayesian analysis.

For example, corresponding to a population size of 10, the death and birth rates are each .04, and the corresponding multinomial realization is (48,912,40), giving .48 and .40 as the usual estimates of the death and birth rates, respectively, when the population is of size 10. I chose 1000 for the realization size because the birth and death rates are very small. I wanted to avoid a transition count of zero! Note that the last row of Table 5.10 does not have an entry (corresponding to the population of size 20, the maximum size), since a birth is impossible, but a death can occur with probability .16.

Using Equation 5.28, the goal of the Bayesian analysis is to estimate the time to extinction assuming a population of size 1.

WinBUGS Code 5.6

```
model;
{
d1~dbeta(1,999)
d2~dbeta(1,999)
d3~dbeta(5,995)
d4~dbeta(4,996)
d5~dbeta(9,991)
d6~dbeta(15,985)
d7~dbeta(18,982)
d8~dbeta(26,974)
d9~dbeta(29,971)
d10~dbeta(48,952)
d11~dbeta(41,959)
d12~dbeta(62,938)
d13~dbeta(74,926)
d14~dbeta(63,937)
d15~dbeta(84,916)
d16~dbeta(93,917)
d17~dmodelbeta(109,891)
d18~dbeta(132,868)
d19~dbeta(139,861)
d20~dbeta(154,846)
b1~dbeta(7,993)
b2~dbeta(3,997)
b3~dbeta(7,993)
b4~dbeta(19,981)
b5~dbeta(27,973)
b6~dbeta(24,976)
b7~dbeta(30,970)
b8~dbeta(31,969)
b9~dbeta(36,964)
b10~dbeta(40,960)
b11~dbeta(41,959)
b12~dbeta(50,950)
b13~dbeta(46,954)
b14~dbeta(55,945)
b15~dbeta(57,943)
b16~dbeta(66,936)
b17~dbeta(67,933)
b18~dbeta(67,933)
b19~dbeta(80,920)

d.12<-d1*d2
d.13<-d.12*d3
d.14<-d.13*d4
d.15<-d.14*d5
d.16<-d.15*d6
```

```
d.17<-d.16*d7
d.18<-d.17*d8
d.19<-d.18*d9
d.110<-d.19*d10
d.111<-d.110*d11
d.112<-d.111*d12
d.113<-d.112*d13
d.114<-d.113*d14
d.115<-d.114*d15
d.116<-d.115*d16
d.117<-d.116*d17
d.118<-d.117*d18
d.119<-d.118*d19
d.120<-d.119*d20

b.12<-b1*b2
b.13<-b.12*b3
b.14<-b.13*b4
b.15<-b.14*b5
b.16<-b.15*b6
b.17<-b.16*b7
b.18<-b.17*b8
b.19<-b.18*b9
b.110<-b.19*b10
b.111<-b.110*b11
b.112<-b.111*b12
b.113<-b.112*b13
b.114<-b.113*b14
b.115<-b.114*b15
b.116<-b.115*b16
b.117<-b.116*b17
b.118<-b.117*b18
b.119<-b.118*b19

mu1<-
1/d1+b1/d.12+b.12/d.13+b.13/d.14+b.13/d.14+b.14/d.15+b.15/d.16+b.16/d.17
+b.17/d.18+b.18/d.19+b.19/d.110+b.110/d.111
+b.111/d.112+b.112/d.113+b.113/d.114+b.114/d.115+b.115/d.116
+b.116/d.117+b.117/d.118+b.118/d.119+b.119/d.120
}
```

I used 55,000 observations for the simulation with a burn-in of 5,000, and the results are reported in Table 5.11.

The posterior distribution of three death and birth rates are portrayed in Table 5.11, while the posterior median to the time to extinction is 919,400 time units. When the population is size 1, note that the probability of a death is .0004; thus, it is not surprising that the median time to extinction is extremely large. The student will be asked to enlarge the posterior analysis to the parameter μ_{10}, the average time to extinction assuming that the population is size 10.

TABLE 5.11

Posterior Analysis of Logistic Growth Model

Parameter	Mean	SD	Error	2 1/2	Median	97 1/2
b_1	.00699	.00263	.000008	.0028	.0066	.0130
b_2	.0030	.0017	.000005	.00062	.0026	.0071
b_3	.0069	.0026	.000007	.0028	.0066	.0130
d_1	.0009	.0009	.000003	.000025	.0006	.0036
d_2	.0010	.00099	.000003	.00002	.0006	.0036
d_3	.0049	.00223	.000006	.00163	.0046	.0102
μ_1	5.37×10^7	3.4×10^9	1.04×10^7	24,200	919,400	1.2×10^8

5.6 Epidemic Processes

5.6.1 Introduction

This section deals with using Markov chains to model the evolution of an epidemic then explores the use of Bayesian techniques to provide inferences for the unknown transition probabilities (which contain the basic parameters that describe the epidemic) of the process. First to be discussed is an explanation of the basic principles of an epidemic; that is, the biological foundation of an epidemic. Next, a deterministic version of a simple epidemic is presented, which lays rudiments of the stochastic version of an epidemic model. Of primary importance in the study of epidemics is to determine the average duration of the epidemic. Various versions which generalize the simple epidemic are the chain binomial models which include the Greenwood model and the Reed–Frost model. The relationship between the epidemic model and the previously discussed birth and death process is elucidated.

The model is referred to as the SIS model because susceptible individuals (S) becomes infected (I) but do not develop immunity after they recover. This is designated by $S \rightarrow I \rightarrow S$, because the infected individuals can again become infected. No latent period is included in the simple epidemic model; thus, people that become infected can pass the disease along to others. Also assumed is no vertical transmission of the disease; that is, the infection cannot be passed on to the mother's offspring. Thus, the offspring immediately become susceptible. Because the number of births and deaths are the same, the total population size is $N = S + I$. Suppose the interval between time n and time $n + 1$ is small enough so that at most 1 event occurs. Therefore, the following can occur: (1) a susceptible person becomes infected, (2) a susceptible person gives birth (and a corresponding death of either a susceptible or infected individual), or (3) an infected person recovers. Suppose the probability of a susceptible individual becoming infected is $\beta I/N$, where β is the number of contacts made by one infectious individual that results in one infection during the interval $(n, n + 1)$; thus, only $\beta S/N$ of these contacts may result in a new infection, and the total number of new infections by the whole class of infected individuals is $\beta SI/N$. Suppose susceptible and infected persons are born or die with probability b and that infected individuals recover with probability γ.

5.6.2 Deterministic Model

Let $I(n)$ and $S(n)$ denote the number of infected and the number of susceptible individuals at time n, respectively, where the dynamics of the chain follow the system of difference equations:

$$\begin{aligned} S(n+1) &= S(n) - \beta S(n)I(n)/N + I(n)(b+\gamma), \\ I(n+1) &= \beta S(n)I(n)/N + I(n)(1-b-\gamma), \end{aligned} \tag{5.32}$$

where $n = 0, 1, 2, \ldots, S(0) > 0, I(0) > 0, S(n) + I(n) = N$.

These equations are interpreted as follows: The number of new susceptible individuals at time $n+1$ is the number of individuals that did not become infected $S(n)[1 - \beta I(n)/N]$, plus the number of infected individuals that recovered, namely, $\gamma I(n)$, and plus the offspring (newborns) from the infected group $bI(n)$. Since the total population size is constant, the number of offspring from the susceptible class is the same as the number of susceptible people that die, namely, $bS(n)$. The restrictions on the unknown parameters is $0 < \beta \le 1$, $0 < b + \gamma \le 1$.

In the second equation of Equation 5.32, $S(n)$ is replaced by $N - I(n)$, giving

$$\begin{aligned} I(n+1) &= I(n)[\beta(N - I(n))/N] + 1 - b - \gamma \\ &= I(n)[1 + \beta - b - \gamma - \beta I(n)/N]. \end{aligned} \tag{5.33}$$

There are two equilibrium solutions; that is, where $I(n+1) = I(n) = E$, which are $E = 0$ and $E = N[1 - (b+\gamma)/\beta]$.

The equilibrium point is a function of the reproduction number

$$R_0 = \beta/(b+\gamma), \tag{5.34}$$

which has an interesting meaning in epidemiology: It is when the whole population is susceptible and one infected and infectious person is introduced into the population, R_0 is the average number β of successful contacts during the period of infectivity, then $1/(b+\gamma)$ will result in a new infected person. Note that if $R_0 > 1$, then one infected individual produces more than one new infection, and if $R_0 < 1$, then one infectious individual produces less than one new infection.

5.6.3 Stochastic Model

Assume the facts about the simple model in the deterministic case, but where $I(n)$ is a random variable denoting the number of infected individuals at time n, where the population is of size $N = I(n) + S(n)$, and $S(n)$ is a random variable denoting the number of susceptible people at time n, $n = 0, 1, 2, \ldots$. Now assume that the time interval from n to $n+1$ is small enough so that there is at most one change over that time period. For the Bayesian analysis, focus will be on the number of infected people $I(n)$ with state space $\{0,1,2,\ldots,N\}$, which has two classes $\{0\}$ and $\{1,2,\ldots,N\}$, where state 0 is absorbing, denoting that the infection dies out. The transition matrix is defined by the following three equations:

$$\begin{aligned} P_{i,i+1} &= \Pr[I(n+1) = i+1 | I(n) = i] = \beta i(N-i)/N = \lambda_i, \\ P_{i,i-1} &= \Pr[I(n-1) = i-1 | I(n) = i] = (b+\gamma)i, \quad \text{and} \\ P_{i,i} &= \Pr[I(n+1) = i | I(n) = i] = 1 - \lambda_i - (b+\gamma)i. \end{aligned} \tag{5.35}$$

Thus, the one-step transition matrix of the number of infected people is

$$P = \begin{pmatrix} 1, 0, \ldots\ldots\ldots\ldots\ldots\ldots\ldots\ldots, 0 \\ b+\gamma, 1-b-\gamma-\lambda_1, \lambda_1, 0, \ldots\ldots\ldots, 0 \\ 0, 2(b+\gamma), 1-2(b+\gamma)-\lambda_2, 0, \ldots\ldots, 0 \\ . \\ . \\ . \\ 0, \ldots\ldots\ldots 0, N(b+\gamma), 1-N(b+\gamma) \end{pmatrix}. \tag{5.36}$$

The goal as a Bayesian is to make inferences about the unknown parameters, the birth rate b, the number of contacts β, and the number of infected people that recover γ. I will begin with estimating the parameters b, γ, and β used as data transition counts for the first three entries of the second row of P, where the transition counts will be realizations generated via the multinomial distribution.

Consider the multinomial distribution with parameter $p = (.005,.9851,.0099)$, then using the R function rmultinom, the following realization was generated with transition counts $(n_{10}, n_{11}, n_{12}) = (3,993,4)$ corresponding to the transition probabilities

$$\begin{aligned} P_{10} &= b+\gamma, \\ P_{11} &= 1-\lambda_1-(b+\gamma), \\ P_{12} &= \lambda_1 = \beta(N-1)/N, \\ N &= 100, \\ b &= \gamma = .0025, \end{aligned} \tag{5.37}$$

and

$$\beta = .01$$

Therefore, the transition probabilities are estimated as $\tilde{P}_{10} = .003$, $\tilde{P}_{11} = .973$, and $\tilde{P}_{12} = .004$. Thus, using Equation 5.37, the estimates for the parameters of the epidemic are .0040404 for β and .0015 for b and γ. What are the Bayesian estimates?

Assume that the prior distribution for the transition probabilities is the improper prior

$$\zeta(P_{10}, P_{11}, P_{12}) = \sum_{j=0}^{j=2} P_{1j}^{-1}, \tag{5.38}$$

where

$$\sum_{j=0}^{j=1} P_{1j} = 1, \quad 0 < P_{1j} < 1, \quad \text{and} \quad j = 1,2,3$$

It can be shown the posterior distribution of the transition probabilities is Dirichlet with parameter $(n_{10}, n_{11}, n_{12}) = (3,993,4)$; thus, the marginal posterior distribution of P_{10} is beta(3,997), that of P_{11} is beta(993,7), and that of P_{12} is beta(4,996).

The Bayesian analysis is executed with WinBUGS Code 5.8 with 35,000 observations for the simulation and a burn-in of 5,000. The step command g2 gives the posterior probability that $R_0 > 1$.

WinBUGS Code 5.7

```
model;
{
p10~dbeta(3,997)
p11~dbeta(993,7)
p12~dbeta(4,996)
bplusgamma<-p10
N<-100
beta<-p12*N/(N-1)
b<-bplusgamma/2
R0<-beta/bplusgamma
g1<-step(R0-.5)
g2<-step(R0-1)
}
```

As a result of the simulation, the Bayesian analysis for the simple epidemic model is reported in Table 5.12.

Note that the posterior probability that $R_0 > 1$ is .663; that is, in symbols, $g_2 = \Pr[R_0 > 1|\text{data}] = .663$. The estimates of the other parameters P_{10}, P_{11}, and P_{12} are similar to the usual estimates given earlier. Recall that the transition probabilities $P_{10} = .005$, $P_{11} = .9851$, $P_{12} = .0099$ were used to generate the multinomial realization $(n_{10}, n_{11}, n_{12}) = (3,993,4)$ for the Bayesian analysis, and these values for the transition probabilities should be compared

TABLE 5.12

Posterior Analysis for Simple Epidemic

Parameter	Mean	SD	Error	2 1/2	Median	97 1/2
R_0	2.029	2.655	0.0097	0.2935	1.389	7.538
b	.00149	.000859	.0000032	.00031	.00133	.00361
$b+\gamma$	.002986	.001718	.0000064	.00062	.00267	.00722
P_{10}	.002986	.001718	.0000064	.00062	.00267	.00722
P_{11}	.993	.00262	.0000096	.987	.9933	.9972
P_{12}	.00399	.001992	.0000083	.0011	.00366	.00879
β	.004037	.002003	.0000112	.00113	.003705	.00886
g_2	.663	.4727	.002646	0	1	0

to the posterior means (.0029,.993,.0039) of Table 5.12. It appears that the posterior means are quite close to these values.

For additional information about the stochastic epidemic model, refer to Allen and Burgin,[10] Daley and Gani,[11] and Allman and Rhodes.[12]

5.6.4 Chain Binomial Epidemic Models

The following description of chain binomial models relies heavily on pages 137–143 of Allen.[1] Let $S(n)$ and $I(n)$ denote discrete random variables for the number of susceptible and infected people at time n, respectively, where the time interval between time n and $n + 1$ is the latent period, that is, the time until individuals become infectious, for $n = 0, 1, 2, \ldots$ Thus, the number of infected individuals $I(n)$ represents newly infected individuals who were latent during the time period $n - 1$ to n, and these people are infectious. They will contact susceptible individuals at time n, who consequently could become infected at time $n + 1$. The infected $I(n)$ individuals are removed or recovered in the next time interval from n to $n + 1$. Births and deaths do not occur, and the newly infected individuals at time $n + 1$ and those susceptible at that time represent all those who were susceptible at time n; thus,

$$S(n+1) \ + I(n+1) = S(n). \tag{5.39}$$

The epidemic terminates when $I(n) = 0$ for some n. The preceding explanation for chain binomial models is somewhat simplified, and the reader is referred to Daley and Gani[11] for additional information about the subject. Let α be the probability of a contact between an infected and a susceptible individual, and suppose that β is the chance the contact results in the susceptible person becoming infected; therefore, the probability that a susceptible person does not become infected is

$$p = 1 - \alpha\beta \tag{5.40}$$

In view of Equation 5.40, $1 - p$ is the probability of a contact not resulting in an infection, and as will be revealed, p is an important parameter in both versions of the chain binomial model.

There are two versions of the chain binomial model: (1) the Greenwood and (2) the Reed–Frost.

First, the Greenwood model is investigated. At time $n + 1$, there are $S(n + 1)$ susceptible people among which $S(n + 1)$ contacts were not successful, and the number of contacts that were successful is $I(n + 1) = S(n + 1) - S(n)$. This implies that the probability of a one-step transition from $S(n + 1)$ to $S(n)$ is

$$P_{S(n),S(n+1)} = \binom{s(n)}{s(n+1)} \left[p^{i(n)}\right]^{s(n+1)} \left[1 - p^{i(n)}\right]^{s(n)-s(n+1)}, \tag{5.41}$$

where $s(n + 1) = 0, 1, 2, \ldots, s(n)$, and $0 < p < 1$. Note that this determines the $s(0) \times s(0)$ transition matrix of the Greenwood model in Equation 5.41 for an epidemic, which is to be initiated with $I(0) > 0$ infectious people.

Note that the state space for the chain is $\{S(n),n = 1,2,\ldots\}$, where $S(0)$ is positive. A realization of the process is depicted by $\{s(0),s(1),\ldots,s(t)\}$, where the number of infected people at time t is zero, that is, $I(t) = 0$, which is equivalent to $S(t - 1) = S(t)$. The duration of the epidemic is t, and the size of the epidemic is the number of susceptible people who became infected during the duration or $S(0) - S(t)$. This type of epidemic is of the chain binomial type because the distribution of $S(n + 1)$ is binomial with parameters $s(n)$ and p. From Equation 5.41, the $(s(0) + 1) \times (s(0) + 1)$ matrix is

$$P = \begin{pmatrix} 1,0,0,\ldots\ldots\ldots\ldots\ldots\ldots\ldots\ldots,0 \\ (1-p),p,0,\ldots\ldots\ldots\ldots\ldots\ldots\ldots\ldots,0 \\ (1-p)^2,2p(1-p),p^2,0,\ldots\ldots\ldots\ldots\ldots\ldots,0 \\ \cdot \\ \cdot \\ \cdot \\ (1-p)^{s(0)},\binom{s(0)}{1}p(1-p)^{s(0)-1},\binom{s(0)}{2}p^2(1-p)^{S(0)-2},\ldots,p^{s(0)} \end{pmatrix}. \quad (5.42)$$

Now let $s(0) = 3$, then the 4×4 matrix P reduces to

$$P = \begin{pmatrix} 1,0,\ldots\ldots\ldots\ldots\ldots\ldots,0 \\ 1-p,p,0,\ldots\ldots\ldots\ldots\ldots,0 \\ (1-p)^2,2p(1-p),p^2,\ldots\ldots\ldots,0 \\ (1-p)^3,3p(1-p)^2,3p^2(1-p),p^3 \end{pmatrix} \quad (5.43)$$

Now let $\beta = .05$, the probability that a susceptible person is infected after contact with an infected person, and suppose that the probability of a contact between a susceptible individual and an infected individual is $p = .025$, which is the probability that a susceptible person is infected. Substitute .025 for p in Equation 5.43 to give as the transition matrix

$$P = \begin{pmatrix} 1,0,0,0 \\ .02500,.97500,0,0 \\ .000625,.04875.95060,0 \\ .0000156,.001821,.07129,.92685 \end{pmatrix}. \quad (5.44)$$

In terms of α and β, the transition matrix is

$$P = \begin{pmatrix} 1,0,0,0 \\ \alpha\beta, 1-\alpha\beta, 0, 0 \\ \alpha^2\beta^2, 2(1-\alpha\beta)\alpha\beta, (1-\alpha\beta)^2, 0 \\ \alpha^3\beta^3, 3(1-\alpha\beta)\alpha^2\beta^2, 3\alpha^2\beta^2(1-\alpha\beta), (1-\alpha\beta))^3 \end{pmatrix}. \tag{5.45}$$

There are two classes {0} and {1, 2, 3} where the latter states are transient and the first is absorbing. Does it have a limiting distribution?

R Code 5.3 generates realizations from the Greenwood process using the transition matrix P (Equation 5.44):

R Code 5.3

```
markov <- function(init,mat,n,labels) {
+ if (missing(labels)) labels <- 1:length(init)
+ simlist <- numeric(n+1)
+ states <- 1:length(init)
+ simlist[1] <- sample(states,1,prob=init)
+ for (i in 2:(n+1))
+ { simlist[i] <- sample(states,1,prob=mat[simlist[i-1],]) }
+ labels[simlist]
+ }
> p<-matrix(c(1,0,0,0,
.02493,.9751,0,0,
.001567,.04672,.9517,0,
.000156,.004232,.004232,.9298),nrow=4,ncol=4,byrow=TRUE)
```

A realization of size 25 is generated by the Markov function markov(init,p,25) with starting value of state 1:

```
init<-c(0,1,0,0)
1,1,1,1,1,1,1,1,1,1,1,1,1,1,1,1,1,1,1,1,1,1,1,1,1 ,
```

with starting value of 2:

```
init<-c(0,0,1,0)
2,2,2,2,2,2,2,2,2,2,2,2,2,2,2, 1,1,1,1,1,1,1, 0,0,0,0,
```

with initial value of 3:

```
init<-c(0,0,0,1)
3,3,3,3,3,3,3,3,3,3,3,3,3,3,3,3,3,3,3,3,3,3,3,3,3,
```

and starting value of 0:

```
init<-c(1,0,0,0)
0,0,0,0,0,0,0,0,0,0,0,0,0,0,0,0,0,0,0,0,0,0,0,0,0.
```

Do these simulations demonstrate the transient behavior of the states 1, 2, and 3 and that 0 is an absorbing state?

For the Bayesian analysis, I used multinomial realizations of size 500 for each row of P (Equation 5.44), which generates the transition count matrix Q (Equation 5.46). The transition counts will serve as the sample information for Bayesian estimation of p (the second entry of the second row of P), the fundamental unknown parameter of the process with transition matrix P (Equation 5.44).

$$Q = \begin{pmatrix} 500,0,0,0 \\ 10,490,0,0 \\ 1,17,482,0 \\ 1,2,2,495 \end{pmatrix}. \tag{5.46}$$

Assuming a uniform prior for p, the posterior density of p is beta (491,11), with posterior mean of 491/502 = .978087649. A more complete analysis is executed with WinBUGS Code 5.8.

WinBUGS Code 5.8

```
model;
{
p~dbeta(491,11)
p10<-1-p
p11<-p
p20<-pow(1-p,2)
p21<-2*p*(1-p)
p22<-pow(p,2)
p30<-pow(1-p,3)
p31<-3*p*pow(1-p,2)
```

```
p32<-3*pow(p,2)*(1-p)
p33<-pow(p,3)
p3219<-p32*p21*p10
q310<-3*p*pow(1-p,2)*(1-p*p)
s3210<-6*pow(p,3)*pow(1-p,3)
}
```

The Bayesian simulation is for the transition probabilities of the transition matrix P (Equation 5.43), whose entries are in terms of the fundamental parameter p. For example, note that $p30$ of the code is the first entry P_{30} of the last row of Equation 5.43. The posterior analysis is executed with 45,000 observations for the simulation and 5,000 for the burn-in. The Bayesian analysis shown in Table 5.13 is quite complete in that the relevant posterior information is shown for all the entries of the transition matrix (Equation 5.43).

Thus, the fundamental parameter p (the probability that a susceptible person does not become infected) is estimated with the posterior mean as .9781 with a 95% credible interval (.9637,.989). Based on the posterior medians, the transition matrix P in Equation 5.43 is estimated as

$$\tilde{P} = \begin{pmatrix} 1,0,0,0 \\ .0212,.9787,0,0 \\ .000454,.04172,.9578,0 \\ .000009..001334,.06125,.937 \end{pmatrix} \tag{5.47}$$

TABLE 5.13

Posterior Analysis for the Greenwood Model

Parameter	Mean	SD	Error	2 1/2	Median	97 1/2
p	.9781	.00648	.000037	.9637	.9787	.989
P_{10}	.0219	.00648	.000037	.01104	.02132	.03631
P_{11}	.9781	.00648	.000037	.9637	.9787	.989
P_{20}	.000522	.000314	.000001	.000122	.000454	.001318
P_{21}	.0428	.01236	.000071	.02184	.04172	.06998
P_{22}	.9567	.01267	.000073	.9287	.9578	.978
P_{30}	.000013	.000012	.0000000	.000001	.000009	.000047
P_{31}	.00152	.000906	.0000052	.000361	.001334	.003812
P_{32}	.0626	.0176	.000101	.0324	.06125	.1012
P_{33}	.9358	.0185	.000107	.895	.9374	.9672
P_{3210}	.000074	.000068	.0000003	.000007	.000054	.00026
s_{3210}	.000074	.000068	.0000003	.000007	.000054	.00026

Note that the last row of the table which shows the posterior distribution of the sample path $\{s(0), s(1), s(2), s(3)\} = \{3, 2, 1, 0\}$ to extinction of the epidemic is estimated as .000074 with the posterior mean and a 95% credible interval (.000007,.00026). Thus, the posterior probability that the epidemic ends after 4 time units along that particular path is extremely small.

Up to this point, Bayesian inference has focused on the estimation of the fundamental parameter p, the probability a susceptible person does not become infected; now the focus will be on predicting future transition counts for the Greenwood model. That is, starting with a given number of susceptible and infected individuals, what is the Bayesian predictive mass function for the future number of infected persons?

Recall that given $S(n) = s(n)$, the distribution of $I(n + 1)$ is binomial with parameters $s(n)$ and $(1 - p)s(n)$; thus, given $s(n)$ susceptible people at time n, the average number of infections is

$$E[I(n+1)|S(n) = s(n)] = (1-p)s(n), \tag{5.48}$$

where $(1 - p)$ is the probability that a susceptible person will be infected.

The predictive mass function of $I(n + 1)$ is easily determined. The conditional mass function of $I(n + 1)$, given $S(n) = s(n)$, is binomial with parameters $s(n)$ and $(1 - p)s(n)$, which is averaged with respect to p over the posterior distribution of p (which is beta with parameters 490 and 11). Recall that of the number of susceptible people, that on the average, $(1 - p)$ will become infected.

A slight variation of the Greenwood model is the Reed–Frost version of a chain binomial epidemic. It is very similar to the Greenwood model but with a slightly different transition matrix:

$$P_{S(n),S(n+1)} = \binom{s(n)}{s(n+1)}\left[p^{i(n)}\right]^{s(n+1)}\left[1-p^{i(n)}\right]^{s(n)-s(n+1)}, \tag{5.49}$$

where $i(n)$ is the number of infected people at time n and the number infected at time n is $I(n) = S(n - 1) - S(n)$ for $n = 1, 2, \ldots$. The primary difference between the Greenwood version (Equation 5.41) and the Reed–Frost one is that the probability p of a susceptible individual being infected depends on the number of infected, namely, the probability $p^{i(n)}$. Let us evaluate the posterior probability of the path $\{s(0),s(1),s(2),s(3)\} = \{3,2,1,0\}$, which has the probability

$$P_{32}P_{21}P_{10} = 3p^2(1-p)2p(1-p)p(1-p) = 6p^3(1-p)^3$$
$$= 6(.975)^3(.025)^3 = 6(.926859)(.000015625) = .000086893$$

With the Bayesian approach, refer to WinBUGS Code 5.9, the statement s_{3210}, and the last row of Table 5.13, which is identical to the next to the last row, which in turn corresponds to the posterior probability of the same sample path. The posterior distribution of the two probabilities P_{3210} and s_{3210} are the same, as they should be. Since the number of infections $E[I(n+1)|S(n) = s(n)] = (1-p)s(n)$, the transition probabilities of the Reed–Frost model (Equation 5.49) are the same as those for the Greenwood model (Equation 5.43). For the two models to differ, $i(n) > 1$, for some n.

Now consider the sample path {s(0),s(10),s(2),s(3)} = {3,1,0,0}, which occurs with probability

$$P_{31}P_{10} = 3p(1-p)^2(1-p^2) = 3(.975)(.275)^2(.049375) = .0109219$$

It is left for the student to do the Bayesian analysis to estimate $P_{31}P_{10}$. Refer to WinBUGS Code 5.8 and the WinBUGS statement s3210<-6*pow(p,3)*pow(1-p,3) which is the probability of the sample path {3,1,0,0} assuming the Reed–Frost model.

5.7 Molecular Evolution

Molecular evolution is another interesting example where DTMCs provide a realistic model for the genetics mechanism involving deoxyribonucleic acid (DNA). A Bayesian analysis (estimation, testing hypotheses, and prediction) will be presented for several versions of molecular evolution.

5.7.1 Simple Model

According to Allman and Rhodes,[12] a simple model of molecular evolution can be described by a Markov chain with four states. The four states are the four DNA bases: adenine (A), guanine (G), cytosine (C), and thymine (T). The transition probabilities represent the probability of a base substitution, and the time scale {1,2,...} represents generations; that is, one models the evolution with the probability of one base substitution per generation occurring along the segment of DNA. Let

$$p(0) = (p_A(0), p_G(0), p_C(0), p_T(0)) \tag{5.50}$$

be the initial distribution denoting the initial proportion of each type base in the DNA segment (expected fraction of sites occupied by each base). The mutation is modeled by assuming that only one base substitution can occur per generation $n \to n+1$. This process of molecular evolution has transition matrix

$$P = \begin{pmatrix} 1-3d, d, d, d \\ d, 1-3d, d, d \\ d, d, 1-3d, d \\ d, d, d, 1-3d \end{pmatrix}, \tag{5.51}$$

therefore,

$$P_{ii} = 1-3d,\ i = 1,2,3,4, \quad P_{ij} = d,\ i \neq j, 0 < d < 1/3$$

It is obvious that each base communicates with the other three with probability d as portrayed in Figure 5.4.

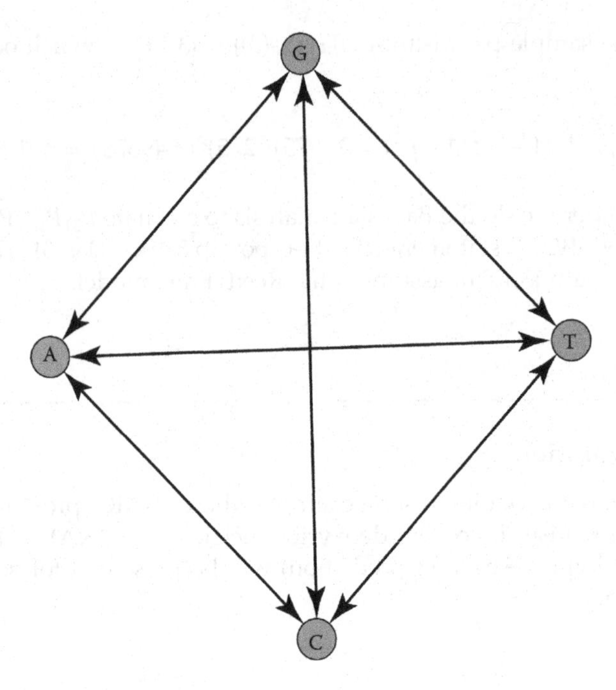

FIGURE 5.4
Directed graph of molecular evolution.

This model is known as the Jukes–Cantor model[13] and page 49 of Allen[1] gives a detailed description of the model. It can be shown that the probability that there has been no substitutions of base i in n generations is

$$P_{ii}^n = \Pr[X(n) = i|X(0) = i] = 1/4 + (3/4)(1 - 4d)^n.$$

Thus, the probability that a base substitution has happened is

$$q = 1 - P_{ii}^n = (3/4)[1 - (1 - 4d)^n] = (3/4)[1 - (1 - 4\alpha/3)^n], \tag{5.52}$$

where $\alpha = 3d$ is the rate of mutation. Note that α is the probability that a particular base is substituted by one of the other three bases. Our main interest is to estimate d and α and to test hypotheses about them. For the Bayesian analysis, let $d = 1/12$, then

$$P = \begin{pmatrix} 3/4, 1/12, 1/12, 1/12 \\ 1/12, 3/4, 1/12, 1/12 \\ 1/12, 1/12, 3/4, 1/2 \\ 1/12, 1/12, 1/12, 3/4 \end{pmatrix} \tag{5.53}$$

is the transition matrix corresponding to a mutation rate of $\alpha = 1/4$.

The transition counts will be generated with the multinomial distribution using the rows of Equation 5.53 as the probability vector. The transition counts are displayed in the following matrix Q;

$$Q = \begin{pmatrix} 37,4,3,6 \\ 5,35,6,4 \\ 5,4,36,5 \\ 1,9,3,37 \end{pmatrix}. \tag{5.54}$$

It is interesting to note that in practice, one would observe Q (not the known transition matrix P [Equation 5.54]), and using Q as sample information one would make inferences about the unknown transition probabilities $P_{ij} : i, j = 1, 2, 3, 4$. Based on proportions, the usual estimates are portrayed in the matrix

$$R = \begin{pmatrix} .74,.08,.06,.12 \\ .10,.70,.12,.08 \\ .10,.08,.72,.10 \\ .02,.18,.06,.74 \end{pmatrix}. \tag{5.55}$$

Thus, consider the first row, which shows what happens to base A over 50 generations, namely, .74 of 50 generations; no substitution occurred for A; but 8% of 50 generation A was substituted with the base G, 6% by C; and 12% by T, with a mutation rate of 26%. That is, A mutates 26% over 50 generations.

A question of interest is on the basis of the estimates R, how are inferences of the parameter d of the matrix in Equation 5.51 determined? Of course, this assumes that one knows the pattern of molecular evolution given by Equation 5.51. Suppose one did not know that the chain has the pattern displayed by Equation 5.51, then could one deduce that pattern based on the sample information (Equation 5.54)? Using a Bayesian approach, both of these problems will be addressed.

The first inference problem to be solved is that of estimating the parameter d (or, equivalently, the mutation rate) of Equation 5.51; that is, one is assuming that molecular evolution occurs according to the pattern in Equation 5.51. The value of d, the off-diagonal term, is estimated by taking the average of the 12 off-diagonal terms, then the diagonal entry is estimated by computing $1 - 3d$, and the analysis is executed with WinBUGS Code 5.9.

The Bayesian approach assumes that the transition probabilities have an improper prior distribution

$$g(P) \propto \prod_{j=1}^{j=4} P_{ij}^{-1}, \tag{5.56}$$

where and $\sum_{j=1}^{j=4} P_{ij} = 1$, $i = 1, 2, 3, 4$.

By the Bayes theorem, the first row of the transition matrix has a posterior Dirichlet distribution with parameter (37,4,3,6); thus, the diagonal element P_{11} has a posterior beta distribution with parameter (37,13).

WinBUGS Code 5.9

```
model;
{
# the elements of the transition matrix have beta distributions
p11~dbeta(37,13)
p12~dbeta(4,46)
p13~dbeta(3,47)
p14~dbeta(6,44)
p21~dbeta(5,45)
p22~dbeta(35,15)
p23~dbeta(6,44)
p24~dbeta(4,46)
p31~dbeta(5,45)
p32~dbeta(4,46)
p33~dbeta(36,14)
p34~dbeta(5,45)
p41~dbeta(1,49)
p42~dbeta(9,41)
p43~dbeta(3,47)
p44~dbeta(37,13)
# d is the average of the 12 off-diagonal terms
d<-(p12+p13+p14+p21+p23+p24+p31+p32+p34+p41+
p42+p43)/12
# diag is the diagonal entry Pii
diag<-1-3*d
}
```

The analysis is executed with 45,000 observations for the simulation and 5,000 for the burn-in. See WinBUGS Code 5.9.

Thus, the posterior analysis estimates the off-diagonal entries with a posterior mean of .0915 and estimates the diagonal entries as .7254. Note that the 95% credible interval for d contains the number $1/12 = .0833$, and that for the diagonal entry, the 95% credible interval contains .75. This suggests that the data support the evolution pattern of matrix (Equation 5.51).

The second phase of the Bayesian analysis will center on a formal Bayesian test of the hypothesis that the molecular evolution pattern of Equation 5.51 is correct.

Consider the transition matrix and the null hypothesis

$$\text{H: } P_{12} = P_{13} = P_{14} = P_{21} = P_{23} = P_{24} = P_{31} = P_{32} = P_{34} = P_{41} = P_{42} = P_{43} \tag{5.57}$$

versus the alternative that H is not true. Note that the null hypothesis states that all off-diagonal entries of the transition matrix are the same, which implies that all the diagonal elements are the same. This is the pattern specified by the matrix in Equation 5.41.

In order to test the hypothesis H, let the prior probability of the null hypothesis be π_o and that of the alternative π_1.

What is required is the joint distribution of the transition counts

$$n_{\text{off}} = \left(n_{12}, n_{13}, n_{14}, n_{21}, n_{23}, n_{24}, n_{31}, n_{32}, n_{34}, P_{41}, n_{42}, n_{43}\right), \tag{5.58}$$

given the transition probabilities

$$P_{\text{off}} = \left(P_{12}, P_{13}, P_{14}, P_{21}, P_{23}, P_{24}, P_{31}, P_{32}, P_{34}, P_{41}, P_{42}, P_{43}\right), \tag{5.59}$$

which has the multinomial mass function

$$g(n_{\text{off}}|P_{\text{off}}) = \left[n!/\prod_{\substack{i=1 \\ i \neq j}}^{i=4} \prod_{j=1}^{j=4} n_{ij}! \right] \prod_{i=1}^{i=4} \prod_{\substack{j=1 \\ j \neq i}}^{j=4} P_{ij}^{n_{ij}}. \tag{5.60}$$

Under the null hypothesis, the likelihood function in Equation 5.60 reduces to

$$g(n_{\text{off}}|\text{H}) = \left[n!/\prod_{\substack{i=1 \\ i \neq j}}^{i=4} \prod_{j=1}^{j=4} n_{ij}! \right] p^{n}, \tag{5.61}$$

where n is the total of the off-diagonal transition counts shown in the matrix Q in Equation 1.53, and p is the common value of the transition probabilities.

$$n = \sum_{i=1}^{i=4} \sum_{\substack{j=1 \\ j \neq i}}^{j=4} n_{ij}.$$

The predictive distribution corresponding to (5.61)

$$g(n_{\text{off}}) = \int_{0}^{1} g(n_{\text{off}}|\text{H})\xi(p)dp = 1/n, \tag{5.62}$$

where $\xi(p)$, the prior density of p, is chosen as uniform.

The posterior probability of the null hypothesis is

$$p_0 = \pi_o g(n_{\text{off}})/[\pi_0 g(n_{\text{off}}) + \pi_1 \lambda_1(n_{\text{off}})], \tag{5.63}$$

where the predictive mass function under the alternative hypothesis is

$$\lambda_1(n_{\text{off}}) = \int \rho_1(P_{\text{off}}) g(n_{\text{off}}|P_{\text{off}}) dP_{\text{off}}, \tag{5.64}$$

and ρ_1 is the prior density of P_{off}, the off-diagonal entries of the transition matrix.

Assuming an improper prior density for ρ_1

$$\rho_1(P_{\text{off}}) \propto \prod_{\substack{i=1 \\ i \neq j}}^{i=4} \prod_{j=1}^{j=4} P_{ij}^{-1},$$

it can be shown that Equation 5.63 reduces to

$$\lambda_1(n_{\text{off}}) = \text{n}!/\prod_{i=1}^{i=4} \prod_{\substack{j=1 \\ j \neq i}}^{j=4} n_{ij}! = .000046296 \tag{5.65}$$

Now let $\pi_0 = \pi_1 = .5$, then the posterior probability of the null hypothesis is $p_0 = (.5/54)/[(.5/54) + .5(.000046296)] = .009259259/.009282407 = .995018719$. Therefore, the data of observed counts Q (Equation 5.53) very strongly support the assertion that molecular evolution follows the pattern specified by the Jukes–Cantor model specified by the matrix in Equation 5.51! This is not a surprising conclusion, since to some extent, it is implied by the corresponding observed transition probabilities portrayed by matrix R (Equation 5.54) and by the Bayesian estimation analysis displayed in Table 5.14.

Kimura[14] proposed another version of molecular evolution given by the transition matrix

$$P = \begin{pmatrix} 1-a-2d, a, d, d \\ a, 1-a-2d, d, d \\ d, d, 1-a-2d, a \\ d, d, a, 1-a-2d \end{pmatrix}, \tag{5.66}$$

where, $0 < a < 1/3$ and $0 < d < 1/3$. Note that when $a = d$, the Kimura chain reduces to the Jukes–Cantor model.

If $a \neq d$, what is the difference between the Jukes–Cantor model (Equation 5.51) and the variation in Equation 5.65? The Kimura model is a two-parameter model which assumes that the bases occur with equal frequency and the rate of all transitions is the same and that the rate of all transversions are the same.

A transvesion refers to the substitution of a (two-ring) purine for a (one-ring) pyrimide, or vice versa. Recall that the two purine bases are adenine and guanine, while the other two bases cytosine and thymine are pyrimidines. See Futuyma[15] for additional information

TABLE 5.14

Posterior Analysis for Molecular Evolution

Parameter	Mean	SD	Error	2 1/2	Median	97 1/2
d	.0915	.0115	.0000551	.07015	.0911	.1154
P_{ii}	.7254	.0347	.000165	.6539	.7267	.7896

about molecular evolution. Thus, there are two types of transitions, A $\leftrightarrow$ G and C $\leftrightarrow$ T, and four transversions: C $\leftrightarrow$ A, C $\leftrightarrow$ G, G $\leftrightarrow$ T, and A $\leftrightarrow$ T.

From an inferential standpoint, it is of interest to test the hypothesis H that $a = d$ versus the alternative A: $a \neq d$.

Consider the Kimura version (Equation 5.66) with $a = .0316$ and $d = .1666$, then the corresponding transition matrix is

$$P = \begin{pmatrix} .7917, .0416, .1666, .1666 \\ .0416, .7917, .1666, .1666 \\ .1666, .1666, .7917, .0416 \\ .1666, .1666, .0416, .7917 \end{pmatrix}, \tag{5.67}$$

and suppose the corresponding distribution of row counts is assumed to be multinomial with a total of 50, then using R to generate the row counts, a possible transition count matrix is

$$Q = \begin{pmatrix} 33, 4, 7, 6 \\ 1, 38, 7, 4 \\ 6, 7, 35, 2 \\ 11, 3, 1, 35 \end{pmatrix}. \tag{5.68}$$

Therefore, with probability .0416, a G base is substituted for an A base, but with probability .1666, a C base is substituted for A. In a similar fashion out of 50 cycles, 33 times out of 50 A did not change, while 4 times out of 50, the C base was substituted in the first position. Note that the usual estimate for no change in the A base in the first position is $33/50 = .66$ compared to the value .79, which was used to generate the count.

Our main concern is to provide Bayesian inferences for the unknown transition probabilities of the process which includes estimating the eight differences between a and d of the Kimura model given by Equation 5.66. For example, the difference $P_{12} - P_{13} = a - d$ in the transition matrix (Equation 5.66).

If an improper prior distribution is used for the transition probabilities, then the rows of the transition counts have a posterior distribution, which is Dirichlet. For example, for the first row of Q, the counts have a Dirichlet (33,4,7,6), and the transition probability P_{11} has posterior distribution which is beta(33,17).

The Bayesian analysis is executed with WinBUGS Code 5.10 using 35,000 observations for the simulation and 5,000 for the burn-in, and the results are reported in Table 5.15:

WinBUGS Code 5.10

```
model;
{
p11~dbeta(33,17)
p12~dbeta(4,46)
p13~dbeta(7,43)
p14~dbeta(6,44)
p21~dbeta(1,49)
```

```
p22~dbeta(38,12)
p23~dbeta(7,43)
p24~dbeta(4,46)
p31~dbeta(6,44)
p32~dbeta(7,43)
p33~dbeta(35,15)
p34~dbeta(2,48)
p41~dbeta(11,39)
p42~dbeta(3,47)
p43~dbeta(1,49)
p44~dbeta(35,15)
# the a-d differences
d1213<-p12-p13
d1214<-p12-p14
d2123<-p21-p23
d2124<-p21-p24
d3431<-p34-p31
d3432<-p34-p32
d4341<-p43-p41
d4342<-p43-p42
```

The Bayesian analysis is quite interesting as exhibited in Table 5.15. For example, the first row shows the posterior distribution of one of the $a - d$ differences specified by $P_{12} - P_{13}$, which has a posterior mean of –.06118 and a posterior median of –.05945. The 95% credible interval for this difference does not contain zero, implying informally that there is no difference in this particular one (of eight) $a - d$ differences. The main emphasis for this analysis is estimating these $a - d$ differences of the Kimura model, not testing hypotheses.

It is left for the student in the exercises to perform a formal test that the Kimura model reduces to the Jukes–Cantor model by testing the null hypothesis specified by

$$H\colon P_{12} = P_{13} = P_{21} = P_{23} = P_{31} = P_{34} = P_{42} = P_{43}. \tag{5.69}$$

TABLE 5.15
Posterior Analysis Kimura Model

Parameter	Mean	SD	Error	2 1/2	Median	97 1/2
$P_{12} - P_{13}$	–.06018	.06172	.000338	–.1851	–.05945	.06048
$P_{12} - P_{14}$	–.0406	.05921	.000337	–.159	–.0397	.0755
$P_{21} - P_{23}$	–.12	.05222	.00029	–.2323	–.1162	–.02737
$P_{21} - P_{24}$	–.05999	.04295	.000257	–.1546	–.05635	.01677
$P_{34} - P_{31}$	–.07982	.05298	.000296	–.1897	–.0773	.02108
$P_{34} - P_{32}$	–.1002	.005579	.000338	–.2161	–.09795	.004984
$P_{43} - P_{41}$	–.2008	.0613	.000388	–.3286	–.1979	–.08915
$P_{43} - P_{42}$	–.0401	.0384	.000232	–.124	–.0365	.0200

5.8 Ehrenfest Model of Cell Diffusion

The Ehrenfest model is an interesting example of the transfer of a model from physics to biology. A quandary arose at the turn of the century, when Boltzman attempted to explain thermodynamics on the basis of kinetic theory. On the one hand, as the system evolves, a conservative system is time reversible and should return infinitely often to the neighborhood of the initial state. On the other hand, based on experimental observation and thermodynamics, starting with any initial state, the system will irreversibly move toward equilibrium. This was stated by Zermelo, namely, that the irreversibility from the theory of thermodynamics and the recurrence property of conservative dynamical conditions cannot be reconciled. See Kannan[16] for further information about the physics of the Ehrenfest model and how it has been adopted by biology for cell diffusion. From the preceding language, one can show that such phenomena can be modeled via stochastic processes. Ehrenfest and Ehrenfest[17] clarified the situation using an urn model to explain the biology of diffusion through a membrane.

Suppose $2N$ molecules are labeled $1,2,\ldots,2N$ and are distributed between two cells, C_1 and C_2, say, x in the first cell and $2N - x$ in the second cell. At time n, an integer is chosen at random among the integers $1,2,\ldots,2N$, and the molecule with that number is identified. This molecule diffuses to the other cell, through the cell membrane. This process of diffusion is continued indefinitely. Let the random variable $X(n)$ denote the number of molecules in C_1 after the nth transfer (molecular diffusion through the cell membrane), then clearly $\{X(n), n \geq 0\}$ is a Markov chain with state space $\{0,1,2,\ldots,2N\}$. Let us compute the transition matrix as follows: Suppose that there are i molecules in C_1 at time n, then the probability that any one of these molecules is selected to diffuse to the other cell is $i/2N$. Also, there are $2N - i$ molecules in C_2; thus, the probability that one of them diffuses through the membrane is $(2N - i)/2$; hence the transition probabilities are

$$P_{i,i-1} = i/2N$$

$$\text{and} \tag{5.70}$$

$$P_{i,i+1} = 1 - i/2N.$$

Also, states 0 and $2N$ are reflecting barriers. If this process is observed for a long time, it is clear that the chain will visit each state infinitely often, where N is an equilibrium value. Assume that the initial state i is far removed from N, then it is also intuitively clear that state i will be visited infinitely often. It is not so obvious that the time between visits to i is so large that state i is not recurrent. For a large number of molecules and one diffusion per second, the expected return time can exceed billions of years, and the chain appears to be irreversible. Irreversibility resolves this contradiction! Bellman and Harris[18] introduced a continuous version of the diffusion process, and this chain will be studied in Chapter 7.

In matrix form, the transition probabilities in Equation 5.70 is given as

$$P = \begin{pmatrix} 0,1,0,0,\ldots\ldots\ldots\ldots\ldots\ldots,0,0 \\ 1/2N,0,(2N-1)/2N,0,0,\ldots\ldots\ldots 0,0 \\ 0,2/N,0,(2N-2)/2N,0,0,\ldots\ldots\ldots 0,0 \\ \cdot \\ \cdot \\ \cdot \\ 0,0,\ldots\ldots\ldots,(2N-1)/2N,0,1/2N \\ 0,0,0,\ldots\ldots\ldots\ldots\ldots\ldots,0,1,0 \end{pmatrix}. \tag{5.71}$$

Note that this matrix is of the order of $2N+1$ corresponding to the state space $\{0,1,2,\ldots,2N\}$. It is clear that the chain is irreducible and the states are recurrent. It is easy to derive the stationary distribution of the Ehrenfest chain as where for state i:

$$\pi(i) = \binom{2N}{i} 2^{-2N}, \quad i = 0,1,\ldots,2N. \tag{5.72}$$

Also, it will be of interest to estimate the average return time to state i:

$$E(i) = 2^{2N}[i!(2N-i)!/(2n)!] < \infty. \tag{5.73}$$

Bayesian inferences will be directed at estimating the stationary distribution and the average return time, given by Equations 5.72 and 5.73, respectively.

Consider the special case $N = 5$; thus, the transition matrix (Equation 5.71) reduces to the 11th order of matrix Q:

$$Q = \begin{pmatrix} 0,1,0,0,\ldots\ldots\ldots\ldots,0 \\ 1/10,0,9/10,0,0,0,0,0,0,0,0 \\ 0,2/10,0,8/10,0,0,0,0,0,0,0 \\ 0,0,3/10,0,7/10,0,0,0,0,0,0 \\ 0,0,0,4/10,0,6/10,0,0,0,0,0 \\ 0,0,0,0,5/10,0,5/10,0,0,0,0 \\ 0,0,0,0,0,6/10,0,4/10,0,0,0 \\ 0,0,0,0,0,0,7/10,0,3/10,0,0 \\ 0,0,0,0,0,0,0,8/10,0,2/10,0 \\ 0,0,0,0,0,0,0,0,9/10,0,1/10 \\ 0,0,0,0,0,\ldots\ldots\ldots,0,1,0 \end{pmatrix}.$$

The matrix Q will be used as sample information for the Bayesian analysis. Using the Q matrix, the multinomial distribution was used to generate transition counts as sample information given by the T matrix.

$$T = \begin{pmatrix} 0,20,0,0,0,0,0,0,0,0,0 \\ 2,0,18,0,0,0,0,0,0,0,0 \\ 0,1,0,19,0,0,0,0,0,0,0 \\ 0,0,4,0,16,0,0,0,0,0,0 \\ 0,0,0,9,0,11,0,0,0,0,0 \\ 0,0,0,0,7,0,13,0,0,0,0 \\ 0,0,0,0,0,11,0,9,0,0,0 \\ 0,0,0,0,0,0,16,0,4,0,0 \\ 0,0,0,0,0,0,0,18,0,2,0 \\ 0,0,0,0,0,0,0,0,16,0,4 \\ 0,0,0,0,0,0,0,0,0,20,0 \end{pmatrix}.$$

If one assumes an improper prior for the transition probabilities, the rows of the transition count matrix follow a Dirichlet distribution; thus, for example, the two nonzero entries of the second row follow a Dirichlet (2,18) distribution. It is suspected that $N = 5$, but not sure. Based on the sample information from matrix T, we will test the hypothesis that $N = 5$. If $N = 5$, then the following null hypothesis H is true:

$$\begin{aligned} &P_{10} = 1/10, \\ &P_{21} = 2/10, \\ &P_{32} = 3/10, \\ &P_{43} = 4/10, \\ \text{H}: \ &P_{54} = 5/10, \\ &P_{65} = 6/10 \\ &P_{76} = 7/10, \\ &P_{87} = 8/10, \\ &P_{98} = 9/10 \end{aligned} \tag{5.74}$$

Note that there are many forms for the alternative. Under the null hypothesis, the order of the ratios $1/10, 2/10$, etc., are quite specific; however, under the alternative, one could have a completely different order of these ratios.

The likelihood for the observed transition counts is

$$f(n|P) = \left[20!/\prod_{i=1}^{i=9}\prod_{j=i-1}^{j=8} n_{ij}!\right]\prod_{i=1}^{i=9}\prod_{j=i-1}^{j=8} P_{ij}^{n_{ij}}\left(1-P_{ij}\right)^{20-n_{ij}}, \tag{5.75}$$

where

$$n = (n_{10}, n_{21}, n_{32}, n_{43}, n_{54}, n_{65}, n_{76}, n_{87}, n_{98}) \tag{5.76}$$

and

$$P = (P_{10}, P_{21}, P_{32}, P_{34}, P_{45}, P_{56}, P_{67}, P_{78}, P_{89}), \tag{5.77}$$

where

$$1 = \sum_{i=1}^{i=9} \sum_{j=i-1}^{j=8} P_{ij}.$$

The transition counts corresponding to the n vector earlier is

$$n_{\text{obs}} = (2, 1, 4, 9, 7, 11, 16, 18, 16).$$

Recall that the posterior probability of the null hypothesis is

$$p_0 = \pi_0 f(n|\text{H}) / [\pi_0 f(n|\text{H}) + \pi_1 \int \rho_1(P) f(n|P) \text{d}P], \tag{5.78}$$

where from Equations 5.73 and 5.74,

$$f(n|\text{H}) = f(n|P_{\text{H}}),$$

and the components of the vector P_{H} are specified by the null hypothesis, namely,

$$P_{\text{H}} = (1/10, 2/10, 3/10, 4/10, 5/10, 6/10, 7/10, 8/10, 9/10).$$

I choose the following as the prior distribution ρ_1:

$$\begin{aligned}
P_{10} &\sim \text{beta}(.25, 2.25), \\
P_{21} &\sim \text{beta}(.5, 2), \\
P_{32} &\sim \text{beta}(.75, 1.75), \\
P_{43} &\sim \text{beta}(1, 1.5), \\
P_{54} &\sim \text{beta}(1.25, 1.25), \\
P_{65} &\sim \text{beta}(1.5, 1), \\
P_{76} &\sim \text{beta}(1.75, .75) \\
P_{87} &\sim \text{beta}(2.25, .5), \\
P_{98} &\sim \text{beta}(2.25, .5) \\
P_{87} &\sim \text{beta}(2.25, .5), \\
P_{98} &\sim \text{beta}(2.25, .5).
\end{aligned} \tag{5.79}$$

The prior beta distributions for the alternative are chosen so that the prior means are the same as the hypothesized values of the transition probabilities under the null hypothesis; however, the prior variances are relatively large so that one is uncertain, a priori, for the values of the transition probabilities under the alternative hypothesis.

Given this information, the posterior distribution of the transition probabilities (Equation 5.77) under the alternative hypothesis is

$$\begin{aligned}
&P_{10} \sim \text{beta}(2.25, 20, 25),\\
&P_{21} \sim \text{beta}(1.5, 21),\\
&P_{32} \sim \text{beta}(4.75, 17.75),\\
&P_{34} \sim \text{beta}(10, 12.5),\\
&P_{54} \sim \text{beta}(8.25, 14.25),\\
&P_{65} \sim \text{beta}(12.5, 10),\\
&P_{76} \sim \text{beta}(17.75, 4.75),\\
&P_{87} \sim \text{beta}(20, 2.5),\\
&\text{and}\\
&P_{98} \sim \text{beta}(18.25, 4.25).
\end{aligned} \tag{5.80}$$

Given this information, one can easily evaluate the integral $\int \rho_1(P)f(n|P)dP$ in Equation 5.78 and, consequently, the posterior probability of the null hypothesis as

$$p_0 = .998$$

Therefore, the data confirm the Ehrenfest model (Equation 5.71) with $N = 5$.

5.9 Comments and Conclusions

This chapter presents the way a Bayesian would make inferences for a large variety of examples from various specializations in biology. Several examples involve genetics including the inbreeding problem, the Wright model that accounts for mutation and genetics forces, and some examples with DNA including the Jukes–Cantor process and a generalization called the Kimura chain. Also presented in this chapter is the birth and death process and several models for epidemics that deal with deterministic and stochastic versions.

The inbreeding problem is a simple one, where the heredity mechanism allows brother–sister mating. The following description of the inbreeding problem closely follows Allen,[1] and the appropriate chain has six states: (1) AA × AA, (2) AA × A*a*, (3) A*a* × A*a*, (4) A*a* × *aa*,

(5) AA × *aa*, and (6) *aa* × *aa*. The laws of inheritance imply the following makeup of the next generation: (1) If the parents are both of type AA, the offspring will be AA individuals, so that the crossing of brother and sister will be of one type, and $P_{11} = 1$; (b) now suppose the parents are type 2, namely, A*a* × A*a*, then the offspring will occur in the following proportions: 1/2 AA and 1/2 Aa; therefore, the crossing of brother and sister will be 1/4 of {AA × AA}, 1/2 of{AA × A*a*}, and 1/4 of {A*a* × A*a*}. (c) Lastly, if the parents are of type A*a* × A*a*, the offspring are in the proportion 1/4 type AA, 1/4 type A*a*, and 1/4 type *aa*; thus, brother and sister mating will give 1/16 type {AA × AA}, 1/4 type {AA × A*a*}, 1/4 type {A*a* × A*a*}, 1/4 type {A*a* × *aa*}, 1/8 type {AA × *aa*}, and 1/16 type {*aa* × *aa*}. This information is sufficient for determining the 6 × 6 transition matrix P in Equation 5.1 of the inbreeding problem. The R routine involving igraph produces the directed graph of the process by displaying the connection between the six alternatives and is followed by the R function Markov, which generates realizations from the inbreeding process. This function depends on the initial configuration of the process. The primary focus for Bayesian inference is testing hypothesis, namely, the laws of inheritance for the inbreeding problem. Testing hypothesis is based on sample information from multinomial realizations (of size 10 for each row) of the transition matrix, then the posterior probability of the null hypothesis is computed as .999, which implies that the laws of inheritance are valid for the inbreeding model.

Next to be considered is the Wright model of genetics, which allows for the effect of mutation and selection on gene frequency. Transition matrix P is shown to depend on the binomial distribution where the two mutation rates are accounted for, and the main emphasis is on estimating the transition probabilities.

The birth and death process is defined by $\{X(n), n = 0, 1, 2, \ldots\}$, where $X(n)$ is the size of the population at time n, and is the next subject presented in this chapter. The transition matrix is a function of the birth and death rates, which give the probabilities of a birth and death, respectively, at each time point. The birth and death rates depend on the present size of the population, and the process is illustrated with birth and death rates $b_i = bi$ and $d_i = di$, where i is the present size of the population and $b = .02$ and $d = .03$. Since the basic death rate is more than that of the birth rate, one would expect the population to become extinct. Of interest for the Bayesian is to estimate the time to extinction. The logistic growth process is a special case of the birth and death process, and fundamental parameters of the process are estimated via Bayesian methods using multinomial realizations for the data and an improper prior distribution for the birth and death rates.

Much of this chapter is devoted to Bayesian inferences for several versions of an epidemic. In the simplest case at a given time point, there are two types of individuals, infected or susceptible, in a fixed population of individuals. Also, infected individuals can again become susceptible. A more realistic version of the epidemic process is presented as an exercise at the end of the chapter. First to be described is the deterministic process, which includes definitions of the basic parameters. In the stochastic version of the epidemic, the one-step transition matrix is defined in terms of the number of the number of infected people in the population. β, b, and γ are the fundamental parameters of the process, and their meaning is made clear in the following explanation: Suppose the interval between time n and time $n + 1$ is small enough so that at most, one event occurs. Therefore, the following can occur: (1) a susceptible person becomes infected, (2) a susceptible person give birth (and a corresponding death of either a susceptible or infected individual), or (3) an infected person recovers. Suppose the probability of a susceptible individual becoming infected is $\beta I/N$, where β is the number of contacts made by one infectious individual that results in one infection during the interval $(n, n+1)$; thus, only $\beta S/N$ of these contacts may result in a new infection, and the total number of new infections by the whole class of infected individuals is $\beta SI/N$. Suppose susceptible and infected persons are born or die with probability b and that infected individuals recover with probability γ. The transition matrix (Equation 5.36) of the process is determined by the three parameters β, b, and γ. The Bayesian posterior analysis is executed with WinBUGS Code 5.8, where the sample information is data generated by multinomial realizations and an improper prior used for prior information. See Table 5.9 for information about the posterior distribution of the fundamental parameters. Another version of the epidemic model is presented and is called the chain binomial model because the transition probabilities are computed according to the binomial distribution. For this model, the parameters are α, β, and p, where α is the probability that a susceptible person comes in contact with an infected person, β is the probability that the contact results in an infection, and the probability a susceptible person is not infected. There are two versions of the chain binomial model: (1) Greenwood and (2) Reed–Frost. For the Greenwood model, the transition matrix is given by the binomial distribution where the probability of success is given by p, whereas for the Reed–Frost model, the probability of success is given by $p^{i(n)}$, where $i(n)$ is the number of infected at time n and the $I(n)$ process satisfies $S(n) + I(n) = S(n-1)$ for $n = 1, 2, \ldots$.

A Bayesian analysis is performed for the special case $s(0) = 3$, and the transition matrix in Equation 5.43 is expressed in terms of p and in terms of α and β with the transition matrix in 5.45. For the case when $p = .025$, multinomial realizations of size 500 are generated for each row of the transition matrix; then assuming an improper prior distribution for p, the Bayesian analysis is executed with 45,000 observations for the simulation and a burn-in of 5,000.

Our attention is turned to a very interesting topic in genetics, namely, that of molecular evolution where the focus is on four sites of a strand of DNA, and each site is occupied by one of the four bases: (1) adenine, (2) guanine, (3) cytosine, and (4) thymine. The corresponding 4×4 one-step transition matrix is quite special, in that, the off-diagonal entries are all the same, namely, probability d, and this arrangement of transition probabilities is called the Jukes–Cantor model. The time unit is one generation; thus, we will be observing the evolution of the four bases over generations. Usually in such situations, the emphasis is on estimating the mutation rate $3d$. Bayesian inferences consist of estimating the probability of a mutation and the mutation rate, and an example is provided with $d = 1/2$. The sample information is generated using the multinomial distribution of size 50 generations, and the prior distribution is assumed to be improper, and the resulting analysis determined that the posterior mean is .0918. Also, included in the Bayesian analysis is a test of the hypothesis that the Jukes–Cantor model is the proper evolutionary model. It was found that the posterior probability that this is true is $p_0 = .995$. Lastly, the so-called Kimura model is a generalization of the Jukes–Cantor model, and the following Bayesian analysis consists of testing the hypothesis that the Kimura model reduces to the Jukes–Cantor model.

Finally, this concludes with a Bayesian analysis of the Ehrenfest model for diffusion through a cell membrane. This example is quite interesting in that the model's origin is in statistical thermodynamics and was later adopted in biology. Finally, there are several more references that will be of interest to the reader. With additional information about epidemics, see Allen[19] and Anderson and May.[20] A good reference for the role mathematics plays in genetics is given by Ewens,[21] and for the role stochastic processes play in epidemics, refer to Gabriel, Lefevre, and Picard.[22]

5.10 Exercises

1. Consider the probability transition matrix of the inbreeding problem in genetics (Equation 5.1).
   a. Show that the chain is irreducible.
   b. What are the three communicating classes?
   c. Refer to R Code 4.1 and use the R function Markov; generate 100 observations with the starting value of 1. The 1 starting value is the AA × AA for the parents.
   d. Using R Code 5.1 along with the igraph package, generate the transition graph corresponding to the transition matrix in Equation 5.1.
   e. Generate five multinomial realizations of size 10 based on the third row of the transition matrix (for the inbreeding problem) using the R function rmultinom(5,10,p), where p<-(1/16,1/4,1/4,1/4,1/8,1/16). Your results should be similar to those in Table 5.1.
2. Using 35,000 observations for the simulation and 5,000 for the burn-in, execute the Bayesian analysis with WinBUGS Code 5.1. The goal of the analysis is to estimate the transition probabilities $P_{31}, P_{32}, P_{33}, P_{34}, P_{35}, P_{36}$ of the third row of Equation 5.1.
   a. What is the posterior mean and 95% credible interval of P_{31}?
   b. What is used for the prior distribution of these parameters?
   c. What is the sample information you used for the Bayesian analysis? Refer to Table 5.1, which reports the five realizations of size 10 for the third row of the transition matrix.
   d. What is the marginal posterior distribution of P_{31}?
   e. Compare your results to Table 5.2.
3. Consider a test of the null hypothesis

$$H\text{: } P_{31} = 1/16, P_{32} = P_{33} = P_{34} = 1/4, P_{35} = 1/8, P_{36} = 1/16,$$

   which are the rules on inheritance for the inbreeding example. As sample information, use the first realization listed in Table 5.1. Also, assume that the likelihood function is given by Equation 5.3.
   a. Verify Equation 5.4 for the posterior probability of the null hypothesis.
   b. Verify Equation 5.6.
   c. Based on Equations 5.3 through 5.6, verify that the posterior probability of the null hypothesis is .999.

4. Refer to the Wright model for genetics described in Section 5.3.
   a. Based on Equation 5.7, show that the probability transition matrix for the Wright model is defined by the probability mass function (Equation 5.8).
   b. Equations 5.9 and 5.10 account for mutations in the Wright model with mutations rates α_1 and α_2. What are the components of the first row of the transition matrix in terms of α_1 and α_2?
   c. Let $\alpha_2 = .07$. What are the components in the first row of the transition matrix?
   d. Based on the probabilities of the first row of the transition matrix and using the R function rmultinom, generate a multinomial realization of size 100.
   e. Assume that the prior distribution for the transition probabilities of the first row is improper, and using the transition counts (68,28,3,1,0,0,0), show that the posterior distribution of the first row of the transition matrix is Dirichlet (68,28,3,1,0,0,0).
   f. Based on WinBUGS Code 5.2, execute the Bayesian analysis for estimating the entries of the first row of the transition matrix. Use 45,000 for the simulation and a burn-in of 5,000.
   g. Verify the posterior analysis reported in Table 5.4.
5. The Wright model is revised to account for selection forces, see Equations 5.14 and 5.15.
   a. What are the transition probabilities for the second row of the transition matrix (Equation 5.16)?
   b. For the selection force, let $s = 1$, and verify that the second row entries are $P_{10} = .1328$, $P_{11} = .3188$, $P_{12} = .3187$, $P_{13} = .0509$, $P_{14} = .00815$, $P_{15} = .00052$.
   c. Generate five multinomial realizations of size 100 based on the preceding transition probabilities of item b. See Table 5.5.
   d. Assume that the prior distribution of the entries of the second row of the transition matrix is improper given by Equation 5.17, and use as sample information one of the multinomial realizations generated in item c. Show the posterior distribution of the cell counts corresponding to the first two to have a Dirichlet distribution.
   e. Use WinBUGS Code 5.3 for the Bayesian posterior analysis of P_{10} and the selection factor s. Use 35,000 observations for the simulation and 5,000 for the burn-in. Your results should be similar to those in Table 5.6.
   f. What is the posterior median of the selection factor s? Are the results consistent with the value $s = 1$ used to generate the multinomial realizations for the data?

6. Refer to the transition probability matrix for the birth and death process given by Equation 5.21, where d_i and b_i are the death and birth rates when the population is size i. Assume that the birth and death rates are given by $b_i = bi$ and $d_i = di$, where $b = .02$ and $d = .03$. Thus, the birth rate is increasing at 2% for an increase of 1 in the population size, while the death rate is increasing at a rate of 3%.
   a. Refer to R Code 5.2 and generate 100 observations for the birth and death process by using the R function Markov with an initial value of 1 person.
   b. Using the R function stationary, determine the stationary distribution of this birth and death process.
   c. Generate five multinomial realizations of size 10 for the birth and death process using the probability vector p<-c(.03,.95,.02). See Table 5.8.
   d. Assume an improper prior distribution for the birth and death rates b_1 and d_1, the birth and death rates when the population consists of only one individual. Use the fifth multinomial realization for the sample information and execute WinBUGS Code 5.5 with 45,000 observations for the simulation and 5,000 for the burn-in.
   e. What is the 95% credible interval for d_1?
   f. Are the posterior distributions for these birth and death rates skewed?
   g. Demonstrate their skewness by plotting the posterior density of the birth and death rates.
7. Refer to the time of extinction of the birth and death given by Equation 5.28.
   a. Generate multinomial realizations of size 100 of the 10 birth and death processes shown in Table 5.17.
   b. Assume an improper prior for the birth and death rates. Execute a Bayesian analysis using WinBUGS Code 5.5 for estimating the 10 birth and death rates b and d (where $b_i = bi$ and $d_i = di$, $i = 1, 2, \ldots, 10$) and time to extinction. Use 35,000 observations for the simulation and 4,000 for the burn-in.
   c. What are the posterior characteristics for the 10 average time to extinction?
   d. What are the 95% credible intervals for d_3 and b_3?
   e. The posterior median of μ_1 is 76.55. Does this seem reasonable to you?

TABLE 5.16

Posterior Analysis for Birth and Death Process: Fifth Realization

Parameter	Mean	SD	Error	2 1/2	Median	97 1/2
b_1	.02006	.01402	.0000821	.002347	.01696	.05521
d_1	.07009	.02538	.000142	.02875	.06727	.1269

TABLE 5.17

Multinomial Realizations for Birth and Death Process

Population Size	d	$1 - d - b$	b	n_1	n_2	n_3
1	.03	.95	.02	2	97	21
2	.06	.90	.04	8	86	6
3	.09	.85	.06	7	91	2
4	.12	.8	.08	15	81	4
5	.15	.75	.10	11	76	13
6	.18	.7	.12	17	66	17
7	.21	.65	.14	22	64	14
8	.24	.6	.16	23	59	18
9	.27	.55	.18	27	57	16
10	.30	.70		28	72	

8. a. Read Section 5.6.1 and describe the SIS epidemic model.
   b. Carefully explain the difference Equation 5.32 for the evolution of an epidemic. What is the solution to the difference equation?
   c. Define the parameters β, b, and γ of the deterministic version.
   d. Describe the transition matrix as defined by Equation 5.35.
   e. Refer to Equation 5.37, which determines the transition probability matrix for the special case $N = 100$, $\gamma = b = .0025$, $\beta = .01$.
   f. Generate multinomial realizations of size 1000 using the probability vector p<-c (.005,.9851,.0099). This should result in the transition count (3,993,4) for the transition counts corresponding to the second row of the transition matrix (Equation 5.36).
   g. Assume an improper prior distribution (Equation 5.38) corresponding to the first three entries of the second row of the transition count vector (3,993,4). The latter serves as data for the Bayesian analysis for estimating the parameters $(b + \gamma)$, b, β, P_{10}, P_{11}, P_{12}. Execute WinBUGS Code 5.7 for the posterior analysis with 35,000 observations for the simulation and 5,000 for burn-in. The results should be similar to those in Table 5.12.
   h. What is the posterior mean and median of b? Is this skewed? The posterior mean of b is .00149. Is this reasonable? Why?

9. a. Refer to Section 5.64 and describe the purpose of the parameters α, β, and $1 - \alpha\beta$.
   b. Verify that the matrix P (Equation 5.42) is the one-step probability transition matrix of the Greenwood model.
   c. Show that Equation 5.42 reduces to the matrix Equation 5.43 when $s(0) = 3$.
   d. As a special case, let $\alpha = .5$, $\beta = .05$, and verify that the probability transition matrix is Equation 5.44.
   e. What states are transient for the special case?
   f. Using the R function Markov, generate 25 observations for the epidemic with an initial value of 2.
   g. For each row of the transition matrix (Equation 5.44), generate multinomial realizations of size 500. The result should be similar to the matrix Q (1.46).
   h. Assuming a uniform prior distribution for the parameter p, show that the posterior distribution of p is beta (491,11).
   i. In order to estimate p and the other entries of the transition matrix in Equation 5.43, execute WinBUGS Code 5.9 with 35,000 observations for the simulation and 5,000 for the burn-in.
   j. What is the 95% credible interval for p? Is this a reasonable value? Explain why.
10. The SIR model is an abbreviation for susceptible people, infected, and removed individuals during the course of an epidemic. A person is susceptible if they have not had the disease, infected if they currently have the disease, and removed if they have had the disease and have since recovered (and are now immune) or have died. Time is measured in discrete steps, and at each step, each individual can infect susceptible individuals or can recover/die, at which point, the infected is removed. Therefore, this version of an epidemic is more realistic than the previous model. See Jones, Maillardet, and Robinson[23] for additional information about the SIR model. Suppose $S(t)$, $I(t)$, and $R(t)$ denote the number of susceptible individuals, infected, and removed at time t, where at each time point, each infected has a probability α of infecting each susceptible (this assumes each person has an equal chance of contacting all susceptible persons). At the end of each step, after having had a chance to infect people, each infected person has probability β of being removed. The initial conditions are $S(0) = N$, $I(0) = 1$, $R(0) = 0$, where the total population is of size $N + 1$ and remains fixed (is not random), that is, $S(t) + I(t) + R(t) = N + 1$, for all $t = 0, 1, 2, \ldots$.

    Note that at each time point, the chance a person remains uninfected is $(1 - \alpha)^{I(t)}$, that is, each infected must fail to pass on the infection to the susceptible individuals. Because each infected has a probability β of being removed,

$$R(t+1) \sim R(t) + bin(I(t), \beta). \tag{5.81}$$

    Also, $S(t + 1)$, given $S(t)$, is binomial

$$\left[S(t), \quad (1-\alpha)^{I(t)}\right]. \tag{5.82}$$

    In addition, that if $S(t + 1)$ and $R(t + 1)$ are known, then

$$I(t+1) = N + 1 - S(t+1) - R(t+1). \tag{5.83}$$

The following R Code for SIR stipulates the instructions for simulating the SIR epidemic if one knows the inputs $a = \alpha$; $b = \beta$; N, the total population size; and T, the number of time points that the epidemic is followed:

R Code for SIR

```
SIR<-function(a,b,N,T){
+ S<-rep(0,T+1)
+ I<-rep(0,T+1)
+ R<-rep(0,T+1)
+ S[1]<-N
+ I[1]<-1
+ R[1]<-0
+ for ( i in 1:T){
+ S[i+1]<-rbinom(1,S[i],(1-a)^I[i])
+ R[i+1]<-R[i]+rbinom(1,I[i],b)
+ I[i+1]<-N+1-R[i+1]-S[i+1]
+ }
+ return(matrix(c(S,I,R),ncol=3))
+ }
> a<-.001
> b<-.1
> N<-1000
> T<-100
```

An epidemic will be simulated with the following inputs: $a = .001$, $b = .1$, $N = 1000$, and $T = 100$. Thus, an infected person has a probability of .001 of infecting a susceptible, while the probability is .1 that an infected is removed. Table 5.18 portrays an SIR

TABLE 5.18

Simulation of SIR Epidemic

Time	S	I	R
1	1000	1	0
2	998	3	0
3	994	7	0
4	982	18	1
5	964	34	3
6	933	63	5
7	886	102	13
8	795	188	18
9	631	341	29
10	435	504	62
11	256	631	114
12	136	690	175
13	67	690	244
14	41	638	322

simulation with the preceding inputs. I used 100 time points, but the table displays on the first 14 time points. It is interesting to observe that by time 14, there are already 638 infected individuals and 322 have been removed. The student should simulate the epidemic for all 100 time points.

Now the interesting aspect of this exercise is to make Bayesian inferences for the α and β parameters of the SIR model, given the data in Table 5.18. Does the information in this table allow one to estimate α and β with reliable values? In order to estimate these two parameters, the following WinBUGS Code for SIR will be used to execute the posterior analysis. In the code, alphamean is the mean of the 13 alphas, and betameans is the mean of the 13 betas. Assuming an improper prior for the alphas and gams, their corresponding posterior distribution is beta. This follows from Equations 5.81 and 5.82. The Bayesian analysis is executed with 70,000 observations for the simulation and 5,000 for the burn-in.

WinBUGS Code for SIR

```
model;
{
beta4~dbeta(2,16)
beta5~dbeta(2,32)
beta6~dbeta(8,45)
beta7~dbeta(5,97)
beta8~dbeta(11,177)
beta9~dbeta(33,308)
beta10~dbeta(52,452)
beta11~dbeta(61,570)
beta12~dbeta(69,621)
meanbeta<-(beta4+beta5+beta6+beta7+beta8+beta9+beta10+beta11+beta12)/9
gam1~dbeta(998,2)
gam2~dbeta(994,4)
gam3~dbeta(982,12)
gam4~dbeta(964,18)
gam5~dbeta(933,31)
gam6~dbeta(886,47)
gam7~dbeta(795,91)
gam8~dbeta(631,164)
gam9~dbeta(435,196)
gam10~dbeta(256,179)
gam11~dbeta(136,120)
gam12~dbeta(67,69)
gam13~dbeta(41,26)
```

```
alpha1<-1-gam1
alpha2<-1-pow(gam2,1/2)
alpha3<-1-pow(gam3,1/3)
alpha4<-1-pow(gam4,1/4)
alpha5<-1-pow(gam5,1/5)
alpha6<-1-pow(gam6,1/6)
alpha7<-1-pow(gam7,1/7)
alpha8<-1-pow(gam8,1/8)
alpha9<-1-pow(gam9,1/9)
alpha10<-1-pow(gam10,1/10)
alpha11<-1-pow(gam11,1/11)
alpha12<-1-pow(gam12,1/12)
alpha13<-1-pow(gam13,1/13)
meanalpha<-(alpha1+alpha2+alpha3+alpha4+alpha5+alpha6+alpha7+alpha8+alpha9
+alpha10+alpha11+alpha12+alpha13)/13
}
```

Note that α is the posterior mean of the average of the 13 alphas, and β is the posterior mean of the mean of the nine betas calculated with WinBUGS Code for SIR. The posterior mean for α, should be compared to the value of $\alpha = .001$ to generate the data for the epidemic. In a similar way, the posterior mean of β should be compared to $\beta = .1$ used to generate the epidemic data of Table 5.18.

a. Using WinBUGS Code for SIR with 70,000 observations for the simulation with a burn-in of 5,000, execute the Bayesian analysis appearing in Table 5.19.
b. What does the posterior distribution for α appearing in Table 5.19 imply about $\alpha = .001$ used to generate the epidemic data of Table 5.18.
c. What does the posterior distribution of β imply about $\beta = .1$, the value used to generate the epidemic portrayed in Table 5.18?

TABLE 5.19

Bayesian Analysis for SIR Model

Parameter	Mean	SD	Error	2 1/2	Median	97 1/2
α	.02421	.000985	.0000038	.02235	.02419	.0262
β	.0916	.01139	.0000413	.07213	.0906	.1167

11. Consider the Jukes–Cantor model of molecular evolution: At four sites of the DNA, there are four different bases:
    a. Adenine
    b. Guanine
    c. Cytosine
    d. Thymine
    The Jukes–Cantor model specifies that the probability transition matrix P in Equation 5.51 should have equal off-diagonal entries and, hence, equal diagonal components. The off-diagonal entry d is the probability that a base will be substituted by a different base, and 3d is the mutation rate.
    a. Derive Equation 5.52, the probability that a base substitution has occurred.
    b. Show that when $d = 1/12$, the transition matrix is Equation 5.53.
    c. Based on the transition matrix (Equation 5.53), use the appropriate multinomial realization of size 50 to generate transition counts for each row. Your result should be similar to the Q matrix (Equation 5.54).
    d. Verify the entries of matrix R (Equation 5.55). Consider the second row. What is the estimated mutation rate?
    e. Using the Q matrix as data and assuming an improper prior distribution (Equation 5.56) for the transition probabilities, use WinBUGS Code 5.9 to execute a Bayesian analysis with 35,000 observations for the simulation and 5,000 for the burn-in.
    f. Your result should be similar to that in Table 5.14 for the posterior analysis.
    g. Based on the posterior mean, what is your estimate of the mutation rate $3d$? Use WinBUGS Code 5.9 to calculate the posterior mean of $3d$.

12. Refer to the Jukes–Cantor model of molecular evolution. This exercise pertains to the Bayesian analysis for the stationary distribution and time reversibility of the Jukes–Cantor model. Also recall the Bayesian analysis of estimating the parameters of time reversibility. Denote the stationary distribution of a k-state chain by $(\pi_1, \pi_2, ..., \pi_k)$ and let the transition probabilities of the process be $P_{ij}, i, j = 1, 2, ..., k$; then for the chain to be time reversible

$$\pi_i P_{ij} = \pi_j P_{ji}, \; i, j = 1, 2, .., k.$$

Of course, the stationary distribution satisfies the system of equations

$$\pi_j = \sum_i \pi_i P_{ij}, \quad \forall j.$$

Refer to the Jukes–Cantor transition matrix (Equation 5.53) in the special case $d = 1/12$.

a. Find the stationary distribution for the transition matrix (Equation 5.53).
b. Use the Q matrix (Equation 5.54) as the sample information. Make an appropriate revision of WinBUGS Code 4.6 and use the modification to estimate the stationary distribution of the Jukes–Cantor model. Note that the stationary distribution has four components.
c. Based on the stationary distribution found in item b, perform the Bayesian analysis for determining the time reversibility. You will need to write WinBUGS Code similar to WinBUGS 4.8, and your results should be presented in the same format as that of Table 4.13.
d. Based on the analysis described in item c, is the Jukes–Cantor time reversible? Explain your answer carefully!

13. a. Refer to Section 5.8 and briefly describe the Ehrenfest cell diffusion model with $2N$ total molecules and state space

$$S = \{0, 1, 2, ..., 2N\}.$$

b. What is the transition probability matrix for the cell diffusion model? See Equation 5.70, and the general form of the transition matrix is given by Equation 5.71.
c. Consider the special case $N = 5$ with transition probability Q and transition counts T. Assume an improper prior distribution for the rows of the transition probability matrix. Note that the Ehrenfest model must satisfy the null hypothesis (Equation 5.74). Using this information and employing Equations 5.75 through 5.79, test the null hypothesis (Equation 5.74). Calculate the posterior probability of the null hypothesis p_0 in Equation 5.78 as .998.

References

1. Allen, J. S. L. 2011. *An Introduction to Stochastic Processes with Applications to Biology, Second Edition.* Boca Raton, FL: CRC Press.
2. Hoppensteadt, R. 1975. *Mathematical Methods of Population Biology.* Cambridge, UK: Cambridge University Press.
3. Bailey, N. T. J. 1990. *The Elements of Stochastic Processes with Applications to the Natural Sciences.* New York: John Wiley & Sons.
4. Feller, W. 1968. *An Introduction to Probability Theory and Its Applications.* New York: John Wiley & Sons.
5. Wright, S. 1932. The roles of mutation, inbreeding, crossbreeding, and selection in evolution, *Proceedings of the 6th International Congress in Genetics* 1:356–366.
6. Fisher, R. A. 1962. *The Genetical Theory of Natural Selection.* New York: Oxford (Clarendon) Press.
7. Karlin, S., and Taylor, H. M. 1975. *A First Course in Stochastic Processes, Second Edition.* New York: Academic Press.
8. DeGroot, M. H. 1970. *Optimal Statistical Decisions.* New York: McGraw-Hill.
9. Dobrow, R. P. 2016. *Introduction to Stochastic Processes with R.* New York: John Wiley & Sons.
10. Allen, L. J. S., and Burgin, A. 2000. Comparison of deterministic and stochastic SIS and SIR models in discrete time, *Mathematical Biosciences* 163:1–33.
11. Daley, D. J., and Gani, J.1999. *Epidemic Modeling: An Introduction. Cambridge Studies in Mathematical Biology 15.* Cambridge, UK: Cambridge University Press.
12. Allman, E. S., and Rhodes, J. A. 2004. *Mathematical Models in Biology, An Introduction.* Cambridge, UK: Cambridge University Press.
13. Jukes, T. H., and Cantor, C. R. 1969. Evolution of protein molecules, in Munro, H. N. (Ed.) *Mammalian Protein Metabolism.* New York: Academic Press.
14. Kimura, M. 1990. A simple method for estimating evolutionary rates of base substitutions through the comparative studies of sequence evolution, *Journal of Molecular Evolution* 16:111–120.
15. Futuyma, D. J. 2013. *Evolution, Third Edition.* Sunderland, MA: Sinauer.
16. Kannan, D. 1979. *An Introduction to Stochastic Processes.* New York: Elsevier (North Holland).
17. Ehrenfest, P., and Ehrenfest, T. 1907. Über zwei bekannte Einwände gegen das Boltzmannsche H-Theorem, *Physikalishce Zeitschrift* 8:311–314.
18. Bellman, R., and Harris, T. E. 1951. Recurrence time for the Ehrenfest model, *Pacific Journal of Mathematics* 1:179–193.
19. Allen, L. J. S. 2000. Some discrete time SI, SIR, and SIS epidemic models, *Mathematical Biosciences* 124:83–105.
20. Anderson, R. M., and May, R. M. 1992. *Infectious Diseases in Humans: Dynamics and Control.* Oxford, UK: Oxford University Press.
21. Ewens, W. J. 1979. *Mathematics Population Genetics.* Berlin: Springer-Verlag.
22. Gabriel, J. P., Lefevre, C., and Picard, P. 1990. (Editors). *Stochastic Processes in Epidemic Theory: Lecture Notes in Biomathematics.* New York: Springer-Verlag.
23. Jones, O., Maillardet, R., and Robinson, A. 2014. *Introduction to Scientific Programming and Simulation Using R.* Boca Raton, FL: Taylor & Francis.

6

Inferences for Markov Chains in Continuous Time

6.1 Introduction

Our main goal for this chapter is to present Bayesian inferences for processes in continuous time. As in earlier chapters, Bayesian ways to estimate parameters, test hypotheses about those parameters, and predict future observations will be developed. There are many examples of Markov chains in continuous times, and the best known is the Poisson process. The Poisson process is a counting process that records many interesting events such as the number of accidents in a given stretch of highway over a selected period, the number of telephone calls at a switchboard, the arrival of customers at a counter, the number of visits to a website, earthquake occurrences in a particular region, etc.

The definition of the Poisson process begins this chapter, which is followed by a description of the arrival and interarrival times, then various generalizations are considered such as the nonhomogeneous Poisson processes, which include compound processes and processes that contain covariates. Of course, R is employed to generate realizations from the several examples of homogeneous and nonhomogeneous Poisson processes.

6.2 Poisson Process

The Poisson process is a counting process, a collection $\{N(t), t \geq 0\}$ of nonnegative integers, where if $0 \leq s \leq t$, $N(s) \leq N(t)$.

Note that a counting process is indexed by a continuum, the set of positive numbers, and with a state space consisting of the nonnegative integers. Now to be described are three equivalent definitions of the Poisson process. For additional information, see pages 224–229 of Dobrow.[1]

- Definition a

  A Poisson process with parameter λ is a counting process $\{N(t), t \geq 0\}$ such that

  1. $N(0) = 0$
  2. For all $t > 0$, $N(t)$ has a Poisson distribution with parameter λt
  3. The process has stationary increments; that is, for all times $s, t > 0$, $N(t + s) - N(s)$ has the same distribution as $N(t)$; thus,

$$\Pr[N(t+s) - N(s)] = \Pr[N(t) = k] = e^{-\lambda t}(\lambda t)^k / k! \tag{6.1}$$

4. The process has independent increments; that is, for $0 \le q < r \le s < t$, $N(t) - N(s)$ and $N(r) - N(q)$ are independent.

Consequently, the distribution of the number of events depends only on the length of the interval. It should be noted that the mean and variance of the process are $E[N(t)] = \text{Var}[N(t)] = \lambda t$ for all $t > 0$. The next definition of a Poisson process is given in terms of the arrival times of the events. The arrival times are the times the event occur while the interarrival times are the times between consecutive events. It will be shown that the interarrival times are independent and have identical exponential distributions.

- Definition b

For a Poisson process with parameter λ, let Y be the time of the first arrival after time 0, then

$$P[Y > 0] = P[N(t) = 0] = e^{-\lambda t}, \quad t > 0 \tag{6.2}$$

Y is distributed exponential with parameter λ (with mean $1/\lambda$). It is therefore true that the interarrival times between all the following events (after the first) follow the same exponential distribution. This is so because after the first event, the succeeding events start the process all over again.

Let $Y_1, Y_2, \ldots$ be a sequence of independent and identically distributed (i.i.d.) exponential random variables with parameter λ and let

$$N(t) = \max\{n : Y_1 + Y_2 + \ldots + Y_n \le t\},$$

where $N(0) = 0$; then $\{N(t), t > 0\}$ is a Poisson process with parameter λ.

The arrival time for the nth event is

$$S_n = Y_1 + Y_2 + \ldots + Y_n.$$

Thus, the interarrival time between event $k - 1$ and the kth one is

$$Y_k = S_k - S_{k-1}, \quad k = 1, 2, \ldots.$$

It can be shown that the distribution of $S_n \sim \text{gamma}(n, \lambda)$ with density

$$f(s) = \lambda^n s^{n-1} \exp -\lambda s/(n-1)!, \quad s > 0, \tag{6.3}$$

and moments

$$E(S_n) = n/\lambda$$

and

$$\text{Var}(S_n) = n/\lambda^2.$$

- Definition c

  A Poisson process with parameter λ is a counting process $\{N(t), t \geq 0\}$, satisfying the following properties:

  1. $N(0) = 0$
  2. The process has stationary and independent increments
  3. $P[N(t) = 0] = 1 - \lambda h + o(h)$
  4. $P[N(t) = 1] = \lambda h + o(h)$ (6.4)
  5. $P[N(t) > 1] = o(h)$

  where $h \to 0$ and $o(h)/h \to 0$.

Restrictions 3–5 ensure that it is impossible for an infinite number of arrivals to occur in a finite interval, and that for very small intervals, there may occur at most one event. It is left to reader in exercise 1 to show that the three definitions for a Poisson processes are equivalent. See pages 234 and 235 of Dobrow.[1] The beginning sections have laid the foundation for the Poisson process, and the following section will introduce Bayesian inferential techniques for the parameter λ.

6.3 Bayesian Inferences for λ

This section begins with an R Code 6.1 for simulating the arrival times of a Poisson process with parameter $\lambda = 1/2$ over the time interval (0,40]:

R Code 6.1

```
> t<-40
> lamda<-1/2
> N<-rpois(1,lamda*t)
> unifs<-runif(N,0,t)
> arrivals<-sort(unifs)
> arrivals

2.976133, 3.248414, 4.721712, 13.677706, 18.775247, 20.919774,
24.250771, 26.427624, 29.565929, 33.931659, 36.650545, 39.282970.
```

Thus, the first event arrives at time 2.976133; the second, at time 3.248414; and the last, the 12th, at time unit 39.28297; consequently, $N(39.28) = 12$.

Consider the following fact about the minimum n of independent exponential random variables $Y_1, Y_2, \ldots, Y_n$ with parameters $\lambda_1, \lambda_2, \ldots, \lambda_n$, respectively.

Let $M = \min(Y_1, Y_2, \ldots, Y_n)$ t and $t > 0$; then

1. $P[M > t] = \exp(-t(\lambda_1 + \lambda_2 + \ldots + \lambda_n))$ (6.5)
2. For $k = 1, 2, \ldots, n$

$$P[M = Y_k] = \lambda_k/(\lambda_1 + \lambda_2 + \ldots + \lambda_n). \tag{6.6}$$

The first assertion earlier implies that M has an exponential distribution with parameter $\lambda_1 + \lambda_2 + \ldots + \lambda_n$.

The statistical problem is to generate realizations from the n Poisson processes and then estimate the probabilities given by 1 and 2 earlier.

R Code 6.1 is used to generate realizations from $n = 3$ processes mentioned in the following interesting problem described on pages 231 and 232 of Dobrow.[1]

A bus station serves three routes labeled 1, 2, and 3. Buses on each route arrive at the bus station according to three independent Poisson processes. Buses on route 1 arrive at the station on the average every 10 minutes, those on route 2 arrive on the average every 15 minutes, while route 3 buses arrive on the average every 20 minutes.

1. When a person arrives at the station, what is the probability that the first bus to arrive is from route 2?
2. On average, how long will the person wait for some train to arrive?
3. The person has been waiting for 20 minutes for a bus on route 3 to arrive and during this time, three route 1 buses arrive at the station. What is the expected additional time the person will have to wait for the arrival of a route 3 bus?

Note that we have three independent Poisson processes with parameters $\lambda_1 = 10$, $\lambda_2 = 15$, and $\lambda_3 = 20$. Since the parameters are known, it is straightforward to calculate these probabilities. Let Y_1, Y_2, and Y_3 denote the waiting times for buses from routes 1, 2, and 3, respectively, then it is known that $Y_1 \sim \exp(1/10)$, $Y_2 \sim \exp(1/15)$, and $Y_3 \sim \exp(1/20)$.

To answer question 1, note that the desired probability is given by Equation 6.6, namely,

$$P[\min(Y_1, Y_2, Y_3) = Y_2] = \lambda_2/(\lambda_1 + \lambda_2 + \lambda_3) = (1/15)/(13/60) = .31 \quad (6.7)$$

Now let us take the role of a statistician where the three Poisson parameters are unknown, but that one has information about the waiting time for passengers at the bus station. One has available observations for the waiting times Y_i, $i = 1, 2, 3$, which will be generated with R Code 6.1, and $t = 120$ minutes.

When $\lambda_1 = 1/10$, there are 14 arrivals with the following waiting times:

```
7.922561, 31.770897, 50.570418, 57.364507, 57.593699, 61.285045,
70.839773, 78.795652, 83.963642, 84.830358, 91.112292,
94.419953, 106.066164, 109.656950
```

When $\lambda_2 = 1/15$, there are 9 arrivals with the following waiting times:

```
4.224859, 19.282435, 27.570154, 31.322879, 58.255511,
75.070503, 87.359219, 93.729471, 115.969410.
```

Lastly, when $\lambda_3 = 1/20$, there are 8 arrivals with corresponding waiting times (minutes) as follows:

```
13.06094, 33.21101, 37.39390, 47.40320, 75.50575, 96.19654,
97.06688, 112.26442.
```

Bayesian inferences for the parameter of a Poisson process are well known, and the reader is referred to Chapter 6 of Dobrow,[1] Chapter 5 of Insua, Ruggeri, and Wiper,[2] and Albert.[3]

Suppose $Y_1, Y_2, \ldots, Y_n$ is a random sample from an exponential population of size n with parameter λ and corresponding observations $y_1, y_2, \ldots, y_n$, then the likelihood function for λ is

$$l(\lambda|\text{data}) \propto \lambda^n \exp\left(-\lambda \sum_{i=1}^{i=n} y_i\right), \quad \lambda > 0. \tag{6.8}$$

Note that $\sum_{i=1}^{i=n} y_i$ is the waiting time to the nth event.

Prior information will be expressed with the gamma distribution with density

$$f(\lambda) \propto \lambda^{\alpha-1} \exp(-\beta\lambda), \lambda > 0. \tag{6.9}$$

Sometimes prior information is expressed with the improper density

$$g(\lambda) \propto 1/\lambda, \quad \lambda > 0. \tag{6.10}$$

In the former case with the gamma prior, the posterior distribution of λ is gamma ($n + \alpha, \beta + \sum_{i=1}^{i=n} y_i$), while for the improper prior, the posterior distribution is gamma ($n, \sum_{i=1}^{i=n} y_i$).

If one uses the improper prior density (Equation 6.10), the posterior density of λ_1 is gamma (15,109.65), implying that the posterior mean is 15/109.65 = .1367, and the posterior variance is 15/(109.65)(109.65) = .001247, and the 95% credible interval is (.1087,.215). Now consider the estimation of λ_2, the per unit average waiting time for the second route, then the posterior mean is 9/115.96 = .0776 and posterior variance is 9/(115.96)(115.96) = .00066 with a 95% credible interval of (.06,.136). Lastly, for the third bus route with parameter λ_3, the posterior mean is 8/112.26 = .0712 with posterior variance .000634 = 8/(112.96)(112.96). The 95% credible interval is (.055,.13).

We now present the Bayesian analysis using WinBUGS Code 6.1 to determine the posterior distribution of

$$P[\min(Y_1, Y_2, Y_3) = Y_2] = \lambda_2/[\lambda_1 + \lambda_2 + \lambda_3]. \tag{6.11}$$

Assuming that the improper prior density (Equation 6.10), recall that the posterior distribution of λ_i, $i = 1, 2, 3$, is known, namely,

$$\lambda_1 \sim \text{gamma}(14{,}109.65),$$

$$\lambda_2 \sim \text{gamma}(9{,}115.96),$$

and

$$\lambda_3 \sim \text{gamma}(8{,}112.26).$$

WinBUGS Code 6.1

```
model;
{
lamda1~dgamma(14,109.65)
lamda2~dgamma(9,115.96)
lamda3~dgamma(8,112.26)
# equation (6.11)
pm<-lamda2/(lamda1+lamda2+lamda3)
}
```

The Bayesian analysis is executed with 35,000 observations for the simulation with a burn-in of 1,000, giving the results in Table 6.1.

The main parameter of interest is $P[M = \lambda_2|\text{data}]$, the posterior probability that the first train to arrive at the station is from bus route 2.

A point estimate of this parameter is given by the posterior mean of .2808 with a 95% credible interval of (.1417,.4477). Note the Monte Carlo Markov chain (MCMC) error is quite small at .00029.

Our next objective is to conduct a formal test of the null hypothesis

$$H:\lambda_1 = .1 \text{ versus the alternative } A:\lambda_1 \neq .1. \tag{6.12}$$

The data used will be the 14 waiting times at the bus station for busses on route 1, namely, 7.922561, 31.770897, 50.570418, 57.364507, 57.593699, 61.285045, 70.839773, 78.795652, 83.963642, 84.830358, 91.112292, 94.419953, 106.066164, 109.656950.

The test will be based on the 14th waiting time S_{14}, which has a gamma distribution with density

$$f(s_{14}|\lambda_1) = \left[\lambda_1^{14} s_{14}^{14-1} \exp(-\lambda_1 s_{14})\right]/\Gamma(14), \quad s_{14} > 0. \tag{6.13}$$

Thus, S_{14} has a gamma distribution (14, λ_1), and the observed value of S_{14} is 109.6569.

Pages 126 and 127 of Lee[4] will be closely followed for details about the Bayesian approach to testing hypotheses (Equation 6.12), a simple null versus a composite alternative hypothesis.

The Bayesian test is based on the posterior probability of the null hypothesis given by

$$p_0 = \pi_0 f(s_{14}|\lambda_1 = .1)/[\pi_0 f(s_{14}|\lambda_1 = .1) + \pi_1 f_1(s_{14})], \tag{6.14}$$

where π_0 is the prior probability of the null hypothesis $\pi_1 = 1 - \pi_0$,

$$f(s_{14}|\lambda_1 = .1) = \left[(.1)^{14} s_{14}^{n-1} \exp(-.1 s_{14})\right]/\Gamma(14), \tag{6.15}$$

TABLE 6.1

Posterior Distribution for Arrivals at Bus Station

Parameter	Mean	SD	Error	2 1/2	Median	97 1/2
λ_1	.1276	.03404	.00019	.06995	.1246	.2021
λ_2	.0774	.02591	.00014	.03526	.07467	.1363
λ_3	.07124	.02523	.00012	.03087	.06824	.1286
$P(M = \lambda_2)$	.2808	.07864	.00029	.1417	.2761	.4477

$$f_1(s_{14}) = \int \rho_1(\lambda_1) f(s_{14}|\lambda_1)\, d\lambda_1, \tag{6.16}$$

where the integral is over the interval $(0, \infty)$, and $\rho_1(\lambda_1)$ is the prior density of λ_1 under the alternative hypothesis. How does one choose $\rho_1(\lambda_1)$? If one chooses this prior as gamma (α,β), one can show

$$f_1(s_{14}) = \beta^{\alpha}\Gamma(14+\alpha)/\Gamma(\alpha)\Gamma(14)(\beta + s_{14})^{14+\alpha}. \tag{6.17}$$

Now the problem is to choose α and β. Remember that the mean of the prior gamma is $E(\lambda_1) = \alpha/\beta$ and the variance is $\text{Var}(\lambda_1) = \alpha/\beta^2$. Suppose $\alpha = .1$ and $\beta = 1$, then $E(\lambda_1) = .1 = \text{Var}(\lambda_1)$.

It can be shown that $f(s_{14}|\lambda_1 = .1) = 9.2589289$ and $f_1(s_{14}) = 0$; thus, with $\pi_0 = \pi_1 = .5$ and based on Equation 6.14, the posterior probability of the null hypothesis is $p_0 = 1$. The evidence provided by the 14th waiting time together with the prior information imply that the null hypothesis is true.

Note the sensitivity of p_0 to the prior probabilities of the null and alternative hypotheses and to the prior distribution of λ_1 under the alternative hypothesis.

The third phase of the Bayesian inference is to develop the Bayesian predictive distribution of S_{15}, the 15th waiting time for the first bus route based on the posterior distribution of λ_1 induced by S_{14}.

$$f(s_{15}|s_{14}) = \int_0^{\infty} f(s_{15}|\lambda_1) f(\lambda_1|s_{14}) d\lambda_1, \tag{6.18}$$

where

$$f(s_{15}|\lambda_1) = \left[\lambda_1^{15} s_{15}^{15-1} \exp(-\lambda_1 s_{15})\right] / \Gamma(15), \quad s_{15} > 0 \tag{6.19}$$

and

$$f(\lambda_1|s_{14} = 109.65) = \left[\lambda_1^{14} \exp(-\lambda_1(109.65))\right] / \Gamma(14), \quad \lambda_1 > 0. \tag{6.20}$$

Based on Equations 6.18 through 6.20, the predictive density of S_{15} is

$$f(s_{15}|s_{14} = 109.65) = \left[s_{15}^{15-1}\Gamma(30)\right] / \Gamma(14)\Gamma(15)(s_{15} + 109.65)^{30}, \quad s_{15} > 0. \tag{6.21}$$

The student will be asked to compute the density in Equation 6.21 by using the following code:

WinBUGS Code 6.2

```
model;
{
# posterior distribution of lamda1
lamda1~dgamma(15,109.65)
# predictive density of s15
s15~dgamma(15,lamda1)
}
```

I executed the analysis with WinBUGS Code 6.2 using 270,000 observations for the simulation and a burn-in of 5,000, where the results are reported in Table 6.2.

Table 6.3 contains 100 values generated from the predictive density (Equation 6.21).

Note that the predictive distribution appears to be skewed to the right with a mean of 117.3 and a median of 109.9. The right skewness is evident in Figure 6.1.

TABLE 6.2

Bayesian Predictive Distribution for Bus Route 1

Parameter	Mean	SD	Error	2 1/2	Median	97 1/2
λ_1	.1369	.03535	.0000715	.0767	.1338	.2143
S_{15}	117.3	45.19	0.08863	52.75	109.5	226.8

TABLE 6.3

Future Values of S_{15}

83.77,89.58,160.6,74.27,87.25,
76.97,94.05,100.0,101.0,80.83,
69.22,77.03,73.96,120.8,133.0,
112.6,91.21,94.86,144.0,131.3,
77.8,103.1,72.66,115.2,129.3,
63.82,74.24,112.5,123.1,92.33,
142.9,93.23,53.3,98.26,105.8,
101.3,81.08,65.44,72.09,111.7,
97.93,62.18,118.9,71.02,102.4,
81.03,103.7,98.24,110.5,107.6,
102.8,96.02,82.43,113.5,88.26,
52.24,61.58,113.2,106.7,90.15,
84.26,110.1,73.8,89.37,93.89,
111.7,62.92,90.31,92.29,94.87,
92.22,156.8,76.08,78.1,96.15,
90.96,127.8,96.16,84.85,71.09,
86.75,106.5,149.2,92.08,74.57,
99.08,85.64,99.83,71.1,76.48,
141.2,51.87,84.59,87.38,73.04,
146.7,105.9,114.1,54.26,110.3

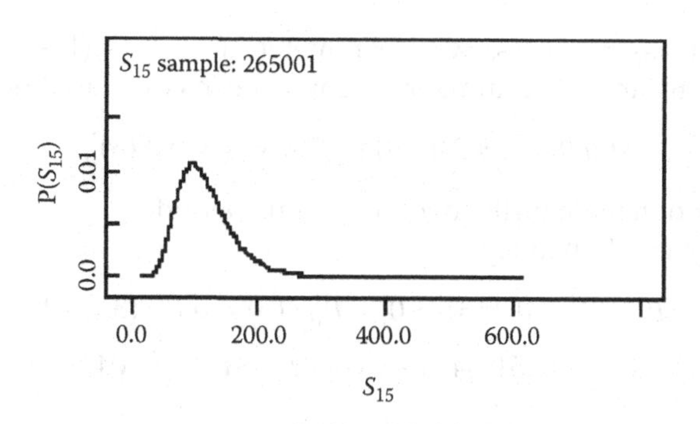

FIGURE 6.1
Predictive density of S_{15}.

6.4 Thinning and Superposition

6.4.1 Birth Rates

The thinning is a Poisson event (arrival) that can be one of several types, each occurring with some nonzero probability. The initial process has a given rate λ, but the subsequent thinned processes have rates smaller than λ induced by the thinning probabilities of the component processes. A good example of this is the birth of humans, where the overall birth rate is, say, λ, and the birth rates for males and females are, say, $p\lambda$ and $(1 - p)\lambda$, respectively, where p is the probability of a male birth.

If the overall birth process follows a Poisson process, it can be shown that the male births follow a Poisson process, and the females, and that both component processes have stationary and independent increments.

More generally, suppose $\{N(t),\ t > 0\}$ is a Poisson process with parameter λ, and assume that each arrival (independent of previous arrivals) is marked by a type k event with probability p_k such that $\sum_{k=1}^{k=n} p_k = 1$; thus, there are n thinned processes. Now let $\{N^k(t),\ t > 0\}$ be the type k component process; then it can be shown that it is a Poisson process with parameter λp_k, where the n processes are independent. See pages 238–240 of Dobrow[1] for additional details.

Returning to the birth Poisson process discussed earlier, according to the United Nations Population Division, the probability of a male birth is $p = .519$, and assume that births at a large municipal hospital is a Poisson process occurring at 2 births per hour. Consider the following three problems:

1. On an 8-hour shift, what is the expectation and standard deviation of the number of female births?
2. Find the probability that only girls were born between 2 p.m. and 5 p.m.
3. Assume that three babies were born at the hospital yesterday. Find the probability that two are female.

Suppose that $\{N(t), t > 0\}$, $\{M(t) > 0\}$ and $\{F(t),\ t > 0\}$ denote the overall, male, and female processes, respectively, and consider the first problem (1) earlier. It is obvious that the females form a Poisson process with parameter $\lambda(1-p) = 2(1-.519) = .962$ births per hour. Therefore, the mean of the number of female births over an 8-hour process is

$$E[F(8)] = 8(2)(.481) = 7.696 = \mathrm{Var}[F(8)], \tag{6.22}$$

the average number of female births over an 8-hour period.

Now consider problem 2, namely,

$$P[M(3) = 0, F(3) > 0] = P[M(3) = 0]P[F(3) > 0]$$
$$= \exp[-2(.519)4][1 - \exp(-2(.481)3)] = .042.$$

Lastly, for problem 3, the desired probability is

$$\binom{3}{2}(.481)^2(.519) = .3602. \tag{6.23}$$

The solution to problems 1, 2, and 3 are based on the "true" values of λ and p; however, in practice, these true values are not available. Instead what is available are observations from the three processes $\{N(t), t > 0\}$, $\{M(t) > 0\}$, and $\{F(t), t > 0\}$. From the overall process, one can estimate λ, and from that, for males, one can estimate $p\lambda$ and, consequently, p. R Codes 6.2 and 6.3 generate observations for the overall births over an 8-hour period with a birth rate of λ per hour and generate male births with 2(.519) = 1.038 births per hour.

R Code 6.2

```
> t<-8
> lamda<-2
> N<-rpois(1,lamda*t)
> unifs<-runif(N,0,t)
> births<-sort(unifs)
> arrivals
1.263526, 1.414724, 1.593349, 2.248500, 2.329316, 2.597407,
2.653645, 2.861839, 3.654373, 3.881924, 4.310339, 4.450808, 5.0
5.008346, 5.085663, 5.315546, 6.850246, 7.320440, 7.524868.
```

R Code 6.3

```
> t<-8
> lamda<-1.038
> N<-rpois(1,lamda*t)
> unifs<-runif(N,0,t)
> malebirths<-sort(unifs)
> malebirths
1.284937, 1.861346, 2.364378, 2.823737, 3.577798, 4.628334,
5.542943, 6.022781, 6.304477, 6.743645, 6.779419.
```

There are 18 overall births with 11 male births over the 8-hour period, which implies a maximum likelihood estimate of 18/8 = 2.25 births per hour for the overall birth rate, while that for the male birth rate, the estimate is 11/8 = 1.375, and consequently, the maximum likelihood estimate of p is 1.375/2.25 = .611. The overall rate of 2.25 should be compared to 2, the value used to generate the data with R Code 6.2. In the same way, the male birth rate estimate of .611 should be compared to the true value of .519.

TABLE 6.4

Bayesian Analysis for Overall and Male Birth Rates

Parameter	Mean	SD	Error	2 1/2	Median	97 1/2
γ	1.623	0.4898	0.00198	0.8108	1.572	2.719
λ	2.391	0.565	0.00212	1.422	2.342	3.618
p	.7187	.2867	.00111	.3033	.672	1.414

What is the Bayesian approach to estimating the overall and male birth rates over an 8-hour period? First, consider estimating the overall birth rate λ, using the conditional distribution of the waiting time S_{18}, given λ as the likelihood function. When this is combined with the improper prior density (Equation 6.10), the posterior distribution of λ is gamma $(18, S_{18})$, where $S_{18} = 7.524868$. In a similar fashion, it can be shown that the posterior distribution of $\gamma = p\lambda$ is gamma (11,6.779419).

WinBUGS Code 6.3

```
model;
{ lamda~dgamma(18,7.5248)
gam~dgamma(11,6.7794)
p<-gam/lamda
}
```

WinBUGS code 6.3 is executed with 70,000 observations for the simulation with a burn-in of 5,000, and the Bayesian analysis is reported in Table 6.4.

The posterior distributions for p and γ are skewed; thus, I recommend the posterior median as the estimate of the parameter. For example, the posterior median of γ is 1.572, which should be compared to $\gamma = p\lambda = 1.038$, the true value of that parameter. In the same manner, the posterior median of λ of 2.343 should be compared to the true value of $\lambda = 2$. Also, the true value of p is .519, which should be compared to the posterior median of .672. These so-called true values are the values used to generate the realizations from the relevant Poisson process via R Codes 6.2 and 6.3. One can see the uncertainty of posterior estimates induced by the sample realizations.

6.4.2 Earthquakes in Italy

The second example of thinning (a marked Poisson process) is taken from Rotondi and Varini,[5] who used a Bayesian approach to model the occurrence and magnitude of earthquakes, and the approach was to analyze data in the Sannio–Matese–Ofanto–Irpinia region of Italy. In a series of important investigations, Vere-Jones,[6] Vere-Jones and Ozaki,[7] and Ogata[8] laid the foundation for the statistical analysis of earthquakes by modeling frequency of earthquakes as point processes.

In particular, our presentation is taken from Ruggeri,[9] who used a Bayesian approach to analyze data from the Sannio–Matese region of southern Italy. Exploratory data analysis by Bivand, Pebesma, and Gomez-Rubio[10] identified three distinct phenomena for the occurrence of earthquakes represented as a time series over the period of 1860–1980, and the data used here are for one strong quake as the main shock for a sequence lasting 1 week where the regions were divided into three geophysical homogenous areas. This is a multivariable model, consisting of three series, X, Y, and Z. X is the interarrival time between major earthquakes (a magnitude of at least 5 on the Richter scale), Y is the magnitude of the earthquake, and Z is the number of minor quakes that occurred in a given area since the previous major quake.

Suppose that the earthquake occurrences follow a Poisson process with parameter λ and that the chance of a major quake is p, and of a minor, $1 - p$; thus, there are two thinned process, one for the major quakes which is a Poisson process with parameter λp, while the Poisson process for the minor quakes has parameter $\lambda(1 - p)$.

It is obvious that the interarrival time X between major shocks has an exponential distribution with mean $1/\lambda p$, while the mean for the interarrival time between minor shocks is $1/\lambda(1 - p)$. Suppose that $X = x$ (the observed interarrival time between major quakes is x), it can be shown that Z is a Poisson process with parameter $x\lambda(1 - p)$ and that the marginal distribution of Z is geometric with parameter p. To see this, consider

$$P(Z = z) = \int_0^\infty P[Z = z|(X = x)]f(x)\,dx = p(1-p)^z, \quad z = 1, 2, \ldots, \tag{6.24}$$

which shows that Z has a geometric distribution with parameter p.

Assuming that Y is independent of X and Z, the distribution of Y is geometric with parameter μ, then assuming $\mu \approx 1$, the joint distribution of the three observations is

$$\begin{aligned} f(x,y,z) &= f(x)P(Y = y)P(Z = z|X = x) \\ &= \lambda e^{-\lambda x}[\lambda(1-p)x]^z \mu(1-\mu)^y / z! \end{aligned} \tag{6.25}$$

Thus, the likelihood function is determined as

$$l(p,\lambda,\mu) \propto \lambda^{n+\sum z_i} e^{-\lambda \sum x_i} p^n (1-p)^{\sum z_i} \mu^n (1-\mu)^{\sum y_i}, \tag{6.26}$$

where $0 < p < 1$, $0 < \mu < 1$, and $\lambda > 0$.

What should be chosen for the prior information about the three unknown parameters p, μ, and λ?

Based on Equation 6.26, the conjugate prior distributions for p, μ, and λ are independent beta, beta, and gamma distributions, respectively. Thus let, a priori,

$$p \sim \text{beta}(a_1, b_1), \quad \mu \sim \text{beta}(a_2, b_2), \quad \text{and} \quad \lambda \sim \text{gamma}(a_3, b_3).$$

Therefore, the corresponding posterior distributions are

$$\begin{aligned} &p|\text{data} \sim \text{beta}\left(n + a_1, b_1 + \sum_i z_i\right), \\ &\mu|\text{data} \sim \text{beta}\left(n + a_2, b_2 + \sum_i y_i\right), \\ &\text{and} \\ &\lambda|\text{data} \sim \text{beta}\left(n + \sum_i z_i + a_3, b_3 + \sum_i x_i\right). \end{aligned} \tag{6.27}$$

The preceding description of the earthquake analysis is from pages 126–130 of Insua, Ruggeri, and Wiper,[2] and in order to complete the analysis, the data in Table 6.5 are used.

TABLE 6.5

Data for Three Areas

	n	$\sum_{i=1}^{i=n} x_i$	$\sum_{i=1}^{i=n} y_i$	$\sum_{i=1}^{i=n} z_i$
Area 1	3	50.1306	9	713
Area 2	16	81.6832	53	1034
Area 3	14	118.9500	100	812

TABLE 6.6

Posterior Distributions for the Earthquakes in Italy

	p ~ **beta(,)**	μ ~ **beta(,)**	λ ~ **gamma(,)**
Area 1	5, 721	11, 11	718, 54.1036
Area 2	18, 1042	24, 55	1052, 85.6832
Area 3	16, 820	22, 102	828, 122.9500

How should the hyperparameters be assigned? According to page 129 of Insua, Ruggeri, and Wiper,[2] the hyperparameters are chosen on the basis of the following argument: A major earthquake occurs with probability p, and since major quakes are less likely than minor earthquakes, p is close to zero, consequently, $a_1 = 2$ and $b_1 = 8$, giving a prior mean of $E(p) = 1/5$. It is approximately true that major earthquakes occur every 10 years, and the investigators choose $b_3 = 4$ and $a_3 = 2$ to reflect that prior information. Now consider μ and remember that it is very close to the number 1; the corresponding hyperparameters are chosen as $a_2 = 8$ and $b_2 = 2$, implying that the prior mean is 4/5. When this prior information is combined with the posterior distributions (Equation 6.27), one can verify the posterior distribution of the three parameters for the three areas given by Table 6.6.

A more detailed analysis is provided with WinBUGS Code 6.4, and the code lists the posterior distribution of the three parameters of the three areas, and the analysis is executed with 45,000 observations for the simulation and a burn-in of 5,000:

WinBUGS Code 6.4

```
model;
{
p1~dbeta(5,721)
p2~dbeta(18,1042)
p3~dbeta(16,820)
mu1~dbeta(11,11)
mu2~dbeta(24,55)
mu3~dbeta(22,102)
lamda1~dgamma(718,54.1036)
lamda2~dgamma(1052,85.6832)
lamda3~dgamma(828,122.9500)
}
```

See Table 6.7 for the posterior analysis.

TABLE 6.7

Bayesian Analysis for Earthquakes in Italy

Parameter	Mean	SD	Error	2 1/2	Median	97 1/2
λ_1	13.27	0.493	0.00232	12.34	13.26	14.26
λ_2	12.28	0.3777	0.00193	11.55	12.27	13.02
λ_3	6.732	0.2352	0.00116	6.282	6.73	7.2
μ_1	.5003	.1041	.000541	.2974	.5006	.7008
μ_2	.3038	.05169	.000252	.2075	.3002	.4098
μ_3	.1775	.03412	.000169	.1154	.1758	.2489
p_1	.006895	.003083	.0000144	.002228	.006443	.0141
p_2	.01694	.003958	.0000198	.01006	.01666	.02552
p_3	.01911	.004731	.0000244	.01095	.01872	.02935
τ_1	.0915	.04103	.000191	.02492	.08545	.1869
τ_2	.208	.04898	.000245	.1231	.2042	.3146
τ_3	.1286	.03215	.000165	.0734	.126	.1987
$1/\tau_1$	13.68	7.868	0.03471	5.35	11.7	33.99
$1/\tau_2$	5.091	1.271	0.006428	3.179	4.896	8.124
$1/\tau_3$	8.294	2.222	0.01154	5.033	7.939	13.62
$1/\tau_1 - 1/\tau_2$	8.59	7.968	0.03551	−0.0422	6.722	28.97

Upon comparing Table 6.7 with Table 5.3 of page 130 of Insua, Ruggeri, and Wiper,[2] the agreement is remarkable, but note that Table 6.7 is much more informative. For example, in addition to the posterior mean and standard deviation, the posterior median and the upper and lower posterior of 2½ percentiles are reported. One can assess the symmetry of the posterior distributions, and one sees the first nine posterior distributions appear to be symmetric about their posterior means.

Also analyzed is the parameter of the interarrival time of major earthquakes, namely, $\tau_i = \lambda_i p_i$, $i = 1, 2, 3$; thus, the mean interarrival times between major earthquakes for area i is $\kappa_i = 1/\tau_i$. For example, with area 1, the posterior median is 11.7 years, an estimate of the time between consecutive earthquakes. On the other hand, the posterior median for the time between major earthquakes for area 2 is 4.896. Also, it is apparent that the three posterior distributions for the three times between consecutive earthquakes are skewed to the right. The gamma distribution appears to be a good approximation to these three posterior distributions.

Our next effort is directed toward determining the predictive distributions of the three series of observations X, Y, and Z. Consider the predictive mass function of $Y(n+1)$, the future value of the magnitude of the $(n+1)$st major earthquake; then we know that the conditional mass function is

$$f(y(n+1)|\mu) = \mu(1-\mu)^{y(n+1)}, \quad y(n+1) = 1, 2, 3, \ldots \tag{6.28}$$

and that the posterior distribution of μ is beta(11, 11); thus, the predictive density function is

$$f(y(n+1)|\text{data}) = f_1/f_2, \tag{6.29}$$

where

$$f_1 = (n+11)\Gamma\left(n+11+\sum_{i=1}^{i=n} y_i + 11\right)\Gamma\left(\sum_{i=1}^{i=n} y_i + 11 + y(n+1)\right) \tag{6.30}$$

and

$$f_2 = \Gamma\left(\sum_{i=1}^{i=n} y_i + 11\right)\Gamma\left(n+11+\sum_{i=1}^{i=n} y_i + 11 + y(n+1) + 1\right). \tag{6.31}$$

Note that the predictive mass function is pertinent for area 1 (see Table 6.6); thus, to compute the future mass function value, let $\sum_{i=1}^{i=n} y_i = 9$.

This section is concluded with a formal test of hypothesis about the mean interarrival times $\kappa_i = 1/\tau_i$, $i = 1, 2, 3$, between the three areas. For simplicity, consider the null hypothesis

$$H: \kappa_1 = \kappa_2 \quad \text{versus} \quad A: \kappa_1 \neq \kappa_2. \tag{6.32}$$

That is to say, the null hypothesis is that the interarrival time between major earthquakes between the first and second areas is the same versus the alternative that they are not the same. The probability of the null hypothesis is given by

$$p_0 = \pi_0 \int_0^\infty f(s_3, s_{16}|\kappa)\rho_0(\kappa)d\kappa / D, \tag{6.33}$$

where ρ_0 is the prior density of κ, the common value of κ_1 and κ_2 under the null hypothesis ($\kappa_1 = \kappa_2 = \kappa$), and π_0 is the prior probability of the null hypothesis with $\pi_1 = 1 - \pi_0$. Also S_3 and S_{16} are the waiting times for the 3rd and 16th major earthquakes of area 1 and area 2, respectively.

$$D = \int_0^\infty f(s_3, s_{16}|\kappa)\rho_0(\kappa)d\kappa + \pi_1 f(s_3, s_{16}), \tag{6.34}$$

where

$$f(s_3, s_{16}|\kappa) = \left[\kappa^{n_1+n_2} s_3^{n_1-1} s_{16}^{n_2-1} e^{-\kappa(s_3+s_{15})}\right] / \Gamma(n_1)\Gamma(n_2), \tag{6.35}$$

and the prior beta density under the null hypothesis is

$$\rho_0(k) = \beta^\alpha \kappa^{\alpha-1} e^{-\kappa\beta} / \Gamma(\alpha). \tag{6.36}$$

$$\int_0^\infty f(s_3, s_{16}|\kappa)\rho_0(\kappa)d\kappa = \left[\beta^\alpha \Gamma(n_1 + n_2 + \alpha)s_3^{n_1-1}s_{16}^{n_2-1}\right]/\Gamma(n_1)\Gamma(n_2)\Gamma(\alpha)(s_3 + s_{16})^{n_1+n_2+\alpha},$$

$$f(s_3, s_{16}) = \int_0^\infty \int_0^\infty f(s_3, s_{16}|\kappa_1, \kappa_2)\rho_1(\kappa_1, \kappa_2)\, d\kappa_1\, d\kappa_2, \tag{6.37}$$

where ρ_1 is the prior density of κ_1 and κ_2 under the alternative hypothesis. Note that

$$f(s_3, s_{16}|\kappa_1, \kappa_2) = \left[\kappa_1^{n_1} s_3^{n_1-1} e^{-s_3\kappa_1}/\Gamma(n_1)\right]\left[\kappa_2^{n_2} s_{16}^{n_2-1} e^{-s_{16}\kappa_2}/\Gamma(n_2)\right], \tag{6.38}$$

and the prior beta density under the alternative is

$$\rho_1(\kappa_1, \kappa_2) = \left[\beta_1^{\alpha_1}\kappa_1^{\alpha_1-1}e^{-\kappa_1\beta_1}/\Gamma(\alpha_1)\right]\left[\beta_2^{\alpha_2}\kappa_2^{\alpha_2-1}e^{-\kappa_2\beta_2}/\Gamma(\alpha_2)\right]. \tag{6.39}$$

Thus,

$$f(s_3, s_{16}) = \left[\beta_1^{\alpha_1}\beta_2^{\alpha_2}\Gamma(n_1 + \alpha_1)\Gamma(n_2 + 2)\right]/\Gamma(\alpha_1)\Gamma(\alpha_2)\Gamma(n_1)\Gamma(n_2)(s_3 + \beta_1)^{(n_1+\alpha_1)}(s_{16} + \beta_2)^{(n_2+\alpha_2)}$$

The preceding expressions (Equations 6.34 through 6.39) depend on known quantities, namely, known parameters of prior information or of sample information, and these are now listed.

The sample information consists of observing three major earthquakes in area 1 where the third occurred after 9 years, while for the second area, the waiting time for 16 quakes was 53 years,

$$S_3 = 9, \quad S_{16} = 53, \quad n_1 = 3, \quad \text{and} \quad n_2 = 16.$$

The following is prior information for the average interarrival times between earthquakes of area 1, $\alpha_1 = .11$ and $\beta_1 = .01$.

Note the prior mean for κ_1 is $\alpha_1/\beta_1 = 11$ with prior variance $\alpha_1/\beta_1^2 = 1100$.

The following is prior information for κ_2, the average time between major earthquakes of area 2. $\beta_2 = .01$ and $\alpha_2 = .1$.

Thus, $E(\kappa_2) = \alpha_2/\beta = .05/.01 = 5$, and $\mathrm{Var}(\kappa_2) = \alpha_2/\beta_2^2 = .05/(.01)^2 = 500$.

Lastly, the common values of the average interarrival time between major earthquakes are $\alpha = .07$, $\beta = .01$, $E(\kappa) = \alpha/\beta = .07/(.01) = 7$, and $\mathrm{Var}(\kappa) = \alpha/\beta^2 = .07/(.01) = 700$.

One must assign a value to π_0 which I take to be .5 because I am not sure which hypothesis, the null or alternative, is the true. Using this information, it is left to the reader to calculate the posterior probability of the null hypothesis p_0 (Equation 6.33).

6.5 Spatial Poisson Process

The spatial Poisson process is a generalization of the one-dimensional process studied in Section 6.4 to two or higher dimensions, and there are many examples: models of location of trees in a forest, the distribution of galaxies in the universe, and clusters of disease epidemics. Let the dimension $d > 1$ and the subset $A \subset R^d$. Suppose that the random variable $N(A)$ counts the number of points in subset A and denote $|A|$ as the size of A (in one dimension, $|A|$ would be a length; in two dimensions, an area; etc.); then the spatial Poisson process $\{N(A), A \subset R^d\}$ is defined as follows:

1. For each set A, $\{N(A), A \subset R^d\}$ is a spatial process with parameter $\lambda|A|$.
2. Whenever A and B are disjoint sets, $N(A)$ and $N(B)$ are independent random $\lambda|A|$.

Note how properties 1 and 2 generalize the Poisson process to higher dimensions, where 1 is the generalization of stationary increments and 2 is a generalization of independent increments.

Consider the following problem, in two dimensions with parameter $\lambda = 1/3$, then what is the probability that a circle of radius 2 centered at (3,4) contains five points?

Let C denote the circle, where $|C| = \pi r^2 = 4\pi$; then

$$P[N(C) = 5] = e^{-\lambda|C|}(\lambda|C|)^5/5! = e^{-(1/3)(4\pi)}(4\pi/3)^5/5! = .1629.$$

Recall how the uniform distribution arises with a Poisson process in one dimension, and one sees that it generalizes to the spatial case. Suppose $N(A) = n$, then the locations of the points in A are uniformly distributed in A. In order to simulate a spatial Poisson process, first simulate a Poisson process with parameter $\lambda|A|$, then generate n points uniformly distributed in A. R Code 6.4 generates a realization from a Poisson process with parameter $\lambda = 100$ on the unit square. Circle C inside the square is centered at (.7, .7) with radius $r = .2$. The simulation was repeated 100,000 times, counting the number of points in the circle at each location. Figure 6.2 depicts one realization of the simulation, and Tables 6.8 and 6.9 list the x and y coordinates of the points appearing in the square. Refer to Table 6.10 for the results of the simulation.

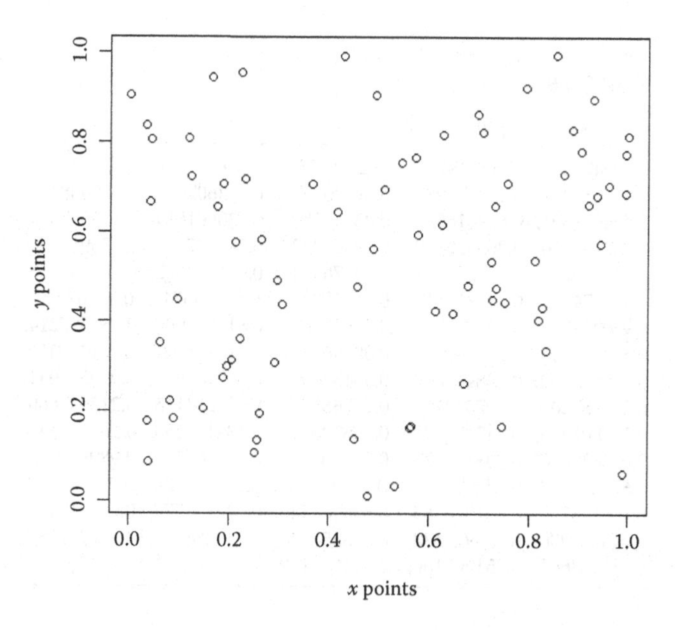

FIGURE 6.2
Spatial Poisson process $\lambda = 100$.

R Code 6.4

```
lamda<-100
> squarearea<-1
> trials<-100000
> simlist<-numeric(trials)

> for( i in 1:trials){
+ N<-rpois(1,lamda*squarearea)
+ xpoints<-runif(N,0,1)
+ ypoints<-runif(N,0,1)
+ ct<-sum((xpoints-0.7)^2+(ypoints-0.7)^2<=0.2^2)
+ simlist[i]<-ct}
> mean(simlist)
[1] 12.56054
> var(simlist)
[1] 12.58442
> xpoints
> ypoints

> lamda*pi*(0.2)^2
[1] 12.56637
> plot(xpoints,ypoints)
```

TABLE 6.8

Abscissa for Spatial Poisson

0.69947304,	0.752098439,	0.123763154,	0.296506146,	0.928317846,
0.429585261,	0.488405587,	0.697309565,	0.993607834,	0.532608426,
0.756079691,	0.826716592,	0.034607809,	0.730601819,	0.118687329,
0.852127976,	0.263830818,	0.793649223,	0.745337202,	0.176954050,
0.452190885,	0.038280711,	0.734769342,	0.901897682,	0.042760698,
0.493274997,	0.367916513,	0,81977386,	0.624886743,	0.867079208,
0.960224926,	0.725773072,	0.225712035,	0.818893997,	0.884597214,
0.577080285,	0.572040944,	0.08996827,	0.195316027,	0.189498151,
0.095972550,	0.038441649,	0.061324936,	0.307401637,	0.999510041,
0.257187202,	0.707372153,	0.477685743,	0.993814115,	0.213965860,
0.147804266,	0.045082209,	0.166208690,	0.188101258,	0.264188643,
0.613630638,	0.003161575,	0.810800111,	0.563908210,	0.989670485,
0.676237988,	0415593890,	0.545226286,	0.565639704,	0.723113632,
0.205693715,	0.291846363,	0.833905545,	0.222310579,	0.627019410,
0.918520986,	0.233912573,	0.933558643,	0.943388963,	0.647354387,
0.455108665,	0.251903168,	0.511865139		

TABLE 6.9

Ordinate Spatial Poisson

0.26124186, 0.44396077, 0.72392878, 0.49111736, 0.89497749, 0.99227313, 0.56237175, 0.86285194, 0.77451082, 0.03185146, 0.71030768, 0.43206807, 0.83520226, 0.65923869, 0.80783829, 0.99454636, 0.19521338, 0.92224793, 0.16692210, 0.65440660, 0.13786445, 0.17622265, 0.47331905, 0.78017173, 0.66753798, 0.90558875, 0.70708760, 0.22333211, 0.61457419, 0.72894909, 0.70386713, 0.44872583, 0.95518757, 0.40273958, 0.82697592, 0.59400583, 0.76483304, 0.18288699, 0.30049257, 0.27454940, 0.44961878, 0.08830324, 0.35170958, 0.43822516, 0.81478120, 0.13588946, 0.82128662, 0.01149270, 0.68743833, 0.57614086, 0.20674913, 0.80565632, 0.94538263, 0.70551711, 0.58287165, 0.42483037, 0.90379620, 0.53658911, 0.16208827, 0.06086748, 0.47875291 0.64352514, 0.75371784, 0.16645801, 0.53436726, 0.31414601, 0.30729664, 0.33627592, 0.36142669, 0.81547725, 0.66096427, 0.71732857, 0.67960241, 0.57335401, 0.41865694, 0.47824777, 0.10624367, 0.69410526

TABLE 6.10

Number of Points in Target Circle C

Counts	0–4	5–9	10–14	15–19	20–24	25–29
	522	19,200	50,028	24,975	3,135	106

The reader should refer to pages 249–252 of Dobrow[1] for additional information about this simulation.

Of course, spatial statistics is an active area of research, and for additional details, refer to pages 249–252 of Bivand, Pebesma, and Gomez-Rubio;[10] and for a Bayesian perspective, to Blangiardo and Camdelli.[11] The R package spatstat is very useful for additional ideas about the simulation of spatial processes similar to the Poisson. See the technical report from the University of California, Los Angeles (http://scc.stat.ucla.edu/page_attachments/0000/0094/spatial_R_1_09S.pdf.).

For example, R Code 6.5 is another way to simulate a spatial Poisson process with intensity function $f(x, y) = 50(x^2 + y^2)$. One must employ the R package spatstat.

A plot of the simulation is portrayed in Figure 6.3.

R Code 6.5

```
pp1<-rpoispp(function(x,y) {50*(x^2+y^2) )})
plot(pp1)
```

From a Bayesian viewpoint, how would one estimate the parameter λ of a spatial Poisson process with parameter λ with probability mass function

$$\Pr[N(A) = k] = (\lambda|A|)^k \exp(-\lambda|A|)/k!, \tag{6.40}$$

where A is a subset of R^n and the known $|A|$ is the "size" of A.

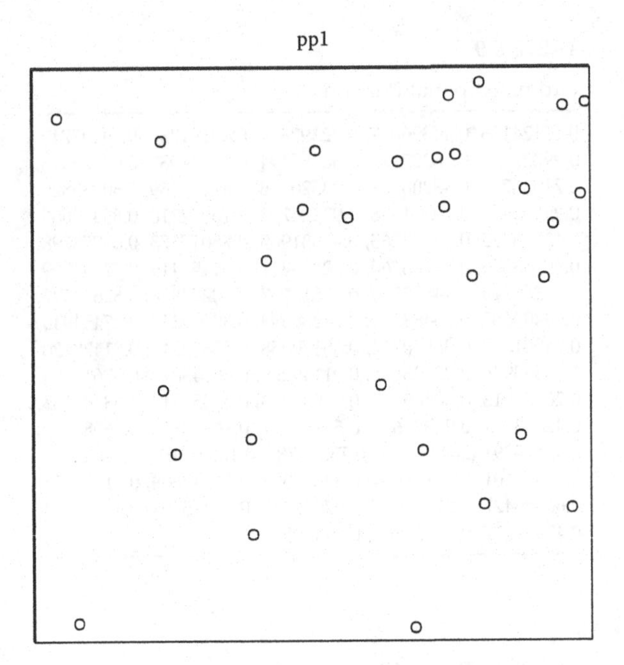

FIGURE 6.3
Spatial Poisson process with intensity $f(x,y) = 50(x^2 + y^2)$.

If one combines the improper prior density

$$g(\lambda) = 1/\lambda, \quad \lambda > 0 \tag{6.41}$$

with the likelihood function (Equation 6.40), then via the Bayes theorem, the posterior distribution of λ is gamma with parameters k (the observed number of points appearing in subset A) and the size of A, namely, $|A|$. Recall the previous simulation where $\lambda = 100$ and A is the circle centered at (.7,.7) with radius A; thus, $|A| = \pi r^2 = .12566$.

I used WinBUGS Code 6.5 for the Bayesian estimation of λ with 45,000 observations for the simulation and a burn-in of 5,000. Referring to Table 6.10, the number of points in A is varied from 1 to 20.

WinBUGS Code 6.5

```
model;
{
for ( k in 1:20){

delta[k]~dgamma(k,.125)}
}
```

Table 6.11 reports the posterior distribution of λ for various values of k, the observed number of points appearing in the subset A.

As k varies from 1 to 20, the posterior median of λ varies from 5.544 to 157.5. Recall that the value of λ used for the simulation is 100; thus if in fact, the observed value of k is 13; the

TABLE 6.11

Posterior Distribution of the Poisson Rate Parameter λ

k	Mean	SD	Error	2 1/2	Median	97 1/2
1	8	8.028	0.04148	0.202	5.544	29.65
2	16.09	11.45	0.05355	1.954	13.44	45.06
3	23.93	13.85	0.06663	4.884	21.3	57.77
4	31.98	15.98	0.07236	8.732	29.35	70.13
5	39.92	17.86	0.09785	12.95	37.27	82.21
6	48.01	19.64	0.09033	17.51	45.37	93.82
7	55.96	21.22	0.1043	22.52	53.23	104.6
8	63.94	22.51	0.1128	27.74	61.44	115.1
9	72.12	23.97	0.135	33.16	69.41	126.4
10	79.89	25.12	0.1241	38.66	77.31	136.4
11	88.08	26.44	0.135	44.56	85.43	147.1
12	95.98	27.82	0.1291	49.54	93.19	157.8
13	104.1	28.91	0.133	55.07	101.6	167.7
14	112.2	29.9	0.1393	61.64	109.6	177.6
15	120.2	31.1	0.1507	67.41	117.4	188.8
16	127.8	31.9	0.1634	73.3	125.1	198
17	135.9	33.08	0.1586	78.86	133.4	207.8
18	143.9	33.81	0.1684	85.08	141.2	216.7
19	151.8	34.81	0.1787	91.46	149.2	227
20	160.1	35.77	0.1647	96.98	157.5	236.8

posterior mean is 101.6, a "good" estimate of λ. On the other hand, if the observed value of k is 3, the posterior median is 21.3, somewhat far away from the true value $\lambda = 100$. This shows how the sample variation affects the posterior median on the observed number of points in the circle! Remember when $\lambda = 100$, and the preceding spatial Poisson process has parameter $|A| = .1256$ per unit area.

The last stage for Bayesian inferences for spatial Poisson processes is to test the hypothesis

$$\mathrm{H}:\lambda = 100 \quad \text{versus} \quad \mathrm{A}:\lambda \neq 100 \tag{6.42}$$

Recall the case discussed previously where simulations were generated from a spatial Poisson process with $\lambda = 100$. See Figure 6.2 for a graph of the simulations, Table 6.10 for the counts of the number of points in the target circle of radius .2, and, finally, the estimation of λ depicted in Table 6.11. All this information is needed to implement the Bayesian test of H versus A.

The posterior probability of the null hypothesis is

$$p_0 = \pi_0 g(k|\lambda = 100)/[\pi_0 g(k|\lambda = 100) + \pi_1 g_1(k)], \tag{6.43}$$

where

$$g(k|\lambda = 100) = \left[(100|A|)^k \exp(-100|A|)\right]/k!, \tag{6.44}$$

$$g_1(k) = \int_0^\infty \zeta_1(\lambda) g(k|\lambda)\, d\lambda, \tag{6.45}$$

$$g(k|\lambda) = \left[(\lambda|A|)^k \exp(-\lambda|A|)\right]/k!, \tag{6.46}$$

and

$$\zeta_1(\lambda) = \beta^\alpha \lambda^{\alpha-1} e^{-\lambda\beta}/\Gamma(\alpha)$$

is the prior density of λ under the alternative hypothesis ($\lambda \neq 100$). Thus, one must choose the hyperparameters α and β. If one chooses an improper prior for, λ, namely, $\zeta_1(\lambda) = 1/\lambda$, one can show

$$g_1(k) = 1/k. \tag{6.47}$$

When $k = 13$, it can be shown that

$$g(13|\lambda = 100) = .109127284 \tag{6.48}$$

and

$$g_1(13) = .076923072; \tag{6.49}$$

thus,

$$p_0 = .5865, \tag{6.50}$$

which implies that the null hypothesis is indeed plausible. In the exercises, the reader will be asked to compute p_0 for values of $k = 1, 2, \ldots, 20$.

This concludes the presentation of Bayesian methods for estimating the parameter of a spatial Poisson process.

6.6 Concomitant Poisson Processes

Three versions of concomitant Poisson processes are considered: (1) independence, (2) complete similarity, and (3) partial similarity.

Suppose k Poisson processes $N_i(t)$ with parameters λ_i, $i = 1, 2, \ldots, k$, where n_i events are observed over the interval $(0, t_i]$. The processes could be related. For example, consider the case where the different processes correspond to different intersections in a large city, where for the ith intersection, $N_i(t)$ counts the number of accidents at intersection i, and the average number of accidents per day (over a 24-hour period) is denoted by λ_i. And finally consider the case where enough is known about the network of intersections, such as their proximity to each other and their location to busy businesses, etc. In large cities traffic communication centers continuously monitor the accidents at each intersection in the networks. Thus, there are k homogeneous Poisson processes, and we present Bayesian inferences about the parameters λ_i for the cases, namely, of independence of and complete and partial similarity between the k processes.

6.6.1 Independence

This is the simplest case where the intersections are far enough away from each other, and one would not expect the accidents at intersection i to affect those at intersection j. Thus, the usual inferential procedures would be repeated for each of the k intersections; that is, if one uses a gamma prior for λ_i, the posterior is also gamma. See Section 6.3.

6.6.2 Complete Similarity

Suppose the k processes have the same parameter λ, then the likelihood function is

$$L(\lambda|\text{data}) \propto \prod_{i=1}^{i=k} (\lambda t_i)^{n_i} \exp(-\lambda t_i) = \lambda^{\sum_{i=1}^{i=k} n_i} \exp\left(-\lambda \sum_{i=1}^{i=k} t_i\right), \quad \lambda > 0.$$

Therefore, if the prior distribution for λ is gamma (α,β), the posterior distribution of λ is also gamma $(\sum_{i=1}^{i=k} n_i + \alpha, \sum_{i=1}^{i=k} t_i + \beta)$. With an improper prior for λ, its posterior distribution is gamma $(\sum_{i=1}^{i=k} n_i, \sum_{i=1}^{i=k} t_i)$ and a posterior mean of $E(\lambda|\text{data}) = \sum_{i=1}^{i=k} n_i / \sum_{i=1}^{i=k} t_i$.

Suppose that over a 24-hour period, one expects to see on the average of seven accidents at six intersections of the city. Let us generate data for the six intersections over a 10-day period.

For example, R Code 6.6 generates observations for an intersection over a 10-day period.

R Code 6.6

```
> t<-10
> lamda<-7
> N<-rpois(1,lamda*t)
> unifs<-runif(N,0,t)
> arrivals<-sort(unifs)
> arrivals
```

See the following for the hourly arrival times for 10 days:

- `Day 1: 0.1917592, 0.2455052, 0.4200395, 0.6920433, 0.7440332, 0.7820917, 0.8121036`
- `Day 2: 1.1804281, 1.2044852, 1.2753514`
- `Day 3: 2.0234064, 2.1679194, 2.3384486, 2.3742640, 2.5719163, 2.6123924, 2.6495179, 2.6508241, 2.8024459`
- `Day 4: 3.4194264, 3.5928215, 3.9179758`
- `Day 5: 4.0691570, 4.6938118, 4.7803756, 4.7994749, 4.9597395`

- Day 6: 5.1070871, 5.2299435, 5.5519447, 5.6570565, 5.9559313
- Day 7: 6.0626929, 6.4611704, 6.6069061, 6.7669566, 6.8498875, 6.9969702
- Day 8: 7.0691965, 7.3914822, 7.3918613, 7.4236556, 7.8239601. 7.8324599. 7.8683294
- Day 9: 8.3143937, 8.7440850, 8.747061
- Day 10: 9.1380792, 9.1626362, 9.1643527, 9.3776940, 9.5955433, 9.8207425, 9.908292

Thus, for day 10, there were seven accidents with the first accident occurring at .13 (of a 24-hour day), and the last, at .908 of a 24-hour day = 21.79 hours. Over the 10-day period, there were five accidents where the last accident occurred at hour 9.9 of day 10; thus, the usual estimate of λ is 55/9.9 = 5.6 accidents per hour.

If one uses an improper prior for λ, the posterior distribution for λ is gamma (56,9.9).

The Bayesian analysis is executed with 35,000 observations for the simulation with a burn-in of 5,000, and the results are reported in Table 6.12.

Recall that the accident count was generated with λ = 12, but the posterior median is 5.515; however, the 95% credible interval (4.182,7.124) does contain 7.

6.6.3 Partial Similarity

Now suppose the intersections have different accident rates, but the rates share a common prior distribution, which is a typical case of partial similarity. Such a situation is easily analyzed by the Bayesian approach. The approach on pages 115 and 116 of Insua, Ruggeri, and Wiper[2] is employed, which is as follows:

$$\begin{aligned} &N_i(t_i) \sim \text{Poisson}(\lambda_i t_i), \\ &\lambda_i | \alpha, \beta \sim \text{gamma}(\alpha, \beta),\ i = 1, 2, .., k, \end{aligned} \tag{6.51}$$

and

$$\begin{aligned} &\alpha \sim \text{gamma}(.001, .001), \\ &\beta \sim \text{gamma}(.001, .001). \end{aligned}$$

The Bayesian analysis will be based on observations generated directly from a Poisson process. WinBUGS Code 6.6 considers five intersections with accident rates of 5.5, 6, 6.5, 7, and 7.5 accidents per day. I chose these rates because the intersection are all on the same street and follow one after the other. The intersection with the largest accident rate is an area that has the most business activity. Also, the accidents were followed for 10 consecutive days.

TABLE 6.12

Posterior Distribution for the Accident Rate

Parameter	Mean	SD	Error	2 1/2	Median	97 1/2
λ	5.554	0.7511	0.003515	4.182	5.515	7.124

WinBUGS Code 6.6

```
model;
{
for ( i in 1:10){

n1[i]~dpois(t1[i])
n7[i]~dpois(t7[i])
n7.5[i]~dpois(t7.5[i])
n5.5[i]~dpois(t5.5[i])
n6[i]~dpois(t6[i])
n6.5[i]~dpois(t6.5[i])
t5.5[i]<-5.5*t1[i]
t6[i]<-6*t1[i]
t6.5[i]<-6.5*t1[i]
t7[i]<-7*t1[i]
t7.5[i]<-7.5*t1[i]

}}

list(t1=c(1,2,3,4,5,6,7,8,9,10))
```

In the following are the daily accidents for 10 days at five intersections:

- Intersection 1 with an average of $\lambda = 5.5$ accidents per day:
  2.0,13.0,19.0,23.0,34.0,
  29.0,39.0,38.0,54.0,61.0.
  Thus, for intersection 1, on day 1, there are 2 accidents, while on day 10, the total number is 61. The usual estimate is $61/10 = 6.1$.

- Intersection 2 with $\lambda = 6$ accidents per day:
  8.0,19.0,19.0,27.0,29.0,
  34.0,45.0,53.0,51.0,60.0.

- Intersection 3 with $\lambda = 6.5$:
  6.0,17.0,27.0,27.0,23.0,
  52.0,37.0,57.0,56.0,71.0.

- Intersection 4 with $\lambda = 7$:
  8.0,8.0,16.0,29.0,35.0,
  31.0,46.0,59.0,62.0,65.0.

- Intersection 5 with $\lambda = 7.5$:
  7.0,15.0,21.0,36.0,43.0,
  35.0,62.0,76.0,61.0,70.0.

Therefore, at day 1, there are 7 accidents, while at day 10, the total at intersection 5 is 70, providing the usual estimate of $70/10 = 7$ compared to the value of 7.5 used to generate the observations.

The posterior analysis is implemented with WinBUGS Code 6.7 using 35,000 observations for the simulation with a burn-in of 2,000. Consider intersection 1; then the week 10 observation is 61, and the corresponding mass function is

$$P[N_1(10) = 61|\lambda_{5.5}] = \left(\lambda_{5.5}^{10}/10!\right)\exp(-\lambda_{5,5}10), \quad \lambda_{5.5} > 0. \tag{6.52}$$

Then this is combined with the improper prior

$$g(\lambda_{5.5}) = 1/\lambda_{5.5}, \quad \lambda_{5.5} > 0. \tag{6.53}$$

The resulting posterior distribution is gamma (61,10). See the following code, which specifies the five posterior distributions corresponding to the five intersections:

WinBUGS Code 6.7

```
model;
{

lamda5.5~dgamma(61,10)
lamda6~dgamma(60,10)
lamda6.5~dgamma(71,10)
lamda7~dgamma(65,10)
lamda7.5~dgamma(70,10)
}
```

The Bayesian posterior analysis is reported in Table 6.13.

For example, when the 10 Poisson observations are generated with rate $\lambda = 7.5$ (corresponding to intersection 5), the posterior median estimate of the rate is 6.659 accidents per day, with a 95% credible interval of (5.471,8.752). This estimate should be compared to $\lambda = 7.5$.

TABLE 6.13

Posterior Distribution for Five Intersections for 10 Days

Parameter	Mean	SD	Error	2 1/2	Median	97 1/2
$\lambda = 5.5$	6.098	0.7837	0.005202	4.688	6.062	7.724
$\lambda = 6$	6.002	0.7742	0.00533	4.585	5.967	7.622
$\lambda = 6.5$	7.096	0.8455	0.0059	5.522	7.062	8.858
$\lambda = 7$	6.491	0.8092	0.005514	5.014	6.453	8.173
$\lambda = 7.5$	7.001	0.8378	0.0059	5.471	6.659	8.752

6.6.4 Poisson Process with Covariates

Covariates can be incorporated into the Poisson model in a variety of ways, but only two are considered here, namely, (1) the direct approach and (2) as part of the prior distribution of the rate λ.

The goal is to determine relationships between several Poisson processes via their covariates. For example, consider the previous example concerning traffic accidents at several intersections. We may want to see what factors (covariates) affect the accident rates between intersections. If the two intersections share the same covariates, one would compare the two accident rates by comparing the effects of the covariates on the accident rates. A simple case is presented: consider two intersections 1 and 2 with a common covariate, say, the population density of the neighborhoods, which encompass those neighborhoods (an area surrounding the intersections). Let $\{N_1(t), t > 0\}$ and $\{N_2(t), t > 0\}$ be the Poisson processes that counts the number of accidents for intersections 1 and 2 with accident rates λ and $\lambda\mu$, respectively. Thus, the parameter μ modifies the accident rate λ, and one would think that λ and $\lambda\mu$ reflect the population density of intersections 1 and 2, respectively. Suppose that n_1 accidents occur in intersection 1 over the time interval $(0, t_1)$ and n_2 accidents over $(0, t_2)$; then the likelihood function is

$$L(\lambda, \mu | n_1, n_2) \propto (\lambda t_1)^{n_1} e^{-\lambda t_1} (\lambda \mu t_2)^{n_2} e^{-\lambda \mu t_2}. \tag{6.54}$$

For the Bayesian approach, it is convenient to employ gamma(α, β) and gamma(δ, γ) prior distributions which implies then following conditional posterior distributions:

$$\begin{aligned} &\lambda | \mu, \text{data} \sim \text{gamma}(\alpha + n_1 + n_2, \beta + t_1 + \mu t_2) \\ &\text{and} \\ &\mu | \lambda, \text{data} \sim \text{gamma}(\delta + n_2, \gamma + \lambda t_2). \end{aligned} \tag{6.55}$$

I will use information about the accident rates for intersections 1 and 5, where for intersection 1 there are $n_1 = 61$ over a $t_1 = 10$ day period, and for the other intersection, there are $n_2 = 70$ over a $t_2 = 10$ day period. For prior information, I used the vague gamma prior with $\alpha = \beta = \delta = \gamma = .01$.

The posterior distributions are now determined by Equation 6.55, and WinBUGS Code 6.8 will implement the Bayesian analysis for estimating λ, μ, and $\lambda\mu$.

WinBUGS Code 6.8

```
model;
{
lamda~dgamma(z1,z2)
z1<-131.01
z2<-10.01+mu*10
mu~dgamma(v1,v2)
v1<-70.01
v2<-.01+lamda*10
}
list(lamda =6,mu=2)
```

TABLE 6.14
Posterior Distribution with Covariates

Parameter	Mean	SD	Error	2 1/2	Median	97 1/2
λ	6.103	0.7751	0.007056	4.669	6.071	7.711
μ	1.165	0.2051	0.00184	0.8157	1.147	1.619
$\lambda\mu$	6.995	0.8376	0.0034	5.445	6.598	8.735

The Bayesian analysis, via WinBUGS Code 6.8, is executed with 45,000 observations for the simulation and a burn-in of 5,000, and the Bayesian analysis is reported in Table 6.14.

The effect of the covariate (population density) is to increase the daily accident rate 6.103 (using the posterior mean) of the first intersection by an amount of 1.165 (using the posterior mean), resulting in a daily accident rate of 6.995 (using the posterior mean) for the other intersection. Are these estimates plausible?

Now another way to include covariates into the Poisson process is with the prior distribution for the rate λ. The model described on page 117 of Insua, Ruggeri, and Wiper[2] considers k Poisson processes $\{N_i(t), t > 0\}$, $i = 1,2,\ldots,k$, where the ith has m covariates $X_i = (X_{i1}, X_{i2}, \ldots, X_{im})$, then the covariates are included in the model as follows:

$$\begin{aligned} N_i(t_i)|\lambda_i &\sim \text{Pois}(\lambda_i t_i),\ i = 1,2,..,k, \\ \lambda_i|\alpha,\beta &\sim \text{gamma}(\alpha\exp(X_i\beta,\alpha)), \end{aligned} \tag{6.56}$$

where α is a scalar and β is an $m \times 1$ unknown vector of regression parameters. Of course, this form of the prior for the λ_i is adopted because the prior mean is

$$E(\lambda_i) = \exp(X_i\beta). \tag{6.57}$$

Our goal is to determine the posterior distribution of α and β. For simplicity, let

$$E(\lambda_i) = \exp(\beta_0 + \beta_1 x_i), \quad i = 1,2; \tag{6.58}$$

that is, there are two intersections, and λ_i is the daily accident rate for intersection i, and x_i is the corresponding population total of the neighborhood (a circle with center at the intersection with a radius of a quarter mile) of intersection. I assume that for intersection 1, the population in the neighborhood is 12,000 persons and 36,000 for the second intersection.

Based on WinBUGS Code 6.9, the student will be asked to verify the Bayesian analysis for estimating β_0 and β_1. I assume for the first exercise that at day 10, there were a total of 61 accidents, while for intersection 2, there were a total of 80 accidents over a 10-day period.

WinBUGS Code 6.9

```
model;
{
# posterior distribution of the two accident rates
lamda1~dgamma(61,10)
lamda2~dgamma(80,10)
# see equations (6.57) and (6.58) for solutions for beta0 and beta1
beta1<-(log(lamda2)-log(lamda1))/24000
beta0<-log(lamda1)-beta1*12000
}
```

TABLE 6.15

Posterior Distribution for β_0 and β_1

Parameter	Mean	SD	Error	2 1/2	Median	97 1/2
β_0	1.633	0.200089	0.000948	1.259	1.667	2.048
β_1	.0000113	.0000071	10^{-8}	10^{-6}	.0000132	.000025

One would conclude that the slope is 0; that is, that the population density does not affect the accident rates at the two intersections (Table 6.15).

6.7 Nonhomogeneous Processes

6.7.1 Intensity Function

To understand nonhomogeneous Poisson process, one must understand the concept of its intensity function, which is defined as

$$\lambda(t) = \lim P(N(t, t+\Delta t) \geq 1)/\Delta t, \tag{6.59}$$

where the limit is as $\Delta t \to 0$, and $N(t)$ has a Poisson distribution. It can be shown that

$$P[N(s,t)] = n] = \left[\left(\int_s^t \lambda(x)\mathrm{d}x\right)^n \exp\left(-\int_s^t \lambda(x)\mathrm{d}x\right)\right]/n!. \tag{6.60}$$

Also, note that the mean value function of the process is

$$E[N(t)] = m(t)] = \int_0^t \lambda(x)\mathrm{d}x, \quad t > 0 \tag{6.61}$$

Consequently,

$$E[N(s,t)] = m(s,t)] = \int_s^t \lambda(x)\mathrm{d}x, \quad s < t,\ s,t > 0, \tag{6.62}$$

and if m is differentiable, it follows from Equation 6.62 that

$$\mathrm{d}m(t)/\mathrm{d}t = \lambda(t). \tag{6.63}$$

Since the intensity function and the mean value function can vary over time, in general, the nonhomogeneous Poisson process does not have stationary increments; however, it does have independent increments, and the superposition (the sum of independent

Poisson process has a Poison distribution) and valid coloring theorem (when the events occur with a set of multinomial probabilities) are valid.

The goal of this section is to develop Bayesian inferences for the parameters of the mean value function or, equivalently, the intensity function. Of special interest is the intensity function

$$\lambda(t) = M\beta t^{\beta-1}, \tag{6.64}$$

where M and β are unknown positive parameters.

Note that if $M = 1$ and $\beta = 1$, the Poisson process is homogenous, but the intensity function is quite flexible in that it can represent increasing, decreasing, convex, and concave properties.

Corresponding to $\lambda(t)$ is the mean value function

$$m(t) = \int_0^t \lambda(x)dx = \int_0^t M\beta t^{\beta-1} = Mt^{\beta}, \quad t > 0. \tag{6.65}$$

Consider observations taken at times $t_1 < t_2 < \ldots < t_k$ with corresponding counts $N(t_i) = n_i, \quad i = 1, 2, \ldots, k,$ then

$$P[N(t_i) = n_i] = \{[m(t_i)]^{n_i} \exp(-m(t_i))\}/n_i!. \tag{6.66}$$

Now use the mean value function (Equation 6.65); then Equation 6.66 reduces to

$$P[N(t_i) = n_i] = M^{n_i} t_i^{\beta n_i} \exp\left(-Mt_i^{\beta}\right)/n_i!. \tag{6.67}$$

See Ntzoufras[12] for making Bayesian inferences for the parameters of nonstandard distributions, such as the nonhomogenous Poisson process (Equation 6.67) with the power law intensity function. Page 275 of Ntzoufras[12] shows how one can provide Bayesian inferences using an approximation to the likelihood function. This approach to approximating the likelihood function will be demonstrated in the next section.

6.7.2 Choosing the Intensity Function

An example involving reliability will illustrate how one can choose the appropriate intensity function. At different times, many systems are subject to reliability decay or growth, and the intensity function is chosen on the basis of exploratory data analysis and prior information about the study at hand. In order to choose the intensity function and, hence, the corresponding mean value function, it is helpful to assess whether the expected number of failures over time becomes infinite or remains at some asymptotic level. If the mean value function remains finite as time increases, then Cox–Lewis[13] nonhomogeneous Poisson process with mean value function

$$m(t) = M\left(1 - e^{-\beta t}\right)/\beta \tag{6.68}$$

is a possible alternative; however, as stated on page 205 of Insua, Ruggeri, and Wiper,[2] if the mean value function becomes unbounded as time increases, a possible alternative is the Musa–Iannino–Okumoto[14] mean value function:

$$M(t) = M\log(t + \beta). \tag{6.69}$$

As mentioned earlier, a very flexible intensity function is the power law:

$$\lambda(t) = M\beta t^{\beta-1} \tag{6.64}$$

which includes a variety of behavior, including constant, growth, or decaying reliability. For example, when $\beta = 1$, the reliability is constant, and if $\beta > 1$, reliability decays, and, finally, when $0 < \beta < 1$, there is reliability growth.

When $0 < \beta < 1$, the nonhomogeneous Poisson process is useful in detecting software bugs, assuming new bugs are not introduced during testing. For our purpose, an example described on pages 207 and 208 of Insua, Ruggeri, and Wiper[2] is adopted. The intensity function exhibits a bathtub behavior for the reliability, which is modeled as a change point with three stages. The points y_1 and y_2 determine the three intervals $I_1 = (0, y_1]$, $I_2 = (y_1, y_2]$, and $I_3 = (y_2, y]$ with the same parameter M, but different β_1, β_2, and β_3. Assume that there are n observations $T_1, T_2, \ldots, T_n$ in the interval $(0, y]$ where n_i is the count in interval i for $i = 1$, 2, 3. The likelihood function is

$$l(M, \beta_1, \beta_2, \beta_3 | \text{data}) = l_1(M, \beta_1) l_2(M, \beta_2) l_3(M, \beta_3), \tag{6.70}$$

where

$$l_1(M, \beta_1 | \text{data}) = M^{n_1} \beta_1^{n_1} \prod_{i=1}^{i=n_1} T_i^{\beta_1 - 1} \exp\left(-M y_1^{\beta_1}\right), \tag{6.71}$$

$$l_2(M, \beta_2 | \text{data}) = M^{n_2} \beta_2^{n_2} \prod_{i=n_1+1}^{i=n_2} T_i^{\beta_2 - 1} \exp\left(-M y_1^{\left(y_2{}^{\beta_2} - y_1^{\beta_2}\right)}\right), \tag{6.72}$$

and

$$l_3(M, \beta_3 | \text{data}) = M^{n_3} \beta_3^{n_3} \prod_{i=n_1+n_2+1}^{i=n_3} T_i^{\beta_3 - 1} \exp\left(-M y_1^{\left(y_1^{\beta_2} - y_2^{\beta_3}\right)}\right). \tag{6.73}$$

As an example, the following data points are generated according to the posterior distributions given by Equation 6.70 through 6.72 using the following code:

R Code 6.7

```
M<-1
> beta<-.4
> t<-30
> N<-rpois(1,lamda)
> unifs<-runif(N,0,t)
> arrivals<-sort(unifs)
> arrivals
```

When the mean value function is

$$m(t) = t^4, \quad 20 < t < 30, \tag{6.74}$$

there are only two events with arrival times at 20.54698, 22.14968.

R Code 6.8

```
> t<-20
> M<-1
> beta<-1
> lamda<-M*t**beta
> N<-rpois(1,lamda)
> unifs<-runif(N,0,t)
> arrivals<-sort(unifs)
> arrivals
```

When the mean value function is

$$m(t) = t, \quad 10 < t < 20, \tag{6.75}$$

there are 10 events with the following arrival times:

```
11.570835, 12.520865, 12.714158, 13.288865, 15.761193
16.741561, 16.948549, 17.125616, 18.301101, 18.812169.
```

R Code 6.9

```
> t<-10
> M<-1
> beta<-2
> lamda<-M*t**beta
> N<-rpois(1,lamda)
> unifs<-runif(N,0,t)
> arrivals<-sort(unifs)
> arrivals
```

When the mean value function is

$$m(t) = t^2, \quad 0 < t < 10, \tag{6.76}$$

there are 94 events with the following arrival times:

```
0.01929614, 0.16683374, 0.30868937, 0.45783466, 0.45783841,
0.60971629, 0.69073963, 0.72312287, 0.73048354, 0.78726970,
0.87282923, 1.11631418, 1.19893630, 1.24930384, 1.29853977,
1.86708471, 1.99640772, 2.09816973, 2.31948325, 2.31964980,
2.31978450, 2.32668238, 2.38249653, 2.39276180, 2.40731530,
2.51792069, 2.69428169, 2.86724672, 2.93153171, 2.95547242,
3.09350756, 3.11019411, 3.41164040, 3.48050033, 3.51716059,
3.52967131, 3.66419103, 3.70468339, 3.73201358, 3.76651419,
```

```
3.77171997, 3.80131188, 3.83412594, 4.27709175, 4.32951888,
4.33360442, 4.45838177, 4.46452007, 4.67891457, 4.72167716,
4.81387984, 5.00014701, 5.24055764, 5.30405366, 5.45888764,
5.49390880, 5.52179333, 5.66381047, 6.16508507, 6.24032858,
6.24427960, 6.29753330, 6.30779498, 6.31612410, 6.41120937,
6.41448096, 6.43383395, 6.77219317, 6.84154629, 6.90076633,
6.92867827, 6.93095933,7.10138919, 7.26750958, 7.27311966,
7.33515741, 7.50370963, 7.52847685,7.67226142, 7.69301810,
7.73796714, 7.78828417, 8.08008341, 8.42653468,
8.42955573, 8.43619858, 8.50564013, 8.70426517 8.74478266
8.80415160, 9.0581363, 9.27984186, 9.28731328, 9.35034508
9.51775680, 9.74506902, 9.96410355, 9.98975439.
```

Thus, this is the bathtub sort of intensity function with three stages over the three intervals (0,10], (11,20], and (21,30], where there are 94 observations in the first, 11 in the second, and 2 in the third. See Ntzoufras[12] for making Bayesian inferences for the parameters of nonstandard distributions given by the likelihood function (Equation 6.70). The Bayesian estimation for M, β_1, β_2, and β_3 is based on WinBUGS Code 6.10 with the code that closely follows Equations 6.70 through 6.73.

I executed the analysis with 45,000 observations for the simulation and 5,000 for the burn-in. See Table 6.16 for the posterior analysis.

WinBUGS Code 6.10

```
model;
{
am<-1
bm<-1
n<-108
#see (6.70)
M~dgamma(c1,c2)
c1<-n+am
c2<- bm+pow(10,beta1)+
pow(20,beta2)-pow(10,beta2)+pow(30,beta3)-pow(20,beta3)
# the prior distributions for the regression coefficients
beta1~dgamma(.5,1)
beta2~dgamma(1,1)
beta3~dgamma(2,1)

a1<-1
b1<-1
n1<-94
# see (6.71)
# this is the log likelihood function M and beta1
for ( i in 1:95){
l1[i]<-(a1+n1-1)*log(beta1)-b1*beta1-
M*pow(10,beta1)+beta1*log(T[i])}
n2<-11
a2<-1
```

```
b2<-1
# see (6.72)
# this is the log likelihood function for M and beta2
for( i in 96:107){
l2[i]<-(a2+n2-1)*log(beta2)-b2*beta2-M*(pow(20,beta2)-
pow(10,beta2))+beta2*log(T[i])}
a3<-1
b3<-1
# see (6.73)
# this is the likelihood function for M and beta3
for ( i in 108:110){
l3[i]<-(a3+n-n1-n2-1)*log(beta3)-b3*beta3-M*(pow(30,beta3)-
pow(20,beta3))+beta3*log(T[i])}
}
list(T= c(0.01929614, 0.16683374, 0.30868937, 0.45783466,
0.45783841, 0.60971629, 0.69073963, 0.72312287, 0.73048354,
0.78726970, 0.87282923, 1.11631418, 1.19893630, 1.24930384,
1.29853977, 1.86708471, 1.99640772, 2.09816973, 2.31948325,
2.31964980, 2.31978450, 2.32668238, 2.38249653, 2.39276180,
2.40731530, 2.51792069, 2.69428169, 2.86724672, 2.93153171,
2.95547242, 3.09350756, 3.11019411, 3.41164040,3.48050033,
3.51716059, 3.52967131, 3.66419103, 3.70468339, 3.73201358,
3.76651419, 3.77171997, 3.80131188, 3.83412594, 4.27709175,
4.32951888, 4.33360442, 4.45838177, 4.46452007, 4.67891457,
4.72167716, 4.81387984, 5.00014701, 5.24055764,5.30405366,
5.45888764, 5.49390880, 5.52179333, 5.66381047, 6.16508507,
6.24032858, 6.2442796, 6.29753330, 6.30779498, 6.31612410,
6.41120937, 6.41448096, 6.43383395, 6.77219317, 6.84154629,
6.90076633, 6.92867827, 6.93095933, 7.10138919, 7.26750958,
7.27311966, 7.33515741, 7.50370963, 7.52847685, 7.67226142,
7.6930181, 7.73796714, 7.78828417, 8.08008341, 8.42653468,
8.42955573,8.43619858, 8.50564013, 8.70426517, 8.74478266,
8.80415160, 9.0581363, 9.27984186, 9.28731328, 9.35034508,
9.51775680, 9.74506902, 9.96410355, 9.98975439, 11.127021,
11.570835, 12.520865, 12.714158, 13.288865, 15.761193, 16.741561,
16.948549, 17.125616, 18.301101, 18.812169, 20.54698, 22.14968))

list(beta1=.2, beta2=1,beta3=3,M=1)
```

TABLE 6.16

Posterior Distribution for the Bathtub Mean Value Function

Parameter	Mean	SD	Error	2 1/2	Median	97 1/2
M	3.309	7.196	0.0101	0.00000	0.231	27.33
β_1	0.4991	0.7065	0.001028	0.000493	0.2277	2.5
β_2	1.001	0.9969	0.001381	0.02528	0.6952	3.696
β_3	1.999	1.411	0.00197	0.2431	1.681	5.554

Recall that the true values for β_1, β_2, and β_3 are .5, 1, and 2, respectively, and that of M is 1. The posterior means are very close to these beta values, but the posterior mean of M is 3.309 and the posterior median is .231. Also, note that the posterior distributions for β_1, β_2, and β_3 are skewed to the right. As an exercise, the student will be asked to study the sensitivity of the posterior distributions to their priors. The following example is similar to that earlier for the washtub-shaped mean value function, but instead, the mean value function is the reverse to that of the washtub shape.

This section on nonhomogeneous Poisson processes is concluded with an example presented on page 253 of Dobrow[1] in the form of a problem. Students arrive at a cafeteria for lunch according to a Poisson process, where the rate of arrival varies in a linear way from 100 to 200 students over the time interval from 11:00 a.m. to noon, but the rate stays constant over the next 2 hours (from noon to 2:00 p.m.) and then decreases linearly to 100 from 2:00 to 3:00 p.m. Find the probability that there are at least 400 people in the cafeteria between 11:30 a.m. and 1:30 p.m. The intensity function is given by

$$\begin{aligned}\lambda(t) &= 100 + 100t, \quad 0 < t < 1,\\ &= 200, \quad 1 < t < 3\\ &= 500 - 100t, \quad 3 < t < 4.\end{aligned} \tag{6.77}$$

It is easy to see that the answer to the question is as follows: The mean value function is

$$E[N(2.5) - N(0.5)] = \int_{.5}^{1} (100 + 100t)\mathrm{d}t + \int_{1}^{2.5} 200t\,\mathrm{d}t = 387.5 \tag{6.78}$$

Thus,

$$P[N(2.5) - N(.5) \geq 400] = 1 - \sum_{k=0}^{k=399} (387.5)^k \exp(-387.5)/k! = .269.$$

Our approach will take more of a statistical approach by generating observations from the nonhomogeneous Poisson process with intensity function (Equation 6.77), then based on that information, estimate the expected value of the process and, consequently, estimate the required probability.

R Code 6.10 for generating the interarrival times for the process with intensity function (Equation 6.77) over the range $0 < t < 1$.

R Code 6.10

```
t<-1
> lamda<-200*t
> N<-rpois(1,lamda)
> unifs<-runif(N,0,t)
> arrivals<-sort(unifs)
> arrivals
```

There are 129 arrivals corresponding to the intensity function $\lambda(t) = 100 + 100t$, $0 < t < 1$:

```
0.002514378, 0.019175922, 0.021953457, 0.024550520, 0.031322105,
0.035207160, 0.042003951, 0.066021339, 0.069204332, 0.070603115,
0.074403316, 0.078209166, 0.081210360, 0.089738117, 0.108841185,
0.113504869, 0.118042807, 0.120448525, 0.127535137, 0.148708492,
0.155505013, 0.160686959, 0.165564084, 0.171232042, 0.185665243,
0.194464218, 0.202340639, 0.212150896, 0.216791938, 0.229751281,
0.233844860, 0.237426403, 0.240561157, 0.257191631, 0.259427338,
0.261023988, 0.261239244, 0.264757474, 0.264951792, 0.265082411,
0.276758405, 0.280244587, 0.311615852, 0.324892686, 0.341942644,
0.347802898, 0.350671225, 0.359282152, 0.372455910, 0.375604996,
0.391797577, 0.395026651, 0.398594826, 0.406915699, 0.417825938,
0.421420146, 0.435862032, 0.436892793, 0.445679733, 0.469381181,
0.478037560, 0.479947491, 0.485462589, 0.494364918, 0.495973951,
0.510708708, 0.516524876, 0.522994348, 0.555194473, 0.565705654,
0.590331443, 0.595593126, 0.606269287, 0.625587527, 0.629214609,
0.633505706, 0.639148909, 0.639599133, 0.646117045, 0.656630432,
0.660690608, 0.676695661, 0.679830155, 0.684899308, 0.684988749,
0.699697015, 0.706919646, 0.716615237, 0.727993494, 0.738322579,
0.739148218, 0.739186125, 0.742365563, 0.745413160, 0.747257771,
0.750100336, 0.755538156, 0.757015456, 0.757025317, 0.759269100,
0.779462443, 0.780866463, 0.781078923, 0.782396013, 0.783245994,
0.784139053, 0.786832943, 0.796762091, 0.801637811, 0.808890634,
0.822337339, 0.823737507, 0.831439372, 0.848291475, 0.874408497,
0.874706832, 0.883884702, 0.888856504, 0.913807917, 0.916263616,
0.916435269, 0.935536873, 0.937345495, 0.937769405, 0.950301845,
0.959554330, 0.966411747, 0.982074249, 0.990829274
```

When the intensity function is $\lambda(t) = 200t$, there are 375 arrivals:

```
1.013929345, 1.017299016, 1.020494032, 1.023492119, 1.029293662,
1.032535843, 1.034444443, 1.037337014, 1.044150100, 1.047903597,
1.048080343, 1.048804146, 1.055148176, 1.058901393, 1.062717973,
1.064533587, 1.069731262, 1.073189439, 1.081210021, 1.084091396,
1.099257308, 1.102238394, 1.111405018, 1.113974762, 1.115141771,
1.119604075, 1.122310411, 1.128649801, 1.129954257, 1.131515992,
1.134076646, 1.139261891, 1.140393564, 1.140692617, 1.145096267,
1.150237783, 1.163802895, 1.164433796, 1.165833532, 1.165907854,
1.176382628, 1.181937235, 1.182783321, 1.184585128, 1.196833924,
1.215449107, 1.227184298, 1.229106881, 1.242416283, 1.252645860,
1.252659581, 1.256543911, 1.258500341, 1.263674010, 1.269127304,
1.283127524, 1.286516622, 1.289281019, 1.290950422, 1.298855664,
1.300081327, 1.311437985, 1.329287803, 1.333032885, 1.337514530,
1.339356022, 1.340879801, 1.340931009, 1.341674382, 1.352022462,
1.357608653, 1.370389704, 1.373695941, 1.377188421, 1.403674372,
1.412190809, 1.416503148, 1.416678626, 1.418442181, 1.422087779,
1.424918116, 1.429967697, 1.444163951, 1.447379901, 1.452092366,
1.455721355, 1.458417613, 1.481371435, 1.485535010, 1.500044102,
```

```
1.500944988, 1.508367766, 1.516415729, 1.517187150, 1.521979619,
1.547682925, 1.566067752, 1.572167292, 1.577183020, 1.582464693,
1.590667617, 1.591216097, 1.593355230, 1.603135045, 1.615538036,
1.618061811, 1.626994546, 1.628422921, 1.630330539, 1.632926330,
1.635977235, 1.637666293, 1.645006708, 1.647880308, 1.648163049,
1.648172640, 1.652674647, 1.653589362, 1.656538000, 1.657236307,
1.661943821, 1.669053135, 1.699143140, 1.708281809, 1.709955084,
1.713104594, 1.714725234, 1.719333043, 1.724069923, 1.731021449,
1.735625281, 1.744656268, 1.746371425, 1.752927465, 1.753558599,
1.758681826, 1.759282306, 1.762545862, 1.767443548, 1.769469178,
1.773373658, 1.774687876, 1.778749374, 1.788461942, 1.790885404,
1.792485123, 1.795371861, 1.818204767, 1.827249561, 1.827614513,
1.834176804, 1.837328507, 1.839253480, 1.844222442, 1.847155063,
1.849525522, 1.850213237, 1.872098574, 1.873283888, 1.878129812,
1.889259991, 1.891425377, 1.892338493, 1.894636884, 1.894837229,
1.897565121, 1.907123635, 1.908636525, 1.910457973, 1.923362812,
1.924344287, 1.930150185, 1.949453704, 1.949761146, 1.953641534,
1.963653137, 1.963705628, 1.965964503, 1.977688008, 1.980482935,
1.985755443, 1.993329713, 1.999674787, 2.004291961, 2.007671704,
2.012298288, 2.017793437, 2.017909379, 2.028398239, 2.031657952,
2.033711193, 2.033917471, 2.047056607, 2.050281285, 2.052463886,
2.053530491, 2.054419905, 2.054703643, 2.059919972, 2.063083232,
2.063497233, 2.069296796, 2.070229900, 2.074059980, 2.078603481,
2.079287798, 2.083572979, 2.084623537, 2.093311061, 2.095123729,
2.101438373, 2.101883553, 2.109193094, 2.119495949, 2.130416757,
2.136031440, 2.136091095, 2.139714617, 2.140818849, 2.140982116,
2.143215024, 2.143642017, 2.147456241, 2.151059430, 2.159486218,
2.165581206, 2.171049164, 2.173861756, 2.174151290, 2.176017579,
2.180252875, 2.181935899, 2.184849538, 2.200547223, 2.208170009,
2.212822450, 2.213719575, 2.220661806, 2.239305554, 2.241199394,
2.243685161, 2.251112889, 2.254011817, 2.258543056, 2.275763491,
2.279060312, 2.285134647, 2.288586070, 2.289982291, 2.294147501,
2.296366955, 2.298688979, 2.301678425, 2.307905431, 2.321390142,
2.321833524, 2.323235367, 2.325190695, 2.336485252, 2.345230376,
2.363177724, 2.364178989, 2.365743879, 2.372871519, 2.373774794,
2.375236578, 2.375773515, 2.376371675, 2.391577682, 2.400035776,
2.411478949, 2.424025024, 2.427385656, 2.432976392, 2.434047745,
2.442873146, 2.446176723, 2.472320898, 2.483721139, 2.486104882,
2.490810028, 2.495618162, 2.510926709, 2.511234161, 2.513839507,
2.519147823, 2.520375510, 2.526825013, 2.527719625, 2.527960403,
2.528866718, 2.530859574, 2.531298494, 2.535594637, 2.542282276,
2.547550235, 2.551692040, 2.555513599, 2.568842423, 2.570009794,
2.573882751, 2.578535879, 2.583370734, 2.591475289, 2.592686206,
2.599859480, 2.611279551, 2.619419517, 2.620431709, 2.623434798,
2.623813957, 2.627428332, 2.629607411, 2.641245481, 2.654289036,
2.655992781, 2.669850807, 2.676764303, 2.678075310, 2.688720737,
2.698684317, 2.703161751, 2.717440890, 2.719578907, 2.724078030,
2.728939681, 2.735276548, 2.735783146, 2.745158123, 2.745165184,
2.746823335, 2.760853430, 2.761141808, 2.761796885, 2.769162985,
```

2.769795155, 2.777490208, 2.783952558, 2.786193984, 2.792533542,
2.793089024, 2.798631219, 2.800673852, 2.802583046, 2.805103524,
2.821825363, 2.832271894, 2.836535176, 2.837648671, 2.846627899,
2.850492782, 2.855327041, 2.878247633, 2.882553231, 2.883212790,
2.889528822, 2.892304451, 2.911982264, 2.913785207, 2.923520705,
2.926015311, 2.939286368, 2.943335389, 2.945615221, 2.946840695,
2.956556404, 2.958435072, 2.970019833, 2.970869889, 2.978547289,
2.979987915, 2.987539444, 2.989231064, 2.995278765, 2.996926318

When the intensity function is $m(t) = 500t - 50t^2$, there are 292 events:

3.0394055042, 3.0395649960, 3.0497750333, 3.0515539590,
3.0534064593, 3.0550988168, 3.0603179419, 3.0611753678,
3.0638260543, 3.0684684142, 3.0690515153, 3.0701252520,
3.0710977903, 3.0761675294, 3.0789520564, 3.0870140810,
3.0901728366, 3.0940281730, 3.1025335584, 3.1041963296,
3.1046456881, 3.106,436313, 3.1156511167, 3.1157595599,
3.1171774771, 3.1178429807, 3.1203698972, 3.1262418255,
3.1303598192, 3.1336451545, 3.1356885713, 3.1359373592,
3.1372633446, 3.1387056960, 3.1472093416, 3.1476380508,
3.1481734151, 3.1515996931, 3.1557536889, 3.1578143174,
3.1655712742, 3.1660238905, 3.1675412720, 3.1700455686,
3.1709044836, 3.1710038716, 3.1710768277, 3.1807788387,
3.1844945773, 3.1865160577, 3.1982697193, 3.1989803519,
3.2010379033, 3.2034131940, 3.2041366743, 3.2121709464,
3.2217008546, 3.2264809981, 3.2292034011, 3.2307586530,
3.2316791369, 3.2357311882, 3.2366268234, 3.2477117376,
3.2505444707, 3.2612449862, 3.2624498317, 3.2690377990,
3.2699324042, 3.2726698127, 3.2808823753, 3.2881072043,
3.2917402433, 3.2969507733, 3.3015268594, 3.3015537094,
3.3063119072, 3.3096362194, 3.3164327303, 3.3168281792,
3.3212339133, 3.3241597982, 3.3273486616, 3.3274049992,
3.3323849887, 3.3333767094, 3.3339362908, 3.3363375543,
3.3369981498, 3.3470437285, 3.3504107334, 3.3604161320,
3.3611478573, 3.3624880472, 3.3654973106, 3.3666472705,
3.3799138367, 3.3843323132, 3.3963465840, 3.3980996637,
3.4005952124, 3.4013850382, 3.4036674043, 3.4044647310,
3.4091130989, 3.4127464471, 3.4152737586, 3.4190298095,
3.4272268470, 3.4378765291, 3.4381807251, 3.4381879056,
3.4405560624, 3.4420253634, 3.4503981983, 3.4537759321,
3.4647708368, 3.4727968536, 3.4743528450, 3.4746357612,
3.4767445046, 3.4786015302, 3.4810818359, 3.4830563124,
3.4834156064, 3.4838489853, 3.4844182320, 3.4893770060,
3.4913737085, 3.4926125128, 3.4944544798, 3.4947529892,
3.4955945443, 3.4957072623, 3.5017801821, 3.5019984478,
3.5088554090, 3.5098619508, 3.5190729275, 3.5214515720,
3.5337872561, 3.5338723445, 3.5349942641, 3.5373605257,
3.5374128018, 3.5415375559, 3.5452064471, 3.5490878643,
3.5529638240, 3.5564432358, 3.5567915970, 3.5569886006,

```
3.5691233557, 3.5804139767, 3.5806436213, 3.5812419420,
3.5821914040, 3.5834934516, 3.5839783000, 3.5847572675,
3.5939939115, 3.5965053625, 3.5973652303, 3.5987382149,
3.5989597440, 3.6125720618, 3.6135758925, 3.6139502143,
3.6165282140, 3.6193393739, 3.6278351685, 3.6300896127,
3.6335888784, 3.6348160263, 3.6383879464, 3.6394935939,
3.6402793005, 3.6429974539, 3.6486793263, 3.6525471816,
3.6569040315, 3.6666116016, 3.6678881897, 3.6687063370,
3.6730838716, 3.6790721621, 3.6885159444, 3.6895445548,
3.6923138006, 3.6925381618, 3.6974107707, 3.7003768077,
3.7019655583, 3.7107408307, 3.7119259490, 3.7147472426,
3.7158508776, 3.7180384109, 3.7188526653, 3.7221041368,
3.7243337510, 3.7297047498, 3.7357104449, 3.7379124304,
3.7393667521, 3.7396834819, 3.7536196075, 3.7553359307,
3.7658056961, 3.7690140931, 3.7693173233, 3.7724120421,
3.7737177517, 3.7764852755, 3.7872026497, 3.7891286425,
3.7931585815, 3.7973917332, 3.7999806516, 3.8075277386,
3.8079434456, 3.8130274843, 3.8147484818, 3.8303364394,
3.8325500693, 3.8331372524, 3.8335735751, 3.8363890983,
3.8367080148, 3.8419249961, 3.8456002781, 3.8456870727,
3.8456912525, 3.8466627011, 3.8488095850, 3.8496039398,
3.8507244056, 3.8529330473, 3.8531479556, 3.8593404116,
3.8616266381, 3.8682547659, 3.8685721802, 3.8690067939,
3.8702020561, 3.8746858975, 3.8793189283, 3.8806014806,
3.8872348368, 3.8886372708, 3.8891721042, 3.8899864918,
3.8920450564, 3.8939412264, 3.8948968519, 3.8973098779,
3.8977870010, 3.9011845272, 3.9041249128, 3.9141902262,
3.9167675348, 3.9190505901, 3.9190733116, 3.9214201123,
3.9217549330, 3.9334586877, 3.9381146226, 3.9382127598,
3.9390543317, 3.9406094989, 3.9510613997, 3.9525732268,
3.9560717000, 3.9623804362, 3.9641936077, 3.9660449354,
3.9664281784, 3.9665969238, 3.9751119539, 3.9754377482,
3.9763981393, 3.9774438757, 3.9774594707, 3.9805419631,
3.9838282093, 3.9859077316, 3.9871207457, 3.9879421517,
3.9906606367, 3.9927617311, 3.9960442316, 3.9980143020
```

WinBUGS Code 6.11

```
model;
{

for( i in 1:129){

n1[i]~dexp(delta1[i])
lamda1[i]<-beta11*i+beta12*i*i
delta1[i]<-1/lamda1[i]}
beta11~dnorm(100,1)
beta12~dnorm(50,1)
```

```
for ( i in 1:375){
n2[i]~dexp(delta2[i])
lamda2[i]<-beta2*i
delta2[i]<-1/lamda2[i]}
beta2~dnorm(200,1)

for( i in 1:292){
n3[i]~dexp(delta3[i])
lamda3[i]<-abs(beta31*i-beta32*i*i)
delta3[i]<-1/lamda3[i]}
beta31~dnorm(500,1)
beta32~dnorm(-50,1)
E<-beta11/2+3*beta12/8+1.5*beta2
}
```

```
list(n1=c(0.002514378,  0.019175922,  0.021953457,  0.024550520,  0.031322105,  0.035207160,
0.042003951, 0.066021339, 0.069204332, 0.070603115, 0.074403316, 0.078209166, 0.081210360,
0.089738117, 0.108841185, 0.113504869, 0.118042807, 0.120448525, 0.127535137, 0.148708492,
0.155505013, 0.160686959, 0.165564084, 0.171232042, 0.185665243, 0.194464218, 0.202340639,
0.212150896, 0.216791938, 0.229751281, 0.233844860, 0.237426403, 0.240561157, 0.257191631,
0.259427338, 0.261023988, 0.261239244, 0.264757474, 0.264951792, 0.265082411, 0.276758405,
0.280244587, 0.311615852, 0.324892686, 0.341942644, 0.347802898, 0.350671225, 0.359282152,
0.372455910, 0.375604996, 0.391797577, 0.395026651, 0.398594826, 0.406915699, 0.417825938,
0.421420146, 0.435862032, 0.436892793, 0.445679733, 0.469381181, 0.478037560, 0.479947491,
0.485462589, 0.494364918, 0.495973951, 0.510708708, 0.516524876, 0.522994348, 0.555194473,
0.565705654, 0.590331443, 0.595593126, 0.606269287, 0.625587527, 0.629214609, 0.633505706,
0.639148909, 0.639599133, 0.646117045, 0.656630432, 0.660690608, 0.676695661, 0.679830155,
0.684899308, 0.684988749, 0.699697015, 0.706919646, 0.716615237, 0.727993494, 0.738322579,
0.739148218, 0.739186125, 0.742365563, 0.745413160, 0.747257771, 0.750100336, 0.755538156,
0.757015456, 0.757025317, 0.759269100, 0.779462443, 0.780866463, 0.781078923, 0.782396013,
0.783245994, 0.784139053, 0.786832943, 0.796762091, 0.801637811, 0.808890634, 0.822337339,
0.823737507, 0.831439372, 0.848291475, 0.874408497, 0.874706832, 0.883884702, 0.888856504,
0.913807917, 0.916263616, 0.916435269, 0.935536873, 0.937345495, 0.937769405, 0.950301845,
0.959554330, 0.966411747, 0.982074249, 0.990829274),

n2=c(1.013929345, 1.017299016, 1.020494032, 1.023492119, 1.029293662, 1.032535843,
1.034444443, 1.037337014, 1.044150100, 1.047903597, 1.048080343, 1.048804146, 1.055148176,
1.058901393, 1.062717973, 1.064533587, 1.069731262, 1.073189439, 1.081210021, 1.084091396,
1.099257308, 1.102238394, 1.111405018, 1.113974762, 1.115141771, 1.119604075, 1.122310411,
1.128649801, 1.129954257, 1.131515992, 1.134076646, 1.139261891, 1.140393564, 1.140692617,
1.145096267, 1.150237783, 1.163802895, 1.164433796, 1.165833532, 1.165907854, 1.176382628,
1.181937235, 1.182783321, 1.184585128, 1.196833924, 1.215449107, 1.227184298, 1.229106881,
1.242416283, 1.252645860, 1.252659581, 1.256543911, 1.258500341, 1.263674010, 1.269127304,
```

1.283127524, 1.286516622, 1.289281019, 1.290950422, 1.298855664, 1.300081327, 1.311437985,
1.329287803, 1.333032885, 1.337514530, 1.339356022, 1.340879801, 1.340931009, 1.341674382,
1.352022462, 1.357608653, 1.370389704, 1.373695941, 1.377188421, 1.403674372, 1.412190809,
1.416503148, 1.416678626, 1.418442181, 1.422087779, 1.424918116, 1.429967697, 1.444163951,
1.447379901, 1.452092366, 1.455721355, 1.458417613, 1.481371435, 1.485535010, 1.500044102,
1.500944988, 1.508367766, 1.516415729, 1.517187150, 1.521979619, 1.547682925, 1.566067752,
1.572167292, 1.577183020, 1.582464693, 1.590667617, 1.591216097, 1.593355230, 1.603135045,
1.615538036, 1.618061811, 1.626994546, 1.628422921, 1.630330539, 1.632926330, 1.635977235,
1.637666293, 1.645006708, 1.647880308, 1.648163049, 1.648172640, 1.652674647, 1.653589362,
1.656538000, 1.657236307, 1.661943821, 1.669053135, 1.699143140, 1.708281809, 1.709955084,
1.713104594, 1.714725234, 1.719333043, 1.724069923, 1.731021449, 1.735625281, 1.744656268,
1.746371425, 1.752927465, 1.753558599, 1.758681826, 1.759282306, 1.762545862, 1.767443548,
1.769469178, 1.773373658, 1.774687876, 1.778749374, 1.788461942, 1.790885404, 1.792485123,
1.795371861, 1.818204767, 1.827249561, 1.827614513, 1.834176804, 1.837328507, 1.839253480,
1.844222442, 1.847155063, 1.849525522, 1.850213237, 1.872098574, 1.873283888, 1.878129812,
1.889259991, 1.891425377, 1.892338493, 1.894636884, 1.894837229, 1.897565121, 1.907123635,
1.908636525, 1.910457973, 1.923362812, 1.924344287, 1.930150185, 1.949453704, 1.949761146,
1.953641534, 1.963653137, 1.963705628, 1.965964503, 1.977688008, 1.980482935, 1.985755443,
1.993329713, 1.999674787, 2.004291961, 2.007671704, 2.012298288, 2.017793437, 2.017909379,
2.028398239, 2.031657952, 2.033711193, 2.033917471, 2.047056607, 2.050281285, 2.052463886,
2.053530491, 2.054419905, 2.054703643, 2.059919972, 2.063083232, 2.063497233, 2.069296796,
2.070229900, 2.074059980, 2.078603481, 2.079287798, 2.083572979, 2.084623537, 2.093311061,
2.095123729, 2.101438373, 2.101883553, 2.109193094, 2.119495949, 2.130416757, 2.136031440,
2.136091095, 2.139714617, 2.140818849, 2.140982116, 2.143215024, 2.143642017, 2.147456241,
2.151059430, 2.159486218, 2.165581206, 2.171049164, 2.173861756, 2.174151290, 2.176017579,
2.180252875, 2.181935899, 2.184849538, 2.200547223, 2.208170009, 2.212822450, 2.213719575,
2.220661806, 2.239305554, 2.241199394, 2.243685161, 2.251112889, 2.254011817, 2.258543056,
2.275763491, 2.279060312, 2.285134647, 2.288586070, 2.289982291, 2.294147501, 2.296366955,
2.298688979, 2.301678425, 2.307905431, 2.321390142, 2.321833524, 2.323235367, 2.325190695,
2.336485252, 2.345230376, 2.363177724, 2.364178989, 2.365743879, 2.372871519, 2.373774794,
2.375236578, 2.375773515, 2.376371675, 2.391577682, 2.400035776, 2.411478949, 2.424025024,
2.427385656, 2.432976392, 2.434047745, 2.442873146, 2.446176723, 2.472320898, 2.483721139,
2.486104882, 2.490810028, 2.495618162, 2.510926709, 2.511234161, 2.513839507, 2.519147823,
2.520375510, 2.526825013, 2.527719625, 2.527960403, 2.528866718, 2.530859574, 2.531298494,
2.535594637, 2.542282276, 2.547550235, 2.551692040, 2.555513599, 2.568842423, 2.570009794,
2.573882751, 2.578535879, 2.583370734, 2.591475289, 2.592686206, 2.599859480, 2.611279551,
2.619419517, 2.620431709, 2.623434798, 2.623813957, 2.627428332, 2.629607411, 2.641245481,
2.654289036, 2.655992781, 2.669850807, 2.676764303, 2.678075310, 2.688720737, 2.698684317,
2.703161751, 2.717440890, 2.719578907, 2.724078030, 2.728939681, 2.735276548, 2.735783146,
2.745158123, 2.745165184, 2.746823335, 2.760853430, 2.761141808, 2.761796885, 2.769162985,
2.769795155, 2.777490208, 2.783952558, 2.786193984, 2.792533542, 2.793089024, 2.798631219,
2.800673852, 2.802583046, 2.805103524, 2.821825363, 2.832271894, 2.836535176, 2.837648671,
2.846627899, 2.850492782, 2.855327041, 2.878247633, 2.882553231, 2.883212790,

```
2.889528822, 2.892304451, 2.911982264, 2.913785207, 2.923520705, 2.926015311, 2.939286368,
2.943335389, 2.945615221, 2.946840695, 2.956556404, 2.958435072, 2.970019833, 2.970869889,
2.978547289, 2.979987915, 2.987539444, 2.989231064, 2.995278765, 2.996926318),

n3=c(3.0394055042, 3.0395649960, 3.0497750333, 3.0515539590, 3.0534064593, 3.0550988168,
3.0603179419, 3.0611753678, 3.0638260543, 3.0684684142, 3.0690515153, 3.0701252520,
3.0710977903, 3.0761675294, 3.0789520564, 3.0870140810, 3.0901728366, 3.0940281730,
3.1025335584, 3.1041963296, 3.1046456881, 3.106,436313, 3.1156511167, 3.1157595599,
3.1171774771, 3.1178429807, 3.1203698972, 3.1262418255, 3.1303598192, 3.1336451545,
3.1356885713, 3.1359373592, 3.1372633446, 3.1387056960, 3.1472093416, 3.1476380508,
3.1481734151, 3.1515996931, 3.1557536889, 3.1578143174, 3.1655712742, 3.1660238905,
3.1675412720, 3.1700455686, 3.1709044836, 3.1710038716, 3.1710768277, 3.1807788387,
3.1844945773, 3.1865160577, 3.1982697193, 3.1989803519, 3.2010379033, 3.2034131940,
3.2041366743, 3.2121709464, 3.2217008546, 3.2264809981, 3.2292034011, 3.2307586530,
3.2316791369, 3.2357311882, 3.2366268234, 3.2477117376, 3.2505444707, 3.2612449862,
3.2624498317, 3.2690377990, 3.2699324042, 3.2726698127, 3.2808823753, 3.2881072043,
3.2917402433, 3.2969507733, 3.3015268594, 3.3015537094, 3.3063119072, 3.3096362194,
3.3164327303, 3.3168281792, 3.3212339133, 3.3241597982, 3.3273486616, 3.3274049992,
3.3323849887, 3.3333767094, 3.3339362908, 3.3363375543, 3.3369981498, 3.3470437285,
3.3504107334, 3.3604161320, 3.3611478573, 3.3624880472, 3.3654973106, 3.3666472705,
3.3799138367, 3.3843323132, 3.3963465840, 3.3980996637, 3.4005952124, 3.4013850382,
3.4036674043, 3.4044647310, 3.4091130989, 3.4127464471, 3.4152737586, 3.4190298095,
3.4272268470, 3.4378765291, 3.4381807251, 3.4381879056, 3.4405560624, 3.4420253634,
3.4503981983, 3.4537759321, 3.4647708368, 3.4727968536, 3.4743528450, 3.4746357612,
3.4767445046, 3.4786015302, 3.4810818359, 3.4830563124, 3.4834156064, 3.4838489853,
3.4844182320, 3.4893770060, 3.4913737085, 3.4926125128, 3.4944544798, 3.4947529892,
3.4955945443, 3.4957072623, 3.5017801821, 3.5019984478, 3.5088554090, 3.5098619508,
3.5190729275, 3.5214515720, 3.5337872561, 3.5338723445, 3.5349942641, 3.5373605257,
3.5374128018, 3.5415375559, 3.5452064471, 3.5490878643, 3.5529638240, 3.5564432358,
3.5567915970, 3.5569886006, 3.5691233557, 3.5804139767, 3.5806436213, 3.5812419420,
3.5821914040, 3.5834934516, 3.5839783000, 3.5847572675, 3.5939939115, 3.5965053625,
3.5973652303, 3.5987382149, 3.5989597440, 3.6125720618, 3.6135758925, 3.6139502143,
3.6165282140, 3.6193393739, 3.6278351685, 3.6300896127, 3.6335888784, 3.6348160263,
3.6383879464, 3.6394935939, 3.6402793005, 3.6429974539, 3.6486793263, 3.6525471816,
3.6569040315, 3.6666116016, 3.6678881897, 3.6687063370, 3.6730838716, 3.6790721621,
3.6885159444, 3.6895445548, 3.6923138006, 3.6925381618, 3.6974107707, 3.7003768077,
3.7019655583, 3.7107408307, 3.7119259490, 3.7147472426, 3.7158508776, 3.7180384109,
3.7188526653, 3.7221041368, 3.7243337510, 3.7297047498, 3.7357104449, 3.7379124304,
3.7393667521, 3.7396834819, 3.7536196075, 3.7553359307, 3.7658056961, 3.7690140931,
3.7693173233, 3.7724120421, 3.7737177517, 3.7764852755, 3.7872026497, 3.7891286425,
3.7931585815, 3.7973917332, 3.7999806516, 3.8075277386, 3.8079434456, 3.8130274843,
3.8147484818, 3.8303364394, 3.8325500693, 3.8331372524, 3.8335735751, 3.8363890983,
3.8367080148, 3.8419249961, 3.8456002781, 3.8456870727, 3.8456912525, 3.8466627011,
3.8488095850, 3.8496039398, 3.8507244056, 3.8529330473, 3.8531479556, 3.8593404116,
3.8616266381, 3.8682547659, 3.8685721802, 3.8690067939, 3.8702020561, 3.8746858975,
3.8793189283, 3.8806014806, 3.8872348368, 3.8886372708, 3.8891721042, 3.8899864918,
```

TABLE 6.17

Posterior Distribution for Regression Parameters Nonhomogeneous Process

Parameter	Mean	SD	Error	2 1/2	Median	97 1/2
β_{11}	99.91	0.9997	0.01157	97.96	99.9	101.9
β_{12}	47.44	1.031	0.01306	45.42	47.44	49.46
β_2	198.1	1.007	0.01247	196.1	198.1	200.1
β_{31}	499.9	1.01	0.01089	498	499.0	501.9
β_{32}	−44.44	1.064	0.01737	−46.54	−44.46	−42.35
E	364.9	1.631	0.01614	361.7	364.9	368.2

```
3.8920450564, 3.8939412264, 3.8948968519, 3.8973098779, 3.8977870010, 3.9011845272,
3.9041249128, 3.9141902262, 3.9167675348, 3.9190505901, 3.9190733116, 3.9214201123,
3.9217549330, 3.9334586877, 3.9381146226, 3.9382127598, 3.9390543317, 3.9406094989,
3.9510613997, 3.9525732268, 3.9560717000, 3.9623804362, 3.9641936077, 3.9660449354,
3.9664281784, 3.9665969238, 3.9751119539, 3.9754377482, 3.9763981393, 3.9774438757,
3.9774594707, 3.9805419631, 3.9838282093, 3.9859077316, 3.9871207457, 3.9879421517,
3.9906606367, 3.9927617311, 3.9960442316, 3.9980143020))
```

```
list(beta11=100,beta12=50, beta2=200,beta31=500,beta32=-50)
```

Recall that the value of β_{11} used to generate the arrival times for the first stage is 100 and that for β_{12} is 50 and that the corresponding posterior means are very close to these values. It can also be confirmed that this is true for the remaining beta parameters. Also, recall the expectation

$$E[N(2.5) - N(0.5)] = \int_{.5}^{1} (100 + 100t)\mathrm{d}t + \int_{1}^{2.5} 200t\,\mathrm{d}t = 387.5.$$

However, the posterior mean given by Table 6.17 is 364.9! Why the discrepancy?

6.8 General Continuous-Time Markov Chains

In this section, Bayesian inferential procedures are presented for continuous-time Markov chains more general than the Poisson process. The interarrival times still have an exponential distribution, but the parameter of the exponential distribution depends on the state the process currently occupies. Of special interest are homogeneous processes with a finite state space where the system remains for a time with an exponential distribution at each state, and upon departing the state, the system changes with regard to the probabilities that depend solely on the leaving state. The process is completely specified by the initial state, the transition probabilities from the current state to the future state and the parameter of the exponential distribution of the current time the process is in the current state. All phases of inference will be explored, which includes estimation, testing hypotheses, and forecasting of future observations (states or interarrival times).

Let $\{X(t), t > 0\}$ be a continuous-time stochastic process with state space $S = \{1,2,...,k\}$ such that when the process enters state i, it remains in state i according to the exponential distribution with parameter ν_i (with mean $1/\nu_i$). At the end of this period (being in state i), the process transitions to another state say $j \neq i$ with transition probability p_{ij}, where $\sum_{j=1}^{j=k} p_{ij} = 1$. It should be emphasized that for some j, it is possible that p_{ij} is zero. If one ignores time, the process occupies the states, the transition probability matrix of the various states is $P = (p_{ij})$ corresponds to what is referred to as the embedded chain. The birth and death process discussed in the following is an example of a continuous-time Markov chain.

In such a process, the state space is $S = \{0,1,...,k\}$ and represents the size of the population. Suppose the process is in state i, then the population can increase by a single birth with rate λ_i and decrease by one unit with a single death occurring at rate μ_i, and the transition matrix is defined as follows:

$$p_{i,i+1} = \lambda_i/(\lambda_i + \mu_i)$$

and (6.79)

$$p_{i,i-1} = \mu_i/(\lambda_i + \mu_i)$$

Also, note that if the population is 0, the size of the population remains zero, and if the population is size k, it can only decrease by one death and, of course, cannot increase by one birth.

Furthermore, in general, for a continuous time chains with state space $S = \{1,2,...,k\}$, when the process is in state i, it remains in state i according to an exponential distribution with parameter $\nu_i > 0$. The jumping intensity from state i to state j is defined by

$$r_{ij} = \nu_i p_{ij}. \tag{6.80}$$

Note that $-\nu_i = \sum_{j \neq i} r_{ij}$ and the kth order matrix $R = (r_{ij})$ are called the infinitesimal generator of the process and will be used when computing the equilibrium distribution of the process.

In order to understand the future behavior of a continuous time chain, the Kolmogorov system of differential equations involving the transition probability function

$$P_{ij}(t) = P[X(t+s) = j|X(s) = i] = P[X(t) = j|X(0) = i] \tag{6.81}$$

is described. Note that this follows because the process is homogeneous; that is, only the length of the interval is important in determining the transition from i to j.

The formulation of the Kolmogorov equations is explained on pages 81–82 of Insua, Ruggeri, and Wiper[2] and is closely followed here. Let

$$P'_{ij}(t) = \sum_{k \neq j} r_{kj} P_{ik}(t) - \nu_j P_{ij}(t) = \sum_{k} r_{kj} P_{ik}(t), \tag{6.82}$$

where

$$P'(t) = \left((\mathrm{d}/\mathrm{d}t) P_{ij}(t)\right).$$

Or in matrix form as

$$P'(t) = RP(t) \tag{6.83}$$

with

$$I = P(0),$$

where I is the identity matrix and $P(t)$ is the matrix of transition probability functions. It is easy to show that the solution to the Kolmogorov system of equations (Equation 6.83) is the exponential form

$$P(t) = \exp(Rt), \tag{6.84}$$

which according to Moler and Van Loan[15] can be solved for the given t using matrix exponentiation.

We now come to the central focus of this section, namely, Bayesian inference of continuous-time Markov chains. To this end, it is assumed that the transition matrix P and the vector ν of the transition rates are unknown and that P is not a function of ν. Assume that the initial state is $x(0)$, which is followed by n transitions times t_i and corresponding observations $x(i)$, $i = 1, 2, \ldots, n$; then the likelihood function is

$$\begin{aligned} l(P, \nu \mid \text{data}) &= \prod_{i=1}^{i=n} \nu_{i-1} \exp\left(\nu_{i-1}(t_i - t_{i-1}) p_{x(i-1),x(i)}\right) \\ &= \prod_{i=1}^{i=n} \nu_i^{n_i} \exp(-\nu_i t_i) \prod_{i=1}^{i=n} p_{ij}^{n_{ij}}, \end{aligned} \tag{6.85}$$

where n_{ij} is the number of transitions from state i to j, t_i is the time occupying state i (with an exponential distribution with parameter ν_i), p_{ij} is the probability of the transition from state i to state j, and $n_{i.} = \sum_{j=1}^{j=k} n_{ij}$ is the number of transitions out of state i. Referring to the likelihood function (Equation 6.85), it is understood that state j is $x(j), j = 1, 2, \ldots, k$. It is obvious from Equation 6.85 that the likelihood function can be written as

$$l(P, \nu \mid \text{data}) = l_1(\nu \mid \text{data}) l_2(P \mid \text{data}). \tag{6.86}$$

Therefore, inferences can be made separately for P and ν, by first generating transitions n_{ij} from i to j, then generating exponential holding times after the transition to state i. Recall Chapter 4 where R Code 4.1 is used to generate transitions according to the transition probability matrix P and to R Code 6.10 of this chapter to generate exponential interarrival times. Thus, starting in state i, one can generate transitions to state j for some of the $j = 1,2,..,k$. If one knows the transition probabilities $p_{ij}, j = 1,2,...,k$, one could employ the multinomial distribution to generate the transitions, as was done in Section 4.3.

Suppose prior information for the ν_i, $i = 1,2,..,k$ is gamma (a_i, b_i), then the posterior distribution of ν_i is gamma $(n_i + a_i, t_i + b_i)$ with posterior mean $(n_i + a_i)/(t_i + b_i)$. Recall that the posterior mean of the transition time T_i is

$$E\left(\nu_i^{-1}|\text{data}\right) = (t_i + b_i)/(n_i + a_i - 1).$$

Now for the prior distribution of the transition probabilities p_{ij}, because of practical reasons, it is important to remember that some of the p_{ij} are zero, but nevertheless, that one may employ a Dirichlet distribution for the row of P; thus, the posterior analysis is demonstrated with an example.

Suppose that the probability transition matrix of the embedded chain is

$$P = \begin{pmatrix} p_{11}, 0, p_{13}, p_{14}, 0 \\ 0, p_{22}, 0, p_{24}, 0 \\ 0, 0, p_{33}, p_{34}, p_{35} \\ p_{41}, p_{42}, 0, p_{44}, p_{45} \\ 0, p_{52}, p_{53}, 0, p_{55} \end{pmatrix} \tag{6.87}$$

with permanence rates ν_i, $i = 1,2,...,5$, the goal is to provide Bayesian inference for the unknown parameters ν_i, and the nonzero p_{ij} of (Equation 6.87). Inferences will be based on the transition counts given by the matrix

$$c = \begin{pmatrix} n_{11}, 0, n_{13}, n_{14}, 0 \\ 0, n_{22}, 0, n_{24}, 0 \\ 0, 0, n_{33}, n_{34}, n_{35} \\ n_{41}, n_{42}, 0, n_{44}, n_{45} \\ 0, n_{52}, n_{53}, 0, n_{55} \end{pmatrix} \tag{6.88}$$

and the observed times t_i occupied in the states i, $i = 1, 2, 3, 4, 5$.

Suppose the occupation time distributions T_i are exponential with means 2, 7, 1, 4, and 5 corresponding to exponential parameter values $\nu_1 = 1/2$, $\nu_2 = 1/7$, $\nu_3 = 1$, $\nu_4 = 1/4$, and $\nu_5 = 1/5$. I used the R Code rexp(samples, ν_i) with samples =1 to generate the occupation times for the various states and computed $t_1 = 1.3257$, $t_2 = 10.2288$, $t_3 = .11038$, $t_4 = 3.2565$, and

TABLE 6.18

Prior Dirichlet Distribution for Rows of P

Row	Parameters
1	(1, 0, 1, 1, 0)
2	(0, 1, 0, 1, 0)
3	(0, 0, 1, 1, 1)
4	(1, 1, 0, 1, 1)
5	(0, 1, 1, 0, 1)

$t_5 = 5.65137$. Using the multinomial generator in R, the following transition counts represented by the matrix

$$C = \begin{pmatrix} 1,0,17,2,0 \\ 0,10,0,20,0 \\ 0,0,7,4,4 \\ 10,7,0,7,6 \\ 0,6,8,0,16 \end{pmatrix} \tag{6.89}$$

are computed.

Thus, the number of transitions out of the various states are $n_1 = 19$, $n_2 = 20$, $n_3 = 8$, $n_4 = 23$, and $n_5 = 14$. This completes the sample information; therefore, in order to execute the posterior analysis, the prior distributions need to be assigned to the unknown parameters P and ν.

The rows of P are assigned prior Dirichlet distributions as shown in Table 6.18:

In addition, the improper prior

$$g(\nu_i) \propto 1/\nu_i,\ i = 1,2,3,4,5$$

is assigned to the parameter of the exponential distribution of the occupation times of the various states.

The Bayesian analysis is executed with WinBUGS Code 6.12 using 45,000 observations for the simulation and 5,000 for the burn-in.

WinBUGS Code 6.12

```
Model;
{
nu1~dgamma(19,1.3257)
nu2~dgamma(20,10.2288)
nu3~dgamma(8,.11038)
```

```
nu4~dgamma(23,3.2565)
nu5~dgamma(14,5.65132)
p11~dbeta(2,20)
p13~dbeta(17,5)
p14~dbeta(3,19)
p22~dbeta(11,21)
p24~dbeta(21,11)
p33~dbeta(8,10)
p34~dbeta(5,13)
p35~dbeta(5,13)
p41~dbeta(11,23)
p42~dbeta(8,26)
p44~dbeta(8,26)
p45~dbeta(7,25)
p52~dbeta(7,26)
p53~dbeta(9,24)
p55~dbeta(17,16)
mu1<-1/nu1
mu2<-1/nu2
mu3<-1/nu3
mu4<-1/nu4
mu5<-1/nu5
}
```

Table 6.19 reports the Bayesian analysis for the example with five states with five intensity parameters and the probabilities of the transition matrix.

Most of the posterior distributions are symmetric about the posterior mean, and it appears that the MCMC simulation errors are sufficiently small so that one had confidence in these estimates. The reader should display the posterior densities of the parameters of Table 6.19 to show the symmetry of the posterior distributions. This chapter will develop in much greater detail the fundamental ideas of continuous-time Markov chains, which are illustrated with applications in business, biology, and medicine.

TABLE 6.19

Posterior Distribution for Continuous Markov Chain

Parameter	Mean	SD	Error	2 1/2	Median	97 1/2
μ_1	.07364	.01788	.0000533	.04644	.07099	.1157
μ_2	.538	.1272	.000392	.3442	.519	.838
μ_3	.01573	.006453	.0000213	.007651	.01435	.03184
μ_4	.1481	.03239	.000103	.09786	.1437	.2242
μ_5	.4351	.1255	.000432	.2547	.414	.7382
ν_1	14.34	3.282	0.01452	8.67	14.1	21.5
ν_2	1.957	0.4385	0.001947	1.195	1.927	2.903
ν_3	72.64	25.65	0.1173	31.26	69.74	130.5
ν_4	7.058	1.474	0.006392	4.468	6.957	10.23
ν_5	2.471	0.66087	0.00289	1.353	2.409	3.927
p_{11}	.09073	.06004	.000182	.01151	.07846	.2358
p_{13}	.7732	.08701	.000268	.5831	.7817	.9176
p_{14}	.1362	.07161	.000226	.03036	.1249	.3037
p_{22}	.3435	.08266	.000252	.1927	.3401	.513
p_{24}	.6565	.08262	.000243	.4863	.6597	.8083
p_{33}	.4443	.1141	.000356	.2303	.4424	.6706
p_{34}	.2778	.1025	.000320	.1032	.2696	.498
p_{35}	.2772	.1025	.000325	.1032	.2688	.4976
p_{41}	.3239	.07927	.000251	.1802	.3204	.4883
p_{42}	.2356	.07159	.000231	.111	.2302	.3891
p_{44}	.2351	.07148	.000205	.111	.2299	.398
p_{52}	.2123	.07027	.000215	.093	.2063	.3655
p_{53}	.2727	.07656	.00024	.1372	.268	.4347
p_{55}	.5151	.08579	.000277	.3477	.515	.6817

6.9 Summary

This chapter on continuous-time Markov chains begins with three definitions of the Poisson process where the emphasis is on Bayesian inferences for one parameter λ. R Code 6.1 generates arrival times of a Poisson process, and these observations are used in the example of Bayesian inferences about λ. All three phases of inferences are displayed including estimation, testing hypotheses, and prediction of future observations. Next to be considered is the concept of the superposition of Poisson processes, and the ideas are explained in terms of an example involving major and minor earthquakes in Italy. Included with Bayesian inference is testing the hypothesis that the rate of occurrence of major earthquakes is the same as that for minor earthquakes.

Next to be presented is a generalization to nonhomogeneous Poisson processes, where the rate of occurrence of events varies over time. An example involving the Bayesian estimation of the regression parameters that may affect the rate of occurrence of events is explained in detail. A test of the hypothesis that the regression parameters have no effect is also presented. This chapter ends with a Bayesian approach to the general continuous-time Markov chain, where the parameters of interest are the average occupation of a given state and the transition probabilities of moving from one state to the others.

The reader should be aware of the basic references for continuous-time Markov chains, both from a classical and a Bayesian perspective. From a non-Bayesian view, see Guttorp[16] and Ross,[17] while for a Bayesian flavor, refer to Geweke, Marshal, and Zarkin,[18] and lastly, for those interested in reliability, see Cano, Moguerza, and Rios-Insua.[19] In addition, pages 103–105 of Insua, Ruggeri, and Wiper[2] list more relevant references.

6.10 Exercises

1. Define a Poisson process with parameter λ.
2. Refer to Section 6.2 and show that the three definitions a, b, and c of a Poisson process are equivalent.
3. Show that the interarrival times of a Poisson process with parameter λ are i.i.d. exponential with mean $1/\lambda$.
4. Use R Code 6.1 with t<-40 and lamda<-1/2 to generate the interarrival times of a Poisson process with parameter $\lambda = 1/2$. How many interarrival times are generated?
5. Show that the distribution of the waiting time to the nth event is gamma with parameters n and $\sum_{i=1}^{i=n} t_i$, where $\sum_{i=1}^{i=n} t_i$ is the waiting time to the nth event.
6. Assume one has observed n events of a Poisson process with parameter λ and that one employs the improper prior density $g(\lambda) \propto 1/\lambda$, $\lambda > 0$ for λ. What is the posterior distribution of λ?
7. Execute WinBUGS Code 6.1 with 45,000 observations for the simulation and a burn-in of 5,000 and verify the posterior analysis reported in Table 6.1. What is the 95% credible interval for M? Is the posterior distribution of M symmetric about its mean?
8. Derive the predictive distribution of the waiting time S_{15} given by Equation 6.21.
9. Refer to Section 6.4 and explain the idea behind the thinning Poisson processes.
10. Using R Code 6.2, do the following:
    a. Generate the birth times that follow a Poisson process with $\lambda = 2$ over an 8-hour period.
    b. Generate the birth arrival times of males using R Code 6.3 with $\lambda = 1.03$ over an 8-hour period.
    c. Show that the posterior distribution of λ of the overall birth rate is gamma with parameters 18 and $S_{18} = 7.524868$.
    d. Show that the posterior distribution of $\gamma = p\lambda$ is gamma (11, 6.779414).
    e. Verify Table 6.4, the Bayesian analysis for p, λ, and $\gamma = p\lambda$.

11. a. Verify Table 6.6, the posterior analysis of the three parameters p, λ, and μ for the three Italian quake zones. The Bayesian analysis is based on data from Table 6.5. Note that λ is the overall rate of earthquakes, while λp and $\lambda(1-p)$ are the rates for major and minor earthquakes, where p is the probability of a major quake.
    b. Execute WinBUGS Code 6.4 with 45,000 observations and 5,000 for the burn-in, and then verify the posterior analysis reported in Table 6.7, the Bayesian analysis for the three earthquake zones.
12. Derive the predictive density (Equation 6.29) of $Y(n+1)$, the future magnitude of a major earthquake in zone 1.
13. Refer to Section 6.5 for the spatial Poisson process.
    a. Define a spatial Poisson process.
    b. Use R Code 6.4 with $\lambda = 100$ with a square area of 1 (the unit square), using 100,000 trials for generating observations for a spatial process. The goal is to estimate the number of points that fall with a circle of radius $r = .2$ and center (.7,.7). For your answer, refer to Table 6.10.
    c. By referring to R Code 6.4, explain how the plot of Figure 6.2 is conducted.
    d. Execute WinBUGS Code 6.5 with 45,000 observations for the simulation and burn-in 5,000 and verify the posterior analysis reported in Table 6.11.
14. a. Refer to Section 6.6 on concomitant Poisson processes and define the following and how they play a role in such chains: (1) independence, (2) complete similarity, and (3) partial similarity.
    b. Refer to Section 6.6.2 on the complete similarity between several Poisson processes and use WinBUGS Code 6.6 with $t = 10$ and $\lambda = 7$ to generate the times of accidents over a 10-day period. Refer to Table 6.12 and see if your results are similar to those in this table.
    c. With the data generated in item b, assuming an improper prior distribution for λ, show that the posterior distribution of the accident rate λ is gamma (56, 9.9).
15. Partial similarity is defined by Equation 6.51.
    a. Use WinBUGS Code 6.6 with 45,000 observations for the simulation and 5,000 for the burn-in to generate the accident times at five intersections over a 10-day period with daily rates $\lambda_1 = 5.5$, $\lambda_2 = 6.0$, $\lambda_3 = 6.6$, $\lambda_4 = 7.0$, $\lambda_5 = 7.5$ for the five intersections. Your results should be similar to those reported in Table 6.13.
    b. Execute WinBUGS Code 6.7 with 35,000 observations for the simulation and a burn-in of 5,000. The posterior distribution for the five accident rates λ_i, $i = 1, 2, 3, 4, 5$ is reported in Table 6.13. Your results should be similar to those.
    c. Are these five posterior distributions symmetric about their posterior means?
    d. Do the 95% credible intervals include the corresponding true λ_i, $i = 1, 2, 3, 4, 5$ values used to generate the accident rates for the five intersections?

16. a. Execute WinBUGS Code 6.8 with 55,000 observations for the simulation and a burn-in of 5,000. Verify the posterior analysis reported in Table 6.14.
    b. Are the posterior means for λ, μ, and $\lambda\mu$ reasonable estimates? Explain why.
    c. What do the parameters λ, μ, and $\lambda\mu$ represent?
    d. What prior distribution is used for λ and μ?
17. There are several ways to incorporate covariates into the Poisson process.
    a. Execute WinBUGS Code 6.9 with 35,000 observations for the simulation and a burn-in of 5,000. Use your results to verify Table 6.14, the Bayesian analysis for two regression coefficients.
    b. Based on the results of Table 6.15, is $\beta_1 = 0$? Explain your answer.
18. a. Execute R Code 6.8 using $M = 1$ and beta $= 1$; generate the arrival times listed in Equation 6.75. The mean value function for this nonhomogeneous Poisson process is $m(t) = 6, 0 < t < 20$.
    b. Execute R Code 6.9 with the mean value function $m(t) = t^2, 0 < t < 10$; that is, use t<1= and M<-1, and beta<-1 as inputs for the code. Your results should be similar to the 94 arrival times reported below Equation 6.76.
19. a. Execute WinBUGS Code 6.10 with 40,000 observations for the simulation and a burn-in of 3,000 and verify Table 6.16.
    b. Is the estimate of M reasonable? Explain your answer.
    c. Are the estimates of β_1, β_2, and β_3 reasonable? Explain your answer.
    d. What prior distribution is used for β_1, β_2, β_3, and M?
    e. Why are the posterior means of these four parameters so close to the true values?
20. Refer to Section 6.8.
    a. Define a continuous-time Markov chain. What is the Markov property of such a process?
    b. What is the distribution of the interarrival times?
    c. What is the parameter of the interarrival time when the process leaves state i but has yet to enter a different state j?
    d. Describe the transition probability matrix of a continuous-time Markov chain.
    e. What are the jumping intensities of such a process?
21. a. Explain the solution $P_{ij}(t)$ to the Kolmogorov system of differential equations (Equation 6.82).
    b. Show that the solution is given by the exponential form (Equation 6.84).
22. Derive the likelihood function (Equation 6.85) for the arrival time parameters ν_i, $i = 1,2,3,4,5$, and the transition probabilities P_{ij}, $i, j = 1,2,3,4,5$.
23. a. What is the prior distribution of the ν_i, $i = 1,2,3,4,5$?
    b. Show that the posterior distribution of $\nu_i, i = 1,2,3,4,5$ is gamma$(n_i + a_i, t_i + b_i)$, $i = 1,2,3,4,5$.
    c. Show that the posterior distribution of the rows of the transition probabilities is Dirichlet.
    d. What is the prior distribution of the rows of the transition probabilities?

24. a. Execute WinBUGS Code 6.12 with 41,000 observations for the simulation and a burn-in of 2,500.
    b. Verify the Bayesian analysis reported in Table 6.19.
    c. What is the posterior mean of the interarrival time when the process leaves state i?

References

1. Dobrow, R. P. 2016. *Introduction to Stochastic Processes with R*. New York: John Wiley & Sons.
2. Insua, D. R., Ruggeri, F., and Wiper, M. P. 2012. *Bayesian Analysis of Stochastic Process Models*. New York: John Wiley & Sons.
3. Albert, J. 1985. Simultaneous estimation of Poisson means under exchangeable and independent models, *Journal of Statistical Computation and Simulation* 23:1–14.
4. Lee, P. M. 1989. *Bayesian Statistics: An Introduction, Second Edition*. New York: John Wiley & Sons.
5. Rotondi, R., and Varini, E. 2003. Bayesian analysis of a marked point process: Application in seismic hazard assessment, *Statistical Methods & Application* 12:79–92.
6. Vere-Jones, D. 1970. Stochastic models for earthquake occurrence, *Geophysical Journal of the Royal Astronomical Society* 42:811–826.
7. Vere-Jones, D., and Ozaki, T. 1982. Some examples of statistical estimation applied to earthquake data, *Annals of the Institute of Statistical Mathematics* 34:189–207.
8. Ogata, Y. 1988. Statistical methods for earthquake occurrences and residual analysis for point processes, *Journal of the American Statistical Association* 83:9–27.
9. Ruggeri, F. 1993. Bayesian comparison of Italian earthquakes, *Quaderno IAMI*, 93.8, Milano: CNR-IAMI.
10. Bivand, R. S., Pebesma, E. G., and Gomez-Rubio, V. 2013. *Applied Spatial Data Analysis with R*. New York: Springer.
11. Blangiardo, M., and Carmeletti, M. 2015. *Spatial and Spatio-Temperal Bayesian Models with R*. New York: John Wiley & Sons.
12. Ntzoufras, I. 2009. *Bayesian Modeling Using WinBUGS*. New York: John Wiley & Sons.
13. Cox, D. R., and Lewis, I. A. W. 1966. *Statistical Analysis of Series of Events*. London: Methuen.
14. Musa, J. D., Iannio, A., and Okumoto, K. 1987. *Software Reliability Measurement, Prediction, Application*. New York: Mc Graw-Hill.
15. Moler, C., and Van Loan, C. 2003. Nineteen dubious ways to compute the exponential of a matrix, twenty five years later, *Siam Review* 45:3–49.
16. Guttorp, P. 1995. *Stochastic Modeling of Scientific Data*. Boca Raton, FL: Chapman and Hall.
17. Ross, S. M. 2009. *Introduction to Probability Models, 10th Edition*, New York: Academic Press.
18. Geweke, J., Marshal, R., and Zarkin, G. 1986. Mobility indices in continuous time Markov chains, *Econometrica* 54:1407–1423.
19. Cano, J., Moguerza, J., and Insua, D. R. 2010. Bayesian reliability, availability, and maintainability analysis for hardware systems described through continuous time Markov chains, *Technometrics* 52:324–334.

7

Bayesian Inference: Examples of Continuous-Time Markov Chains

7.1 Introduction

In Chapter 6, details of the Poisson process, an important case of continuous-time Markov chains (CTMCs), were presented. In this chapter, Bayesian inferences for general CTMC are presented. First to be considered are the important concepts involved in the study of such processes. For example, the ideas of transition rates, holding times, embedded chain, and transition probabilities are defined and explained. Such concepts are illustrated by using R to compute the transition function and to generate observation from the CTMC. This is followed by a presentation of Bayesian inferences of estimation and testing hypotheses about the unknown parameters of the process. Also developed is the Bayesian predictive distribution for future observations of the CTMC.

As with discrete time chains, the understanding of stationary distributions, absorbing states, mean time to absorption, and time reversibility is an essential part when discussing CTMCs.

The chapter is concluded with many examples, including deoxyribonucleic acid (DNA) evolution, birth and death processes, and queuing.

7.2 Foundation of Continuous-Time Markov Chains

The reader should refer to Chapter 7 of Dobrow[1] for an introduction to the essential elements that explain the properties of CTMC and to Chapter 5 of Insua, Ruggeri, and Wiper[2] for a good account of Bayesian inferential procedures pertaining to CTMCs. These references will be closely followed in what is to follow. For those interested in biological examples of CTMCs, see Chapters 6 and 7 of Allen[3] for an in-depth informative approach to the topic. As we have seen, a CTMC behaves like the discrete version except the time between states is distributed as an exponential random variable and events can occur at any time. Thus, there are two sets of parameters that define the CTMC, the parameters of the exponentially distributed interarrival times and the probability transition parameters. As will be seen, the two sets are related by the transition rate parameters.

7.2.1 Markov Property and Transition Function

Let $\{X(t), t > 0\}$ be a continuous-time stochastic process; then it is called a CTMC if

$$\begin{aligned} &P[X(t+s) = j|X(s) = i, X(u) = x(u), 0 \le u < s] \\ &= P[X(t+s) = j|X(s) = i] \end{aligned} \tag{7.1}$$

for all states i, j, and $x(u)$, $0 \le u < s$. Of course, it is understood that the state space is countable.

Also, a CTMC is time homogenous, that is to say,

$$\begin{aligned} P[X(t+s) = j|X(s) = i] &= P[X(t) = j|X(0) = i] \\ &= P_{ij}(t). \end{aligned} \tag{7.2}$$

Thus, the probabilistic properties of a CTMC over the interval $[s, t+s]$ are the same as that over the interval $[0, t]$. Or to express it in another way, when the chain visits state i, its forward behavior from that time toward the future is the same as if the process started in i at time $t = 0$. Note that the function $P_{ij}(t)$ is called the transition function of the process.

Recall that for a Poisson process, the interarrival times are identically exponentially distributed; however, for the CTMC, there is a difference as follows. Let T_i be the holding time of the process that is the time the process occupies state i before switching to another state, and then it can be shown that T_i has an exponential distribution.

To show that T_i is memoryless, consider

$$\begin{aligned} &P[T_i > s+t|X(0) = i] = \\ &P[T_i > s+t, T_i > s|X(0) = i] = \\ &P[T_i > s+t|X(0) = i, T_i > s]P[T_i > s] = \\ &P[T_i > t|X(0) = i]P[T_i > s|X(0) = i], \end{aligned} \tag{7.3}$$

and the last statement of Equation 7.3 shows that the process is memoryless. Recall that the only continuous distribution that is memoryless is the exponential. The evolution of the process can be described as follows: Starting in state i, the process remains in this state for an exponentially distributed time with parameter q_i (average time in this state is $1/q_i$); then it hits a new state j with probability p_{ij} and remains in state j for a time which has an exponential distribution with mean $1/q_j$; then it hits a new state k, $k \ne j$, with probability p_{jk}; etc.

I am assuming that the process does not hit an absorbing state and that the process is not explosive. See pages 268 and 269 of Dobrow[1] for additional information. There is a connection between the transition probabilities p_{ij} and the exponential parameters q_i of the holding times, and this connection is the transition rate of the process.

7.2.2 Transition Rates, Holding Times, and Transition Probabilities

An alternative to describing a CTMC is by the transition rates between pairs of states. When the process is in state i, there is a chance that it will change to one of the other possible states; that is, state i is paired with state j for all states $j \ne i$. If j can be reached from i, one can associate an alarm that is activated after a time that has an exponential distribution with parameter q_{ij}. When state i is first occupied, the alarms are all started at the same time, and the first alarm that is activated determines the next state to be occupied. If the alarm (i,j) is first activated and the process moves to state j, a new set of alarms are activated with exponential transition rates $q_{j1}, q_{j2}, \ldots$. Thus, to repeat, the first alarm that is activated determines the next state to be occupied, etc. The q_{ij} are called transition rates, and from them, the transition probabilities and holding time parameters can be determined.

Suppose the process starts at i, then the alarms are initiated, and the first one that is activated determines the next transition; therefore, the time of the first alarm is the minimum of independent exponential random variable with parameters $q_{i1}, q_{i2}, \ldots$, which is an exponential random variable with parameter $\sum_k q_{ik}$. Thus, the process remains in state i for a holding time, which has an exponential distribution with parameter $\sum_k q_{ik} = q_i$. From i, the chain moves to state j if the alarm (i,j) is activated first, which occurs with probability

$$p_{ij} = q_{ij}/q_i, \tag{7.4}$$

which is the transition probability of moving from state i to state j of the embedded chain. One sees from Equation 7.4 that the transition probabilities of the chain are completely determined by the transition rates q_{ij}.

As a simple example, consider a four-state chain with the transition rates of the chain given by

$$(q_1, q_2, q_3, q_4) = (q_{12} + q_{13} + q_{14}, q_{21} + q_{23} + q_{24}, q_{31} + q_{32} + q_{34}, q_{41} + q_{42} + q_{43}). \tag{7.5}$$

Note that the embedded chain has transition matrix

$$P = \begin{pmatrix} 0, q_{12}/q_1, q_{13}/q_1, q_{14}/q_1 \\ q_{21}/q_2, 0, q_{23}/q_2, q_{24}/q_2 \\ q_{31}/q_3, q_{32}/q_3, 0, q_{34}/q_3 \\ q_{41}/q_4, q_{42}/q_4, q_{43}/q_4, 0 \end{pmatrix}. \tag{7.6}$$

The transition rates q_{ij} are quite important when studying CTMC. Assume that the CTMC has a differentiable transition function $P(t)$, where $p_{ij}(0) = 1$ if $i = j$; otherwise, its value is 0.

Note that if $X(t) = i$, then the instantaneous transition rate of hitting $j \neq i$ is given by

$$\begin{aligned} &\lim_{h \to 0^+} E(\text{number of transitions to } j \text{ in } (t, t+h]) \\ &= \lim_{h \to 0^+} P[X(t+h) = j \mid X(t) = i] \\ &= \lim_{h \to 0^+} p_{ij}(h)/h \\ &= (d/dt) p_{ij}(0) \\ &= p'_{ij}(0). \end{aligned} \tag{7.7}$$

Consider the matrix $Q = P'(0)$; then the off-diagonal elements of Q are the transition rates q_{ij}; that is, $q_{ij} = Q_{ij}$, $i \neq j$, and the diagonal entries are $-q_i$; thus, each row of Q has sum 0.

As an example, consider the four-state chain in Equation 7.6, where $q_{12} = q_{13} = q_{14} = 1, q_{21} = q_{23} = q_{24} = 2, q_{31} = q_{32} = q_{34} = 3$, and $q_{41} = q_{42} = q_{43} = 4$; thus, $q_1 = 3, q_2 = 6, q_3 = 9$, and $q_4 = 12$. For this four-state chain, the process remains in state 1 for an average of 1/3 hours, in state 2 for 1/6 hours, and in state 3 for 1/9 hours, and in state 4 for an average of 1/12 hours. Consequently, the infinitesimal generator matrix is

$$Q = \begin{pmatrix} -3,1,1,1 \\ 2,-6,2,2 \\ 3,3,-9,3 \\ 4,4,4,-12 \end{pmatrix} \tag{7.8}$$

and $\sum_i \pi_i Q_{ij} = 0, \ \forall j$. Since the transition rates determine the transition probabilities of the embedded chain, which is the matrix

$$P = \begin{pmatrix} 0,1/3,1/3,1/3 \\ 1/3,0,1/3,1/3 \\ 1/3,1/3,0,1/3 \\ 1/3,1/3,1/3,0 \end{pmatrix}. \tag{7.9}$$

This is a very special transition matrix, because I chose the transition rates in such a way that each transition has the same chance of occurring, namely, 1/3.

7.2.3 Kolmogorov Forward and Backward Equations and the Matrix Exponential

We now see the role the equation $p_{ij} = q_{ij}/q_i$ plays in determining the transition probability matrix $P(t)$, which is a solution to the forward Kolmogorov equation

$$P'(t) = P(t)Q,$$

where Q is the infinitesimal matrix and $P'(t)$ is the derivative matrix with respect to t of the transition probability matrix.

This is also expressed as

$$P'_{ij}(t) = \sum_k p_{ik}(t)q_{kj} = -p_{ij}(t)q_j + \sum_{k \neq j} p_{ik}(t)q_{kj}. \tag{7.10}$$

For a proof of Equation 7.10, see page 276 of Dobrow.[1] It is obvious that the solution $P(t)$ to Equation 7.10 is given by the matrix equation

$$P(t) = \exp(tQ), \quad t \geq 0, \tag{7.11}$$

where $P(0) = I$.

Also, the solution can be written as

$$P'(t) = (\mathrm{d}/\mathrm{d}t)e^{tQ} = \sum_{n=0}^{n=\infty}(1/n!)(tQ)^n = I + tQ + t^2Q^2/2 + t^3Q^3/3! + \ldots.$$

The next section will use R to express solution $P(t)$ in matrix form of Equation 7.11.

7.2.4 Computing the Transition Function with R

As an example, consider the four-state CTMC, where Q is given by Equation 7.8. R Code 7.1 computes the probability transition matrix as the solution in Equation 7.11 to the differential equation (the forward Kolmogorov equations) in Equation 7.10. Note that the package "expm" must be loaded in order to execute the matrix exponential operation.

R Code 7.1

```
>install.packages("expm")
>library(expm)
# Q is the infinitesimal generator matrix

> Q<-matrix(c(-3,1,1,1,
+          2,-6,2,2,
+          3,3,-9,3,
+          4,4,4,-12),ncol=4,nrow=4,byrow=TRUE)
# the following is the command that executes the matrix exponential
> P<- function (t){expm(t*Q)}
# P(2) is the probability transition matrix at 2
> P(2)
          [,1]      [,2]      [,3]      [,4]
[1,] 0.4800078 0.2399952 0.1599982 0.1199989
[2,] 0.4799903 0.2400060 0.1600023 0.1200014
[3,] 0.4799945 0.2400034 0.1600013 0.1200008
[4,] 0.4799954 0.2400028 0.1600011 0.1200007
# P(0) is the probability transition matrix at t=0.
# this serves as a check since P(0)=I, the identity matrix
> P(0)
     [,1] [,2] [,3] [,4]
[1,]    1    0    0    0
[2,]    0    1    0    0
[3,]    0    0    1    0
[4,]    0    0    0    1
> P(4)
     [,1] [,2] [,3] [,4]
[1,] 0.48 0.24 0.16 0.12
[2,] 0.48 0.24 0.16 0.12
[3,] 0.48 0.24 0.16 0.12
[4,] 0.48 0.24 0.16 0.12
> P(1.2)
```

```
          [,1]      [,2]      [,3]      [,4]
[1,] 0.4806526 0.2395971 0.1598456 0.1199047
[2,] 0.4791942 0.2405007 0.1601882 0.1201169
[3,] 0.4795368 0.2402823 0.1601124 0.1200685
[4,] 0.4796187 0.2402339 0.1600914 0.1200561
>> P(2)
          [,1]      [,2]      [,3]      [,4]
[1,] 0.4800078 0.2399952 0.1599982 0.1199989
[2,] 0.4799903 0.2400060 0.1600023 0.1200014
[3,] 0.4799945 0.2400034 0.1600013 0.1200008
[4,] 0.4799954 0.2400028 0.1600011 0.1200007
> P(0)
     [,1] [,2] [,3] [,4]
[1,]    1    0    0    0
[2,]    0    1    0    0
[3,]    0    0    1    0
[4,]    0    0    0    1
> P(4)
     [,1] [,2] [,3] [,4]
[1,] 0.48 0.24 0.16 0.12
[2,] 0.48 0.24 0.16 0.12
[3,] 0.48 0.24 0.16 0.12
[4,] 0.48 0.24 0.16 0.12
> P(1.2)
          [,1]      [,2]      [,3]      [,4]
[1,] 0.4806526 0.2395971 0.1598456 0.1199047
[2,] 0.4791942 0.2405007 0.1601882 0.1201169
[3,] 0.4795368 0.2402823 0.1601124 0.1200685
[4,] 0.4796187 0.2402339 0.1600914 0.1200561
>
```

For example, when the time is $t = 1.2$ hours, the transition probability matrix is $P(1.2)$; thus, the probability of changing states from state 1 to 2 over a 1.2-hour interval is .2395, etc.

7.3 Limiting and Stationary Distributions

Limiting and stationary distributions for CTMCs are determined in much the same way as that for discrete-time Markov chains (DTMCs).

Recall the definition of the limiting distribution (as $t \to \infty$) π of a process as

$$\lim P_{ij}(t) = \pi_j, \tag{7.12}$$

where π_j is the jth component of the vector π. The limiting distribution does not have to exist, but if it does, it is the stationary distribution of the chain. Of course, the stationary distribution of a CTMC is defined as the vector π that satisfies

$$\pi = \pi P(t), \quad t \geq 0, \tag{7.13}$$

which is equivalent to

$$\pi_j = \sum_i \pi_i P_{ij}(t), \quad t \geq 0 \tag{7.14}$$

for all states j.

Note that the ideas of accessibility, communicating classes, and irreducibility carry over from the discrete case to the continuous case. However, note that for CTMC, all states are aperiodic. Under what conditions does a CTMC have a stationary distribution?

7.3.1 Basic Limit Theorem

Let $\{X(t), t \geq 0\}$ be a CTMC, which is irreducible with a finite state space; then the process has a unique stationary distribution, given by the limiting distribution $\lim_{t\to\infty} P_{ij}(t) = \pi_j$ for all initial states i or expressed in an equivalent fashion as

$$\lim_{t\to\infty} P(t) = \Pi, \tag{7.15}$$

where each row of Π is the same vector, namely, π.

As example, consider the two-state CTMC with probability transition matrix

$$P(t) = [1/(\lambda+\mu)]\begin{pmatrix} \mu + \lambda e^{-(\lambda+\mu)t}, \lambda - \lambda e^{-(\lambda+\mu)t} \\ \mu - \mu e^{-(\lambda+\mu)t}, \lambda + \mu e^{-(\lambda+\mu)t} \end{pmatrix}. \tag{7.16}$$

Therefore, in the limit, the matrix reduces to

$$P(t) = [1/(\lambda+\mu)]\begin{pmatrix} \mu, \lambda \\ \mu, \lambda \end{pmatrix},$$

and the stationary distribution is

$$\pi = (\mu/(\lambda+\mu), \lambda/(\lambda+\mu)). \tag{7.17}$$

There is a relation between the stationary distribution and the infinitesimal generator matrix Q, given by

$$\pi Q = 0, \tag{7.18}$$

or, in scalar terms,

$$\sum_i \pi_i Q_{ij} = 0, \quad \forall j.$$

The following example is from pages 286 and 287 of Dobrow[1] with the infinitesimal generator matrix

$$Q = \begin{pmatrix} -2.0, 1.0, 1.0 \\ 1/2, -1, 1/2 \\ 0, 1/3, -1/3 \end{pmatrix}. \tag{7.19}$$

This corresponds to a three-state chain with states eat, play, and sleep.

A newborn baby is in one of three states: eat, play, and sleep. The baby eats on the average for 30 minutes, plays for an average of 1 hour, and, on the average, sleeps for approximately 3 hours. After eating, there is a 50–50 chance that he will sleep or play, and after playing, there is a 50–50 chance that he will sleep or eat. Lastly, after sleeping, he will always eat. The corresponding transition matrix of the embedded chain is

$$P = \begin{pmatrix} 0, 1/2, 1/2 \\ 1/2, 0, 1/2 \\ 0.0, 1.0, 0, 0 \end{pmatrix}. \tag{7.20}$$

Then, using the fact that $q_{ij} = q_i p_{ij}$, the generator matrix is given by Equation 7.19.

Instead of using Equation 7.18, let us find the stationary distribution of this process with R Code 7.2. The corresponding code computes the corresponding P matrix and then approximates the limiting distribution $P(100)$:

R Code 7.2

```
> Q<-matrix(c(-2,1,1,
+         1/2,-1,1/2,
+         0,1/3,-1/3),nrow=3,byrow=TRUE)
> P<-function(t){expm(t*Q)
+ }
> P(100)
           [,1]       [,2]      [,3]
[1,] 0.07142857 0.2857143 0.6428571
[2,] 0.07142857 0.2857143 0.6428571
[3,] 0.07142857 0.2857143 0.6428571
```

Thus, it is found that the stationary distribution is the vector $\pi = (.07142, .28571, .64288)$; therefore, regardless of the initial activity of the baby, in the long run, the chance of eating, playing, and sleeping is the same as that given by the rows of $P(100)$.

7.4 Mean Time to Absorption with R

As with discrete-time processes, a CTMC can have absorbing states; thus, suppose that $\{X(t), t > 0\}$ has states $\{1, 2, ..., k\}$, where one of those states, say, a, is absorbing; all the other $k - 1$ states are transient. If the chain starts in a transient state, there is a positive probability that the chain will be absorbed.

Suppose the generator matrix is partitioned as

$$Q = \begin{pmatrix} 0, 0^* \\ *, V \end{pmatrix}, \tag{7.21}$$

where V is the $k - 1$ order submatrix of transient states; then if the chain has initial state i, the mean time to absorption is

$$a_i = \sum_j F_{ij}, \tag{7.22}$$

where F_{ij} is the ijth element of the fundamental matrix

$$F = V^{-1}. \tag{7.23}$$

Consider the matrix

$$Q = \begin{pmatrix} -(q_{12} + q_{13}), q_{12}, q_{13} \\ 0.000000, -q_{23}, q_{23} \\ 0, 00000, 0.00, 0.0 \end{pmatrix}. \tag{7.24}$$

Then according to Bartolomeo, Trerotoli, and Serio,[4] the matrix is the generator matrix corresponding to the progression of liver disease among three states: state 1 corresponds to cirrhosis; state 2, liver cancer; and state 3, death. Of course, state 3 is an absorbing state; thus, it is of interest to determine the average time to death, beginning from cirrhosis and from liver cancer.

The fundamental matrix is

$$F = \begin{pmatrix} -(q_{12} + q_{13}), q_{12} \\ 0.000000, -q_{23} \end{pmatrix}^{-1} = \begin{pmatrix} 1/(q_{12} + q_{13}), q_{12}/q_{23}(q_{12} + q_{13}) \\ 0.00000000000000000, 1/q_{23} \end{pmatrix}, \tag{7.25}$$

and the mean time to absorption for a person who has cirrhosis is

$$a_1 = [1/(q_{12} + q_{13})] + [q_{12}/q_{23}(q_{12} + q_{13})]. \tag{7.26}$$

In a similar way, for those with liver cancer, the mean time to death is

$$a_2 = 1/q_{23}. \tag{7.27}$$

For the statistician, one would have to have sample information about the transition rates q_{12}, q_{13}, and q_{13}. Bartolomeo, Trerotoli, and Serio's study[4] estimated these rates as follows: $\tilde{q}_{12} = .0151$, $\tilde{q}_{13} = .0071$, and $\tilde{q}_{23} = .0284$, and these can be substituted into Equations 7.26 and 7.27 for estimates of the mean time to death for those with cirrhosis of the liver and liver cancer, respectively.

Using these estimates, R Code 7.3 computes an estimate, using the simulation of the average time to death for a patient diagnosed initially with cirrhosis of the liver, where the time unit is months:

R Code 7.3

```
> trials<-10000
> simlist<-numeric(trials)
> init<-1
> for ( i in 1:trials){
+ state<-init
+ t<-0
+ while (TRUE){
+ if (state==1){ q12<-rexp(1,0.0151)
+ q13<-rexp(1,.0071)}
+ if (q12<q13) {t<-t+q12
+ state<-2}
+ else{t<-t+q13
+ break}
+ if (state==2){q23<-rexp(1,.0284)
+ t<-t+q23
+ break}
+ }
+ simlist[i]<-t}
> mean(simlist)
[1] 68.37068
```

Therefore, one's estimate of the average time to death for a person initially diagnosed with cirrhosis of the liver is 68.37068 months!

7.5 Time Reversibility

Similar to the discrete case, time reversibility is the last major concept to be defined for a CTMC. A CTMC with generator matrix Q and stationary distribution π is said to be time reversible if and only if

$$\pi_i q_{ij} = \pi_j q_{ji} \quad \forall i, j. \tag{7.28}$$

Time reversibility is related to the idea of global balance. Let π be the stationary distribution to the CTMC; that is, $\pi Q = 0$ is satisfied, which in turn implies

$$\sum_{i \neq j} \pi_i q_{ij} = \pi_j q_j, \quad \forall j. \tag{7.29}$$

Note that the holding time parameter q_j is the transition rate from j and that π_j is the long-time proportion of the time the process is in state j; thus, the right-hand side of Equation 7.29 is the long-term rate the process leaves j. Also, it is clear that $\pi_i q_{ij}$ is the long-term rate of transition from i to j; thus, the left-hand side of Equation 7.29 is the long-term rate that the process enters state j. This in turn implies that for a stationary process, the rates in and out of any state are the same, and the equations in Equation 7.29 are called the global balance equations. As an example, let

$$P = \begin{pmatrix} 0.0, 1.0, 0.0 \\ 1/3, 0, 2/3 \\ 0.0, 1.0, 0.0 \end{pmatrix} \tag{7.30}$$

be the transition matrix of an embedded chain, where the process remains in state 1 for an average of 5 minutes before moving to state 2, where it remains for an average of 2 minutes before moving to state 3, which is occupied for an average of 4 minutes before moving to state 2. It can easily be shown that the stationary distribution is $\pi = (5/19, 6/19, 8/19)$; thus Equation 7.29 implies

$$(\pi_1 q_1, \pi_2 q_2, \pi_3 q_3) = (1/19, 3/19, 2/19). \tag{7.31}$$

Thus, the global balance for this example shows that every 19 minutes, the process incurs one transition to and from state 1, three transitions to and from state 2, and two transitions to and from state 3.

This concludes the fundamental properties that need to be understood in order to perform Bayesian inferences for CTMCs.

For this part of the chapter, Bayesian inferences will be performed for a variety of examples, including some from biology, physics, and business.

7.5.1 DNA Evolution

Recall that DNA evolution was presented for the discrete-time case in Section 5.7, where the chain had four states, namely, the four base nucleotides: (1) adenine, (2) guanine, (3) cytosine, and (4) thymine. In the Jukes–Cantor model, the transition rates are all the same with infinitesimal generator matrix

$$Q = \begin{pmatrix} -3r, r, r, r \\ r, -3r, r, r \\ r, r, -3r, r \\ r, r, r, -3r \end{pmatrix}, \tag{7.32}$$

where the first row and column correspond to adenine; the second row and column, to guanine; the third row and column, to cytosine; and the last row and column, to thymine. Thus, the corresponding transition matrix is

$$P(t) = \exp(tQ)$$

$$= (1/4) \begin{pmatrix} 1+3e^{-4rt}, 1-e^{-4rt}, 1-e^{-4rt}, 1-e^{-4rt} \\ 1-e^{-4rt}, 1+3e^{-4rt}, 1-e^{-4rt}, 1-e^{-4rt} \\ 1-e^{-4rt}, 1-e^{-4rt}, 1+3e^{-4rt}, 1-e^{-4rt} \\ 1-e^{-4rt}, 1-e^{-4rt}, 1-e^{-4rt}, 1+3e^{-4rt} \end{pmatrix}. \tag{7.33}$$

Thus, at time t, the probability that the DNA base adenine is replaced by quinine (or by cytosine or by thymine) is $(1/4)(1-e^{-4rt})$, etc. On the other hand, the probability that adenine at time t is not replaced by another base is $(1/4)(1+3e^{-4rt})$.

The statistical problem is to make inferences about rate r based on observing the evolution at various times.

Based on Equation 7.32, the infinitesimal rates are

$$q_{ij} = r, \quad i \neq j,\ i,j = 1,2,3,4; \tag{7.34}$$

thus, the holding time exponential parameters are

$$q_i = 3r, \quad i = 1,2,3,4. \tag{7.35}$$

Recall that the transition probabilities are given by

$$p_{ij} = q_{ij}/q_i = r/3r = 1/3, \quad i \neq j,\ i,j = 1,2,3,4. \tag{7.36}$$

If one assigns a value to r, one can make inferences about the holding time parameters in Equation 7.35 and the transition probabilities in Equation 7.36; however, one knows that the transition probabilities are all 1/3; thus, only the holding time exponential parameters will be of interest.

In practice, one would observe the holding times of the various states and then from those observations estimate r. Using WinBUGS, we will assume a value of r and then generate the exponential holding times. Consider the holding time for occupying the first-state adenine and assume that its mean time is 2 time units. The WinBUGS program that follows is based on 50 observations of the holding time for adenine with an average holding time of 2 time units. The main objective is to estimate the average holding time for adenine.

The holding time T_1 for adenine has an exponential distribution with parameter λ, namely,

$$f(t_1) = \lambda \exp(-\lambda t_1), \quad t_1 > 0; \tag{7.37}$$

here the mean holding time is

$$E(T_1|\lambda) = 1/\lambda = \mu. \tag{7.38}$$

Let $\lambda = 1/6$ and generate the 50 observations in the list statement of WinBUGS Code 7.1, which corresponds to a mean holding time of 6 time units. Refer to the generator matrix in Equation 7.32 for the DNA example where the holding time exponential parameter is $3r$; thus I let $3r = 1/6$. It is also assumed that the prior distribution of λ is gamma (.001,.001), a noninformative prior with mean of 1 and variance of 1000.

WinBUGS Code 7.1 is executed with 45,000 observations for the simulation and a burn-in of 5,000. The results of the analysis are reported in Table 7.1.

TABLE 7.1

Posterior Analysis for Holding Time for Adenine

Parameter	Mean	SD	Error	2 1/2	Median	97 1/2
λ	.163	.02314	.0001259	.121	.1618	.2118
μ	6.259	0.9034	0.004947	4.721	6.181	8.264

WinBUGS Code 7.1

```
model;
{
lamda~dgamma(.001,.001)
for( i in 1:50){
T1[i]~dexp(lamda)}
mu<-1/lamda
}
list(
T1 = c(5.968,6.741,9.346,6.269,6.117,
3.056,5.674,1.377,1.864,1.956,
0.1391,14.28,22.06,3.203,6.415,
5.078,5.958,0.9036,7.865,5.154,
1.078,5.025,18.92,8.998,24.05,
2.433,0.7362,9.205,0.9738,3.878,
0.9797,1.961,6.133,5.866,2.523,
14.51,3.125,4.142,7.126,2.256,
20.31,17.88,5.138,2.076,1.866,
9.726,0.5665,0.6489,2.325,2.697))
list(lamda=.16)
```

The parameter of interest is μ, the average holding time for adenine, which has a posterior mean of 6.259 with a 95% credible interval of (4.721,8.264). Note that the credible interval contains the value of 6, which was used to generate the observations in the list statement of WinBUGS Code 7.1.

It appears that the distributions are symmetric about their posterior means and that the simulation errors are reasonably small. In summary, based on the posterior median, our estimate of the mean holding time is 6.181. Clearly, the type of analysis can be repeated for the other three holding times. There is nothing new to be learned because the mean holding time μ is the same for all four holding times. The student will be asked to estimate the holding times for the other three DNA bases as an exercise at the end of this chapter.

Now consider another aspect of Bayesian inference for the holding time for the DNA base adenine. Remember that the value $\lambda = 1/6\,(u = 6)$ is used to generate the adenine holding times listed in the list statement of WinBUGS Code 7.1. It is of paramount interest to test the hypothesis that $\lambda = 1/6\,(u = 6)$ is a plausible value of the mean holding time.

Thus, using the Bayesian approach, consider a test of the null hypothesis

$$H_0\colon \lambda = 1/6 \quad \text{versus} \quad H_1\colon \lambda \neq 1/6. \tag{7.39}$$

The approach presented on pages 126–128 of Lee[5] is employed to test a simple null hypothesis versus the two-sided alternative (Equation 7.39).

Recall that the posterior probability of the null hypothesis is given by

$$p_0 = [\pi_0 f(t|\lambda = 1/6)]/[\pi_0 f(t|\lambda = 1/6) + \pi_1 \int_0^\infty \rho_1(\lambda) f(t|\lambda)\,d\lambda], \tag{7.40}$$

where $\pi_0 = P(\lambda = 1/6)$ is the prior probability of the null hypothesis and $\pi_1 = 1 - \pi_0$ is the prior probability of the alternative hypothesis. Also, the probability density of the n holding times is

$$f(t|\lambda) = \lambda^n e^{-\lambda \sum_{i=1}^{i=n} t_i}, \tag{7.41}$$

where $t = (t_1, t_2, .., t_n)$ is the $n \times 1$ vector of holding times.

Also, it is obvious that

$$f(t|\lambda = 1/6) = (1/6)^n e^{-(1/6)\sum_{i=1}^{i=n} t_i}. \tag{7.42}$$

In addition, $\rho_1(\lambda)$ is the prior density of λ under the alternative hypothesis $\lambda \neq 1/6$. How does one choose $\rho_1(\lambda)$? It seems reasonable to choose the improper prior density; thus, let

$$\rho_1(\lambda) = 1/\lambda, \quad \lambda > 0. \tag{7.43}$$

Since $\sum_{i=1}^{i=50} t_i = 306.32$, there is sufficient information to compute the posterior probability of the null hypothesis given by Equation 7.40.

Equation 7.40 for the posterior probability of the null hypothesis is formulated as

$$p_0 = (1/(1 + ra)), \tag{7.44}$$

where

$$ra = \left[\pi_1 \int_0^\infty \rho_1(\lambda) f(t|\lambda)\, d\lambda \right] / [\pi_0 f(t|\lambda = 1/6)]. \tag{7.45}$$

Thus, the posterior probability of the null hypothesis is expressed in terms of the ratio ra. Note that as the ratio ra approaches 0, p_0 approaches 1.

If one lets $\pi_0 = \pi_1 = 1/2$, one can show that

$$\begin{aligned} ra &= \left[\Gamma(50) 6^{50} e^{51.05444}\right] / \left[(306.32)^{50}\right] \\ &= .7356. \end{aligned} \tag{7.46}$$

I took the natural log of ra, used the log gamma function, and then finally used the natural exponential function to arrive at the value of .7356, the posterior probability of the null hypothesis. Thus, based on the 50 exponential holding times for adenine of DNA evolution, and the prior information, one would conclude that $\lambda = 1/6$, that the null hypothesis is plausible. Recall that the value $\lambda = 1/6$ was used to generate the 50 holding times included in the list statement of WinBUGS Code 7.1.

The next objective is to determine the predictive density for the holding times of the DNA base adenine. Let $t(n + 1)$ be the future observation for the holding time for adenine.

Note that the predictive density for $t(n + 1)$ is

$$f(t(n+1)|\text{data}) = \int_0^\infty f(t(n+1)|\lambda)f(\text{data}|\lambda)\zeta(\lambda)d\lambda, \tag{7.47}$$

where

$$f(t(n+1)|\lambda) = \lambda e^{-\lambda t(n+1)}, \quad t(n+1) > 0, \tag{7.48}$$

$$f(\text{data}|\lambda) = \lambda^n e^{-\lambda \sum_{i=1}^{i=n} t_i}, \quad 0 < t_1 < t_2 < \ldots < t_n, \tag{7.49}$$

where t_i is the ith holding time and $\zeta(\lambda)$ is the prior density for λ.

Choosing the improper prior density

$$\zeta(\lambda) = 1/\lambda, \quad \lambda > 0, \tag{7.50}$$

one may show that the predictive density in Equation 7.47 reduces to

$$f(t(n+1)|\text{data}) = \Gamma(n+1)/\left[t(n+1) + \sum_{i=1}^{i=n} t_i\right]^{n+1}, \quad t(n+1) > 0.$$

The predictive density for $t(n + 1)$ is completely determined when the data $n = 50$ and $\sum_{i=1}^{i=n} t_i = 306.32$ are substituted into Equation 7.50. WinBUGS provides a way to compute future values from the predictive density in Equation 7.51. Refer to WinBUGS Code 7.1 and add the command Z[i]~dexp(lamda) and execute WinBUGS Code 7.1 the usual way for the example when I did the posterior analysis with 55,000 observations for the simulation and a burn-in of 5,000; the first five predicted holding times are reported in Table 7.2.

TABLE 7.2

Posterior Predictive Analysis

Future	Mean	SD	Error	2 1/2	Median	97 1/2
$t(51)$	6.186	6.29	0.03337	0.1463	4.21	23.26
$t(52)$	6.254	6.393	0.03186	0.1532	4.236	23.52
$t(53)$	6.226	6.343	0.03236	0.1541	4.292	23.36
$t(54)$	6.209	6.314	0.03075	0.148	4.264	23.48
$t(55)$	6.18	6.312	0.029	0.1498	4.24	23.34

Thus, the mean of the predictive distribution for the first future holding time for adenine is 6.186 time units and a 95% predictive interval of (0.1463,23.26). I used a gamma prior (.001,.001) for λ WinBUGS Code 7.1, but the formula for the predictive density in Equation 7.51 was the improper prior in Equation 7.50; however, there will be very little difference in the predicted future values. Note the uncertainty in the prediction implied by the long with of the prediction intervals.

We next consider a variation of the Jukes–Cantor model called the Kimura[6] model with infinitesimal rates given by the matrix

$$Q = \begin{pmatrix} -(r+2s), r, s, s \\ r, -(r+2s), s, s \\ s, s, -(r+2s), r \\ s, s, r, -(r+2s) \end{pmatrix}. \tag{7.51}$$

Thus, the evolution remains in the base nucleotide adenine for a time with an exponential parameter $r + 2s$. This model distinguishes between the substitutions $a \leftrightarrow g$ (from purine to purine or from pyrimidine to pyrimidine) and transversions (from purine to pyrimidine or vice versa). Recall Section 5.7, where these two evolutionary models (Jukes–Cantor and Kimura) were analyzed in discrete time. Thus, referring to Equation 7.51, the rate r is the exponential time that the process remains in adenine until guanine is substituted. The rate s is the exponential parameter for the holding time in adenine until cytosine is substituted for adenine, and the same rate s is the exponential parameter for the holding time of adenine until adenine is substituted by thymine.

Note that corresponding to the infinitesimal rate matrix Q in Equation 7.51, the probability transition matrix

$$\begin{aligned} P_{ij}(t) &= \left(1 + e^{-4st} - 2e^{-(r+s)t}\right)/4, \quad (i,j) \in \{ag, ga, ct, tc\}, \\ &= \left(1 - 2e^{-4t}\right)/4, \quad (i,j) \in \{ac, at, gc, gt, ca, cg, ta, tg\}, \\ &= \left(1 + e^{-4st} + 2e^{-2(r+s)t}\right)/4, \quad (i,j) \in \{aa, gg, cc, tt\}. \end{aligned} \tag{7.52}$$

See page 282 of Dobrow[1] for additional information about the derivation of Equation 7.52.

The three phases of Bayesian inference will be presented for the Kimura model of molecular evolution. To that end, one must generate the holding times for the two rates r and s and let $r = 1/2$ and $s = 1/8$; thus, the holding time for adenine until the substitution by guanine has an average of 2 time units, while the holding time for adenine until the substitution by cytosine is 8 time units.

I generated data for the holding times T_{12} and T_{13} of adenine until substituted by guanine and cytosine, respectively. The list statement of WinBUGS Code 7.2 contains the 15 holding times, where the exponential parameter for T_{12} is $\lambda_{12} = 1/2$; and for T_{13}, $\lambda_{13} = 1/8$. WinBUGS Code 7.2 is executed with 35,000 observations for the simulation and 5,000 for the burn-in:

WinBUGS Code 7.2

```
model;
{
#lamda13 is the exponential parameter s
lamda13~dgamma(.001,.001)
for ( i in 1:15){

T13[i]~dexp(lamda13)}
# m13 is the average holding time for adenine before it is substituted by
#cytosine
mu13<-1/lamda13

# lamda12 is the exponential parameter r
lamda12~dgamma(.001,.001)
for ( i in 1:15){

T12[i]~dexp(lamda12)}
# mu12 is the average holding time for adenine before it is substituted by
#guanine
mu12<-1/lamda12
# d23 is the difference in the two holding times for adenine
d23<-mu12-mu13
}

list(T13 = c(
11.98,12.58,7.757,24.86,5.816,
8.416,2.104,18.0,3.069,7.62,
0.01324,27.58,3.004,6.996,1.003),
T12 = c(
2.996,3.145,1.939,6.215,1.454,
2.104,0.5259,4.5,0.7673,1.905,
0.003311,6.896,0.751,1.749,0.2509))

list(lamda13=.125,lamda12=.5)
```

The posterior analysis for the two holding times is reported in Table 7.3.

Using the posterior mean of 2.51 time units as the estimate of the average adenine holding time until adenine is substituted by guanine, and 10.07 time units as the estimate of the adenine holding time until substituted by cytosine, leads to a difference between the two estimates as –7.556 time units, which in turn implies that the two are indeed different (because the 95% credible interval of (–14.29,–3.164) does not include zero). This is an informal inference about the difference in two estimates, and it appears that the Kimura model is indeed appropriate and is to be preferred to the Jukes–Cantor model.

TABLE 7.3

Posterior Analysis for the Kimura Model

Parameter	Mean	SD	Error	2 1/2	Median	97 1/2
d_{23}	−7.556	2.858	.01387	−14.29	−7.14	−3.164
λ_{12}	.427	.1109	.000541	.2394	.4163	.6694
λ_{13}	.1064	.0274	.000127	.05981	.1039	.1668
μ_{12}	2.51	0.6957	0.003384	1.494	2.402	4.178
μ_{13}	10.07	2.776	0.0133	5.996	9.62	16.72

However, a more formal Bayesian test of hypothesis of

$$H: \lambda_{12} = \lambda_{13} \quad \text{versus} \quad A: \lambda_{12} \neq \lambda_{13} \tag{7.53}$$

will be presented.

Consider the posterior probability of the null hypothesis

$$p_0 = 1/(1 + \text{ratio}), \tag{7.54}$$

where

$$\text{ratio} = N/D,$$

$$N = \pi_1 \int_0^\infty \int_0^\infty \rho_1(\lambda_{12}, \lambda_{13}) f(T_{12}|\lambda_{12}) f(T_{13}|\lambda_{13}) d\lambda_{12} d\lambda_{13}, \tag{7.55}$$

$$D = \pi_0 \int_0^\infty \rho_0(\lambda) f(T_{12}, T_{13}|\lambda_{12} = \lambda_{13} = \lambda) d\lambda, \tag{7.56}$$

$$\begin{aligned} &\rho_1(\lambda_{12}, \lambda_{13}) = 1/\lambda_{12}\lambda_{13}, \quad \lambda_{12} > 0, \quad \lambda_{13} > 0, \\ &\rho_0(\lambda) = 1/\lambda, \quad \lambda > 0 \end{aligned} \tag{7.57}$$

In addition,

$$f(T_{12}, T_{13}|\lambda_{12} = \lambda_{13} = \lambda) = \lambda^{n_1+n_2} \exp -\lambda\left(\sum_{i=1}^{i=n_1} T_{12}(i) + \sum_{i=1}^{i=n_2} T_{13}(i)\right), \tag{7.58}$$

$$f(T_{12}|\lambda_{12}) = \lambda_{12}{}^{n_{12}} \exp -\lambda_{12}\left(\sum_{i=1}^{i=n_1} T_{12}(i)\right), \tag{7.59}$$

and

$$f(T_{13}|\lambda_{13}) = \lambda_{13}{}^{n_{13}} \exp -\lambda_{13}\left(\sum_{i=1}^{i=n_2} T_{13}(i)\right). \tag{7.60}$$

Note that $n_1 = n_2 = 15$, $\sum_{i=1}^{i=15} T_{12}(i) = 35.2014$, and $\sum_{i-1}^{i=15} T_{13}(i) = 140.7934$.

From the preceding information and assuming $\pi_0 = \pi_1$, it can be shown that the ratio = 403; thus, from Equation 7.54, the posterior probability that the null hypothesis is true is $p_0 = .016$.

This implies that the null hypothesis is not true and that $\lambda_{12} \neq \lambda_{13}$.

Of course, this is not surprising because the $T_{12}(i)$ data are generated using the exponential distribution with parameter $\lambda_{12} = .5$ and the $T_{13}(i)$ data are generated using the exponential distribution with parameter $\lambda_{13} = .125$. The overall conclusion is that the average holding time for adenine until substituted with guanine is different (much smaller) from the holding time for adenine until substituted with cytosine.

Another generalization of DNA molecular evolution is described by Felsenstein and Churchill[7] with infinitesimal rate matrix

$$Q = \begin{pmatrix} -\alpha(1-p_a), \alpha p_g, \alpha p_c, \alpha p_t \\ \alpha p_a, -\alpha(1-p_g), \alpha p_c, \alpha p_t \\ \alpha p_a, \alpha p_g, -\alpha(1-p_c), \alpha p_t \\ \alpha p_a, \alpha p_g, \alpha p_c, -\alpha(1-p_t) \end{pmatrix}, \tag{7.61}$$

where the total holding time for adenine has an exponential distribution with parameter $\alpha p_g + \alpha p_c + \alpha p_t = \alpha(1-p_a)$, or mean $= 1/\alpha(1-p_a)$ until substituted with guanine, or cytosine, or thymine. It can be shown that the stationary distribution of this chain is $\pi = (p_a, p_g, p_c, p_t)$, where $p_a + p_g + p_c + p_t = 1$. Using Q and π and determining that $P(t) = \exp(tQ)$ provide the probability transition matrix as

$$\begin{aligned} P_{ij}(t) &= \left(1-e^{-\alpha t}\right)p_j, \qquad i \neq j, \\ &= e^{-\alpha t} + \left(1-e^{-\alpha t}\right)p_j, \quad i = j, \end{aligned} \tag{7.62}$$

with $i, j = a, g, c, t$. Note that these probabilities do not depend on the initial state i and that the effect of the scalar is to contract or expand the holding time parameter $= \alpha(1-p_a)$ for adenine; therefore, it is of interest to investigate the value of α. This will be achieved in the special case $\pi = (.292,.207,.207,.292)$, the stationary distribution for humans. The first part of the Bayesian analysis is to estimate α by using observations for the holding time of adenine. Of course, the holding time of any base could have been used to estimate α. I will assume that $\alpha = 1$; thus, the exponential parameter for the holding time of adenine is $(1-p_a) = (1-.292) = .708$, which corresponds to an average holding time of 1.4124 time units. Fifteen observations will be generated from an exponential distribution with parameter .708, and then based on these observations, a Bayesian analysis will be executed in order to estimate α. The Bayesian analysis is executed with 35,000 observations for the simulation and 5,000 for the burn-in.

TABLE 7.4

Posterior Analysis for Felsenstein–Churchill[7] Model

Parameter	Mean	SD	Error	2 1/2	Median	97 1/2
α	0.8528	0.2202	0.001089	0.4789	0.8325	1.337
λ	.6038	.1559	.000770	.339	.5894	.9464
μ	1.774	0.4892	0.002441	1.057	1.697	2.95

The results of the analysis are reported in Table 7.4. The prior distribution for λ is a noninformative gamma distribution, and the 15 data values included in the first list statement are generated with an exponential distribution with parameter .708.

WinBUGS Code 7.3

```
model;
{
#lamda has a gamma prior distribution with parameters .001 and .001.
lamda~dgamma(.001,.001)
for ( i in 1:15){
# the HTa vectors are the 15 holding times for adenine
HTa[i]~dexp(lamda)
}
# mu is the average holding time for adenine
mu<-1/lamda
# alpha is the main parameter of interest
alpha<-lamda/.708
}
   list(
HTa = c(
2.116,2.221,1.37,4.389,1.027,
1.486,0.3714,3.178,0.5419,1.345,
0.002338,4.87,0.5304,1.235,0.1772))
# this is the starting value for lamda
list(lamda=.7)
```

Note that λ is the parameter for the exponential holding time for adenine and α is the main parameter of interest and is the scale factor for the holding time, while μ is the average waiting time for adenine.

The main parameter of interest is α and is estimated as 0.8325 with the posterior median and a 95% credible interval of (0.4789,1.337), which implies that it is plausible and reasonable to believe that $\alpha = 1$. Perhaps a more formal test of the null hypothesis

$$H: \alpha = 1 \quad \text{versus} \quad A: \alpha \neq 1. \tag{7.63}$$

is in order.

Note that the testing problem in terms of λ is

$$H: \lambda = .708 \quad \text{versus} \quad \lambda \neq .708. \tag{7.64}$$

The explanation on pages 126 and 127 of Lee[5] is adopted for the Bayesian approach to testing a point null hypothesis. The posterior probability of the null hypothesis is given by

$$p_0 = 1/(1+\gamma), \tag{7.65}$$

where

$$\gamma = \pi_1 \int_0^\infty \rho_1(\lambda) f(t|\lambda) d\lambda / \pi_0 f(t|\lambda = .708). \tag{7.66}$$

In addition,

$$f(t|\lambda = .708) = f(t|\lambda = .708) = (.708)^n \exp\left(-.708\left(\sum_{i=1}^{i=n} t_i\right)\right), \tag{7.67}$$

where vector t is the $n = 15$ holding times in the list statement and

$$\sum_{i=1}^{i=n} t_i = 24.869239. \tag{7.68}$$

Assuming the improper prior for λ as

$$\rho_1(\lambda) = 1/\lambda, \quad \lambda > 0, \tag{7.69}$$

the numerator of the ratio γ is

$$\int_0^\infty \rho_1(\lambda) f(t|\lambda) d\lambda = \Gamma(n) / \left(\sum_{i=1}^{n} t_i\right)^n = \Gamma(15)/(24.869238)^{15}. \tag{7.70}$$

Now assume that $\pi_0 = \pi_1 = 1/2$; then there is sufficient information to compute $\gamma = .000025419$ and the posterior probability of the null hypothesis as

$$p_0 = 1/(1 + .000025419) = .999974581. \tag{7.71}$$

Consequently, one would not reject the null hypothesis; thus, it is plausible to believe that $\alpha = 1$ and that the holding time for adenine has a parameter of $(1 - .292) = .708$ or, an average, holding time of 1.4124 time units.

Our last example for DNA evolution is a generalization of the Felsenstein–Churchill model to the Hasegawa, Kishino, and Yano[8] version for molecular evolution, with infinitesimal rate matrix

$$Q = \begin{pmatrix} -\left(\alpha p_g + \beta p_r\right), \alpha p_g, \beta p_c, \beta p_t \\ \alpha p_a, -(\alpha p_a + \beta p_r), \beta p_c, \beta p_t \\ \beta p_a, \beta p_g, -(\alpha p_t + \beta p_s), \alpha p_t \\ \beta p_a, \beta p_g, \alpha p_c, -(\alpha p_c + \beta p_s) \end{pmatrix}, \tag{7.72}$$

where

$$p_r = p_c + p_t, \quad p_s = p_a + p_g,$$

and

$$p_a + p_g + p_c + p_t = 1.$$

The parameters α and β are unknown positive parameters, and the stationary distribution of the process is $\pi = (p_a, p_g, p_c, p_t)$.

This model makes a distinction between transitions and transversions, distinguishing between the substitutions $a \leftrightarrow g$, from purine to purine or from pyrimidine to pyrimidine, and transversions, from purine to pyrimidine or vice versa, i.e., the substitutions $c \leftrightarrow t$.

At the time, this approach was based on a new statistical method for estimating divergence dates of species from DNA sequence data by a molecular clock approach is developed. This method takes into account effectively the information contained in a set of DNA sequence data.

The molecular clock of mitochondrial DNA was calibrated by setting the date of divergence between primates and ungulates at the Cretaceous–Tertiary boundary (65 million years ago), when the extinction of dinosaurs occurred.

Our investigation will center on the conjecture that $\alpha = \beta$, that the Hasegawa, Kishino, and Jano model (Equation 7.72) reduces to the Felsenstein–Churchill evolutionary process (Equation 7.61), and that the evolutionary process does not distinguish transversions from transitions. Thus, let the null hypothesis be

$$\text{H: } \alpha = \beta \text{ versus the alternative A: } \alpha \neq \beta. \tag{7.73}$$

Then the null hypothesis supports the Felsenstein–Churchill process, and in order to illustrate Bayesian inferences, the holding time data will be generated that favor the alternative hypothesis, the Hasegawa, Kishino, Jano model. Suppose $\alpha = 2$ and $\beta = 8$ and the stationary distribution for humans, namely, $p_a = .292, p_g = .207, p_c = .207$, and $p_t = .292$, is employed to fix the values of the first row of Q in Equation 7.72. Consider the holding time for adenine until substituted by quinine, which is denoted by αp_g, and the holding time for thymine until substituted by adenine, denoted by βp_a. I used $\alpha = 2$ and $\beta = 8$ to generate the holding times for adenine (until substituted by guanine) and thymine (until substituted by adenine), respectively, and these values appear in the list statement of WinBUGS Code 7.4.

There is sufficient information to execute the Bayesian analysis with 35,000 observations for the simulation and 5,000 for the burn-in.

WinBUGS Code 7.4

```
model;
{
lamdaa~dgamma(.001,.001)
lamdat~dgamma(.001,.001)
for( i in 1:12){
# holding time for adenine
HTa[i]~dexp(lamdaa)
# holding time for adenine
HTt[i]~dexp(lamdat)}
# average holding time for thymine
# average holding time for adenine
mua<-1/lamdaa
# average holding time for thymine
mut<-1/lamdat
alpha<-lamdaa/pg
beta<-lamdat/pt
dat<-alpha-beta
# DNA nucleotide DNA bases for humans
pa<-.292
pg<-.207
pr<-.499
pt<-.292
}
list(
HTa = c(
2.831,0.4571,5.36,0.583,0.02068,
2.604,2.556,0.8819,2.356,1.371,
0.9336,1.711,1.012,0.324,0.2455),
HTt = c(
0.08592,0.01524,0.07607,0.5035,0.01662,
0.3833,1.271,0.714,1.07,0.04821,
0.2838,0.03206,1.259,0.7526,0.1634))
list(lamdaa=.414,lamdat=2.92)
```

The prior distributions for the exponential parameters λ_a and λ_t are noninformative gamma (.001,.001) distributions, the conjugate prior to the exponential distribution; thus, most of the information for the Bayesian analysis is based on the holding time observation HTa and HTt listed in WinBUGS Codes 7.4.

Some of the posterior distributions are skewed; thus, the Bayesian estimates will be based on the posterior median. For example, consider α with an estimate of 2.559, which in turn should be compared with $\alpha = 2$, the value used to generate the data in the list statement of WinBUGS Code 7.4.

Also note that the posterior median of λ_a is .538, which is the exponential parameter for the holding time of adenine (before being substituted by guanine) and should be compared

TABLE 7.5

Posterior Analysis for the Hasegawa, Kishino, and Yano Model

Parameter	Mean	SD	Error	2 1/2	Median	97 1/2
α	2.675	0.7658	0.004055	1.389	2.599	4.379
β	9.135	2.642	0.01419	4.736	8.858	15.04
d_{at}	−6.46	2.748	0.01492	−12.51	−6.221	−1.749
λ_a	.5538	.1585	.000839	.2876	.538	.9064
λ_t	2.667	0.7714	0.004143	1.383	2.587	4.392
μ_a	1.968	0.6191	0.003331	1.103	1.859	3.477
μ_t	.4104	.1298	.000687	.2279	.3877	.7273

to $\lambda_a = .414$, the value used to generate the exponentially distributed holding times. The key parameter is d_{at}, the difference between alpha and beta and its posterior median of −6.221 with a 95% credible interval of (−12.51,−1.747), which implies that $\alpha \neq \beta$. It is left for the student to develop a formal Bayesian test of $\alpha = \beta$. For the Bayesian method of testing hypotheses, see pages 126 and 127 of Lee.[5] Also, as in the previous example of molecular evolution (the Jukes–Cantor model), it is easy to derive the predictive density of a future adenine (or thymine) holding time, and this will be left as an exercise for the student. See Table 7.5.

7.5.2 Birth and Death Processes

The discrete-time version of birth and death processes are presented in Section 5.4, and the continuous-time analogue will be described in this section. Such processes are time-reversible Markov chains that are quite valuable in a variety of scientific disciplines.

Let the present state of the chain be i; then the process can gain one unit corresponding to a birth or decrease one unit corresponding to a death, where $X(t)$, $t > 0$, is the size of the population at time t. Since this is a continuous-time process, the chain is defined in terms of the infinitesimal rate matrix

$$Q = \begin{pmatrix} -\lambda_0, \lambda_0, 0, 0, \ldots\ldots\ldots\ldots \\ \mu_1, -(\lambda_1 + \mu_1), \lambda_1, \ldots\ldots\ldots \\ 0, \mu_2, -(\lambda_2 + \mu_2), \lambda_2, \ldots. \\ 0, 0, \mu_3, -(\lambda_3 + \mu_3), \lambda_3, . \\ \cdots \\ \cdot\cdot \\ \cdot\cdot \end{pmatrix}, \tag{7.74}$$

corresponding to the state space $S = \{0, 1, 2, \ldots..\}$. The state 0 is an absorbing barrier; that is, once the population size reaches 0, it dies out. Thus, if the population size is 1, the holding time distribution of state 1 until the next death is the exponential parameter μ_1, while on the other hand, the holding time distribution of state 1 until a birth has an exponential distribution with parameter λ_1. Note that the birth rates λ_i and death rates μ_i depend on the present size of the population.

Since the process is time reversible, one can derive the stationary distribution via the local balance equations

$$\pi_i\lambda_i = \pi_{i+1}\mu_{i+1}, \quad i = 0, 1, 2, \ldots, \tag{7.75}$$

which implies that

$$\pi_1 = \pi_0\lambda_0/\mu_1$$

and

$$\pi_2 = \pi_1\lambda_1/\mu_2 = \pi_0\lambda_0\lambda_1/\mu_1\mu_2.$$

Thus, the pattern is

$$\begin{aligned}\pi_k &= \pi_0\lambda_0\lambda_1\ldots\lambda_{k-1}/\mu_1\mu_2\ldots\mu_k \\ &= \pi_0\prod_{i=1}^{i=k}\lambda_{i-1}/\mu_i, \quad k = 0, 1, 2, \ldots.\end{aligned} \tag{7.76}$$

Since the stationary distribution is a probability distribution,

$$1 = \sum_{k=0}^{k=\infty}\pi_k = \pi_0\sum_{k=0}^{k=\infty}\prod_{i=0}^{i=k}\lambda_{i-1}/\mu_i, \quad k = 0, 1, 2\ldots. \tag{7.77}$$

However, it is necessary to impose the constraint

$$\sum_{k=0}^{k=\infty}\prod_{i=0}^{i=k}\lambda_{i-1}/\mu_i < \infty, \quad i = 1, 2\ldots$$

so that Equation 7.77 converges; then the unique stationary distribution is determined as

$$\pi_k = \pi_0\prod_{i=1}^{i=k}\lambda_{i-1}/\mu_i, \quad k = 0, 1, 2, \ldots,$$

where

$$\pi_0 = \left(\sum_{k=0}^{k=\infty}\prod_{i=1}^{i=k}\lambda_{i-1}/\mu_i\right)^{-1}. \tag{7.78}$$

In the infinite case, notice all states are transient except the absorbing state 0.

Now consider a finite state space with infinitesimal rate matrix and state space $S = \{0,1,2,3,4\}$:

$$Q = \begin{pmatrix} -\lambda_0, \lambda_0, 0.0, 0.0, 0.00 \\ \mu_1, -(\mu_1 + \lambda_1), \lambda_1, 0, 0 \\ 0, \mu_2, -(\mu_2 + \lambda_2), \lambda_2, 0 \\ 0, 0, \mu_3, -(\mu_3 + \lambda_3), \lambda_3 \\ 0.0, 0.0, 0.0, \mu_4, -\mu_4 \end{pmatrix}. \tag{7.79}$$

For the purpose of illustration, suppose $\mu_1 = 2$, $\lambda_1 = 1$, $\mu_2 = 3$, $\lambda_2 = 2$, $\mu_3 = 4$, and $\lambda_3 = 3$.

Using these values, I generated 20 holding times for six exponential holding times: From state 1 to 0 with parameter 2, from state 1 to 2 with parameter 1, from state 2 to 1 with parameter 3, from state 2 to 2 with parameter 2, from state 3 to 2 with parameter 4, and from state 3 to 4 with parameter 3. These six holding times are reported in the list statement of WinBUGS Code 7.5. Note that the birth rate for a population of size i is $\lambda_i/(\mu_i + \lambda_i)$, and the corresponding death rate is $\mu_i/(\mu_i + \lambda_i)$, $i = 1, 2, 3$. The following Bayesian analysis is executed with 45,000 observations for the simulation and 5,000 for the burn-in:

WinBUGS Code 7.5

```
model;
{
mu1~dgamma(.001,.001)
lamda1~dgamma(.001,.001)
mu2~dgamma(.001,.001)
lamda2~dgamma(.001,.001)
mu3~dgamma(.001,.001)
lamda3~dgamma(.001,.001)

for ( i in 1:20){
# is the vector of holding times when the process is in state 1 until the
process switches to state 0
HT10[i]~dexp(mu1)
HT12[i]~dexp(lamda1)
HT21[i]~dexp(mu2)
HT23[i]~dexp(lamda2)
HT32[i]~dexp(mu3)
HT34[i]~dexp(lamda3)}
#p10 is the death rate when the population size is 1
p10<-mu1/(mu1+lamda1)
# p12 is the birth rate when the population is size 1
p12<-lamda1/(mu1+lamda1)
# p21 is the death rate when the population is size 2
p21<-mu2/(mu2+lamda2)
#p23 is the birth rate when the population is size 2
p23<-lamda2/(mu2+lamda2)
# p32 is the death rate when the population is size 3
```

```
p32<-mu3/(mu3+lamda3)
#p34 is the birth rate when the population is size 3
p34<-lamda3/(mu3+lamda3)
# the following is the average holding time for the process in state 1 until
it switches to state 0
nu1<-1/mu1
#the following is the average holding time for the process in state 1 until
it switches to state 2
vu1<-1/lamda1
nu2<-1/mu2
vu2<-1/lamda2
nu3<-1/mu3
vu3<-1/lamda3
}
list(
HT10 = c(
0.4236,0.0738,0.01755,0.2506,0.1679,
0.1345,0.1335,0.2103,0.5676,0.1256,
1.25,0.08106,0.08959,0.789,0.966,
0.4133,0.01223,0.08178,0.7944,0.2417),
HT12 = c(
0.6452,0.3766,0.4237,0.8919,0.1333,
0.2981,0.4902,0.1167,0.2695,0.7278,
0.2951,3.941,1.342,0.4537,0.9498,
1.354,0.5427,0.5481,0.1861,0.8717),
HT21 = c(
0.3853,0.0437,0.05582,0.1248,0.2109,
0.2065,0.5253,0.9537,0.07334,0.01717,
0.1436,0.2864,0.7099,0.04587,0.4547,
0.2007,0.438,0.5084,0.3366,0.4267),
HT23 = c(
0.2497,0.1747,0.09605,0.08779,0.1217,
0.1621,0.1332,0.255,2.358,1.472,
0.2527,0.6953,1.937,0.191,1.772,
0.2749,0.03768,0.1866,0.41,1.134),
HT32 = c(
0.007235,0.07714,0.02351,0.3239,0.1192,
0.2329,0.3503,0.1093,0.1339,0.5003,
0.2394,0.336,0.08296,0.229,0.06038,
0.07797,0.125,0.035,0.2428,0.4869),
HT34 = c(
1.067,0.2683,0.6207,0.1697,0.1857,
0.1188,0.2984,0.03796,0.2262,0.4814,
0.5348,0.4655,0.3812,0.1125,0.00313,
0.9027,0.0557,0.4312,0.4711,0.1909))
list(mu1=2,mu2=3,mu3=4,lamda1= 1,lamda2=2,lamda3=3)
```

The Bayesian analysis employs the exponential distribution with parameter μ_i as the likelihood function for μ_i and uses the noninformative prior gamma (.001,.001) for μ_i, while for the birth rate parameter λ_i, its likelihood has an exponential distribution with parameter λ_i and prior gamma (.001,.001). Refer to WinBUGS Code 7.5 for the statements corresponding to the likelihood function and prior of λ_i and μ_i, $i = 1,2,3$.

The Bayesian analysis is for the three birth rates λ_i, $i = 1,2,3$, and death rates μ_i, $i = 1,2,3$. The corresponding average holding times for the deaths are given by η_i, $i = 1,2,3$, and by ν_i, $i = 1,2,3$, for the births.

The transition probabilities are denoted by p_{10}, p_{12}, p_{21}, p_{23}, p_{32}, and p_{34}, and 20 exponential holding times for the death and births are generated; thus, the transition probabilities are based on 20 transitions to the neighboring states.

Consider the transition probability p_{12} from a population of 1 to a population of size 2 (because of a birth); then the corresponding posterior mean is .3188 with 95% credible interval (.1968,.4638). On the other hand, the posterior mean of p_{10} is .6812 with 95% credible interval (.5362,.8034), implying that the chances are higher for the population to become extinct than to increase by 1. See Table 7.6.

The next inference to be considered is the estimation of the average time to extinction for the process

$$Q = \begin{pmatrix} -\lambda_0, \lambda_0, 0.0, 0.0, 0.00 \\ \mu_1, -(\mu_1 + \lambda_1), \lambda_1, 0, 0 \\ 0, \mu_2, -(\mu_2 + \lambda_2), \lambda_2, 0 \\ 0, 0, \mu_3, -(\mu_3 + \lambda_3), \lambda_3 \\ 0.0, 0.0, 0.0, \mu_4, -\mu_4 \end{pmatrix}, \tag{7.80}$$

TABLE 7.6

Posterior Analysis for Birth and Death Process

Parameter	Mean	SD	Error	2 1/2	Median	97 1/2
λ_1	1.345	0.3015	0.001345	0.8237	1.322	2
λ_2	1.666	0.3725	0.001656	1.016	1.637	2.475
λ_3	2.852	0.6333	0.002825	1.749	2.806	4.222
μ_1	2.927	0.6571	0.002994	1.787	2.877	4.347
μ_2	3.25	0.7288	0.003365	1.988	3.196	4.825
μ_3	5.268	1.174	0.005562	3.225	5.176	7.089
η_1	.3598	.08517	.0003899	.23	.3476	.5595
η_2	.3239	.0763	.0003522	.2072	.3129	.50321
η_3	.1997	.04686	.000214	.1281	.1932	.31
p_{10}	.6812	.06828	.000299	.5362	.6852	.8032
p_{12}	.3188	.06828	.000299	.1968	.31248	.4638
p_{21}	.6575	.07082	.000335	.5092	.6612	.7852
p_{23}	.3425	.07082	.000335	.2148	.3388	.4908
p_{32}	.6454	.07145	.000327	.498	.6488	.7756
p_{34}	.3546	.07145	.000327	.2244	.3512	.502
ν_1	.7827	.1847	.000854	.5001	.7567	1.214
ν_2	.6317	.1487	.000664	.4041	.6109	.9846
ν_3	.3689	.08655	.000398	.2369	.3564	.5718

where $\mu_1 = 2$, $\lambda_1 = 1$, $\mu_2 = 3$, $\lambda_2 = 2$, $\mu_3 = 4$, $\lambda_3 = 3$, and $\mu_4 = 4$. In the list statement of WinBUGS Code 7.6 appears the 20 observations for the data labeled HT4, the holding times with exponential parameter $\mu_4 = 4$.

The other holding times were used in WinBUGS Code 7.5. Our main goal is the estimation of the average time to extinction using the pattern in Equation 2.79, and the following formula given that parameter.

Consider a birth and death process with state space $\{0, 1, 2, \ldots, k\}$ with one absorbing state 0, where the remaining are transient states, and let V be the $(k - 1)$ by $(k - 1)$ matrix of transient states, and let $F = -V^{-1}$, and then the average time to absorption from state i, $i = 1, 2, \ldots, k$, is

$$a_i = \sum_{i=1}^{i=k} F_{ij}, \tag{7.81}$$

where F_{ij} is the ijth element of F. See pages 288 and 289 of Dobrow[1] for a proof.

WinBUGS Code 7.6 is the WinBUGS Code for estimating the average time to extinction for the birth and death process with matrix Q in Equation 7.79. Noninformative gamma (.001,.001) distributions were used as priors for the parameters: the death rates μ_i, $i = 1, 2, 3, 4$, and birth rates λ_i, $i = 1, 2, 3$. The Bayesian analysis is executed with 57,000 observations and 1,000 for the burn-in.

WinBUGS Code 7.6

```
model;

{
#prior distributions for the parameters

lamda1~dgamma(.001,.001)
mu1~dgamma(.001,.001)
lamda2~dgamma(.001,.001)
mu2~dgamma(.001,.001)
lamda3~dgamma(.001,.001)
mu3~dgamma(.001,.001)

mu4~dgamma(.001,.001)

for ( i in 1:20){
# the seven holding times for (2.79)
HT10[i]~dexp(mu1)

HT12[i]~dexp(lamda1)

HT21[i]~dexp(mu2)

HT23[i]~dexp(lamda2)
HT32[i]~dexp(mu3)
HT34[i]~dexp(lamda3)
HT4[i]~dexp(mu4)}
```

```
# the -V matrix
v[1,1]<-mu1+lamda1
v[1,2]<-(-lamda1)
v[1,3]<-0
v[1,4]<-0
v[2,1]<-(-mu2)
v[2,2]<-mu2+lamda2
v[2,3]<- (-lamda2)
v[2,4]<-0
v[3,1]<-0
v[3,2]<-(-mu3)
v[3,3]<-mu3+lamda3
v[3,4]<-(-lamda3)
v[4,1]<-0
v[4,2]<-0
v[4,3]<-(-mu4)
v[4,4]<-mu4
# F is the inverse of -V.
F[1:4,1:4]<-inverse(v[1:4,1:4])

}

list(
HT10 = c(
0.4236,0.0738,0.01755,0.2506,0.1679,
0.1345,0.1335,0.2103,0.5676,0.1256,
1.25,0.08106,0.08959,0.789,0.966,
0.4133,0.01223,0.08178,0.7944,0.2417),
HT12 = c(
0.6452,0.3766,0.4237,0.8919,0.1333,
0.2981,0.4902,0.1167,0.2695,0.7278,
0.2951,3.941,1.342,0.4537,0.9498,
1.354,0.5427,0.5481,0.1861,0.8717),
HT21 = c(
0.3853,0.0437,0.05582,0.1248,0.2109,
0.2065,0.5253,0.9537,0.07334,0.01717,
0.1436,0.2864,0.7099,0.04587,0.4547,
0.2007,0.438,0.5084,0.3366,0.4267),
HT23 = c(
0.2497,0.1747,0.09605,0.08779,0.1217,
0.1621,0.1332,0.255,2.358,1.472,
0.2527,0.6953,1.937,0.191,1.772, .
0.2749,0.03768,0.1866,0.41,1.134),
HT32 = c(
0.007235,0.07714,0.02351,0.3239,0.1192,
0.2329,0.3503,0.1093,0.1339,0.5003,
```

```
0.2394,0.336,0.08296,0.229,0.06038,
0.07797,0.125,0.035,0.2428,0.4869),
HT34 = c(
1.067,0.2683,0.6207,0.1697,0.1857,
0.1188,0.2984,0.03796,0.2262,0.4814,
0.5348,0.4655,0.3812,0.1125,0.00313,
0.9027,0.0557,0.4312,0.4711,0.1909),

HT4 = c(
0.7442,0.4533,0.07353,0.03283,0.1242,
1.001,0.1311,1.057,0.09183,0.226,
0.3407,0.1795,0.01064,0.003547,0.3841,
0.6416,0.06772,0.06674,0.2898,0.1095)))

list(mu1=2,mu2=3,mu3=4,mu4=4,lamda1= 1,lamda2=2,lamda3=3)
```

The posterior distributions are skewed; thus, the posterior medians should be used as point estimators. For example, when the process is in state 4, the posterior median is .7252 time units, but if the process is in state 1, the posterior median time to extinction is .399 time units. This seems to appear to be reasonable because closer states to the absorbing state 0 have smaller posterior medians. See Table 7.7.

7.5.3 Random Walk

The random walk is a birth and death process with constant birth rate $\lambda_i = \lambda$ and death rate $\mu_i = \mu$; that is, the birth and death rates do not depend on the population size. From what is known about the general birth and death process (see Equation 7.79), the stationary distribution is

$$\pi_k = \pi_0 \prod_{i=1}^{i=k} (\lambda/\mu) = \pi_0 (\lambda/\mu)^k, \quad k = 0, 1, \ldots, \tag{7.82}$$

where $\lambda < \mu$ and

$$\pi_0 = \left(\sum_{k=0}^{k=\infty} (\lambda/\mu)^k \right)^{-1} = 1 - \lambda/\mu. \tag{7.83}$$

TABLE 7.7

Posterior Analysis for Extinction

Parameter	Mean	SD	Error	2 1/2	Median	97 1/2
a_1	0.4248	0.4461	0.001472	0.2738	0.399	0.6533
a_2	0.5676	1.36	.004491	0.3172	0.493	10.028
a_3	0.6699	5.77	.01083	0.2741	0.4882	1.638
a_4	1.063	5.77	.01919	0.3704	0.7253	2.844

Thus, the stationary distribution is geometric, namely,

$$\pi_k = (1 - \lambda/\mu)(\lambda/\mu)^k, \quad k = 0,1,2,\ldots. \tag{7.84}$$

Bayesian inferences will focus on estimating the rates μ and λ and testing the hypothesis that $\lambda < \mu$. Recall that for this process, the infinitesimal rate matrix is

$$Q = \begin{pmatrix} -\lambda, \lambda, 0, 0, \ldots\ldots\ldots\ldots \\ \mu, -(\mu+\lambda), \lambda, 0, 0, \ldots. \\ 0, \mu, -(\mu+\lambda), \lambda, 0, \ldots. \\ 0, 0, \mu, -(\mu+\lambda), \lambda, 0, . \\ . \\ . \\ . \end{pmatrix}. \tag{7.85}$$

Observations for the holding times with exponential parameters λ and μ will be generated, and then based on those observations, Bayesian inferences will be presented. Suppose that $\lambda = 2 < \mu = 5$. The likelihood functions for λ and μ are determined by the exponential density and the 17 holding time observations denoted by HT2 for λ and HT5 for μ. Noninformative gamma (.001,.001) distributions are assumed for the two unknown parameters. The objective of the Bayesian analysis is to estimate the stationary distribution of the random walk and to estimate $\Pr(\lambda > \mu|\text{data})$, thus testing the hypothesis H: $\lambda > \mu$.

WinBUGS Code 7.7 is executed with 35,000 observations for the simulation and a burn-in of 5,000. See Table 7.8 for the posterior analysis.

WinBUGS Code 7.7

```
model;
{
# the prior distributions for lamda and mu
lamda~dgamma(.001,.001)
mu~dgamma(.001,.001)
for ( i in 1:17){
HT2[i]~dexp(lamda)
HT5[i]~dexp(mu)}

m2<-1/lamda
m5<-1/mu
# the pi[k] are the stationary probabilities
for( k in 0:5){
pi[k]<-(1-lamda/mu)*pow(lamda/mu,k)}

diff<-lamda-mu
#prob is the    Pr(λ > μ)
```

```
prob<-step(diff)
}

list(
HT2 = c(
1.155,0.1775,1.513,0.1776,0.9627,
0.602,0.05385,0.2169,0.1129,0.3904,
0.7396,0.06482,0.09624,0.1167,0.5177,
0.511,0.3539),
HT5 = c(
0.09529,0.1754,0.4323,0.04315,0.01035,
0.05834,0.05996,0.09225,0.07383,0.1131,
0.1176,0.1822,0.1276,0.2809,0.001542,
0.01506,0.2859))

list( mu=5,lamda=2)
```

The posterior distributions for the parameters of the random walk appear to be symmetric about the posterior mean, and it also appears that $\lambda < \mu$ because the 95% credible interval for the difference $\lambda - \mu$ excludes zero and has a negative posterior mean of −5.66. It should be remembered that the 17 holding times were generated with the values $\lambda = 2 < \mu = 5$. With such a small sample size, it is not surprising that the posterior means for λ and μ are not very close to their hypothetical values.

The hypothesis $\lambda > \mu$ has posterior probability of 0, as measured by the posterior median; thus, one would conclude that $\lambda < \mu$, further implying that the stationary distribution is well defined in that the geometric series in Equation 7.84 converges!

7.5.4 Yule Process

The Yule process is a pure birth process where each individual has the same rate λ of a birth; thus, if the population size is i, the rate probability of a birth is λ_i, $i = 1, 2, 3, \ldots$, with the infinitesimal rate matrix

TABLE 7.8

Posterior Analysis for Random Walk

Parameter	Mean	SD	Error	2 1/2	Median	97 1/2
diff	−5.66	1.985	0.01111	−9.939	−5.506	−2.185
λ	2.19	0.526	0.003034	1.286	2.148	3.332
m_2	.4846	.1241	.000729	.3002	.4656	.7778
m_5	.1345	.03501	.000207	.08321	.1304	.219
μ	7.85	1.915	0.01085	4.567	7.671	12.01
π_1	.1974	.03646	.000126	.1209	.2013	.2495
π_2	.06812	.03107	.000106	.01713	.05645	.1354
π_3	.02151	.01953	.000066	.00242	.01581	.07544
π_4	.008273	.01373	.000046	.000341	.00442	.04181
π_5	.003485	.01363	.000046	.000482	.00124	.02316
$\Pr(\lambda > \mu)$	.0001333	.01155	.000053	0	0	0

$$Q = \begin{pmatrix} -\lambda, \lambda, 0, 0, 0, 0, 0, 0, 0, 0..... \\ 0, -2\lambda, 2\lambda, 0, 0, 0, 0, 0....... \\ 0, 0, -3\lambda, 3\lambda, 0, 0, 0, 0, 0.... \\ . \\ 0, 0, 0, 0, 0, -i\lambda, i\lambda, 0, 0, 0, .. \\ . \\ . \end{pmatrix}. \tag{7.86}$$

Note for such a process, all states are transient and the limiting probability does not exist. It is easily shown that the transition function is

$$P_{ij}(t) = \binom{j-1}{i-1} e^{-i\lambda t} \left(1 - e^{-\lambda t}\right)^{j-1}, \quad i \le j,\ t > 0, \tag{7.87}$$

which is the negative binomial distribution. In particular, starting with a population of size 1, the probability the size remains at 1 at time t is given by

$$P_{11}(t) = \exp(-\lambda t). \tag{7.88}$$

Of concern to the Bayesian is the estimation of λ and of the transition probability in Equation 7.87. This is easily accomplished by generating holding times with an exponential distribution with parameter λ.

Using the 20 observations for the holding times with exponential parameter $\lambda = 3$, the Bayesian analysis consists of estimating λ and the probability function in Equation 7.88, the probability that the population remains at one individual for times $t = 1, 2, \ldots$. The analysis is executed with 35,000 observations for the simulation and a burn-in of 5,000.

WinBUGS Code 7.8

```
model;

{
lamda~dgamma(.01,.01)
for ( i in 1:20){

HT[i]~dexp(lamda)}
for ( t in 1:10){
p11[t]<-exp(-lamda*t)}
}

list(
HT = c(
0.9923,0.6045,0.09804,0.04378,0.1655,
1.335,0.1748,1.409,0.1224,0.3013,
0.4542,0.2393,0.01419,0.004729,0.5122,
0.8555,0.09029,0.08898,0.3865,0.146))

list(lamda=3)
```

TABLE 7.9

Posterior Analysis for the Yule Process

Parameter	Mean	SD	Error	2 1/2	Median	97 1/2
λ	2.486	.5562	.003134	1.525	2.441	3.691
$P_{11}(1)$	.096	.05049	.0002882	.02494	.08711	.2176
$P_{11}(2)$	.01177	.01308	.0000755	.0006221	.007589	.04733
$P_{11}(3)$	.001757	.003381	.0000201	.0000155	.000661	.0103

The results of the Bayesian analysis are reported in Table 7.9. Note that the posterior mean of λ is 2.4876 with a 95% credible interval of (1.525,3.691), and the posterior mean of $P_{11}(1)$ is .096 and that the posterior mean of $P_{11}(t), t = 2,3$, is decreasing as it should.

It is straightforward to test hypotheses about λ and to derive the formula for future holding times with exponential parameter λ. The student will be asked to perform these inferences as problems at the end of the chapter.

7.5.5 Birth and Death with Immigration

From the least complex, such as the Yule process, to the more complicated, we now study the so-called birth and death process with immigration.

Suppose the immigration with rate ν is included in the simple birth and death process; then the infinitesimal rate matrix is

$$Q = \begin{pmatrix} -\nu, \nu, 0, 0, 0, 0, 0, 0, 0, \ldots\ldots\ldots\ldots\ldots\ldots \\ \mu, -(\nu + \lambda + \mu), \nu + \lambda, 0, 0, 0, 0, \ldots\ldots\ldots \\ 0, 2\mu, -(\nu + 2(\lambda + \mu)), \nu + 2\lambda, 0, 0, 0, 0, \ldots\ldots \\ 0, 0, 3\mu, -(\nu + 3(\lambda + \mu)), \nu + 3\lambda, 0, 0, 0\ldots\ldots \\ . \\ . \\ . \end{pmatrix}, \quad (7.89)$$

where the rates μ, ν, and λ are positive and are the parameters of the relevant holding times, which have an exponential distribution.

Referring to Equation 7.89, it is apparent that if the population is 0, it can increase by one person if one person immigrates to the population. Our goal is to estimate the parameters and to test hypotheses about those parameters. In order to perform a Bayesian analysis, let $\mu = 1/2$, $\nu = 1/3$, and $\lambda = 1$; that is, if the population is size i (i = 1,2,...), the average number of immigrants per day is 3, the average number of births is 1, and the average number of deaths is 2 per day. Note the birth and death rates depend on the present population size, but the immigration rate does not. The Bayesian analysis is relatively straightforward and is executed with WinBUGS Code 7.9 with noninformative prior gamma (.01,.01) distributions and using 35,000 observations for the simulation and a burn-in of 5,000:

WinBUGS Code 7.9

```
model;

{
lamda~dgamma(.01,.01)
mu~dgamma(.01,.01)
nu~dgamma(.01,.01)
for( i in 1:16){

HTmu[i]~dexp(mu)
HTvu[i]~dexp(nu)
HTlamda[i]~dexp(lamda)}
mua<-1/mu
nua<-1/nu
lamdaa<-1/lamda
}

list(
HTlamda = c(
1.159,1.833,0.2687,0.1907,0.2907,
0.2291,0.6853,0.5028,0.6184,2.253,
0.3,3.233,0.9386,0.9576,1.274,
0.2392),
HTmu = c(
0.1633,0.3263,0.7803,0.5374,0.9227,
0.2175,0.6166,0.9843,7.635,0.8423,
0.2366,0.7924,3.172,0.6005,0.4124,
0.08125),
HTvu = c(
2.049,1.409,3.333,4.758,5.905,
0.03409,6.001,1.666,0.281,1.086,
3.418,1.319,2.271,0.4253,6.108,
5.066))

list(lamda=1,vu=.333,mu=.5)
```

The Bayesian analysis for the immigration model is portrayed in Table 7.10.

TABLE 7.10

Posterior Distributions for Birth and Death with Immigration

Parameter	Mean	SD	Error	2 1/2	Median	97 1/2
λ	1.07	0.2679	0.001294	0.6131	1.047	1.661
$E(\lambda)$	0.9971	0.2663	0.001282	0.6002	0.9551	1.631
μ	0.8717	0.2181	0.001046	0.4999	0.8513	1.35
$E(\mu)$	1.223	0.3249	0.001632	0.7408	1.175	2.001
ν	.3544	.08832	.000458	.2033	.3468	.5478
$E(\nu)$	3.008	0.7985	0.004228	1.825	2.884	4.919

The main parameters of interest are the immigration rate and the average number of immigrants per day.

With regard to the immigration rate, the posterior mean is .3544 with a 95% credible interval (.2003,.5478), while for the average number of immigrants per day, the posterior median is 2.884 with posterior standard deviation of 0.7985. Note the asymmetry in the posterior distribution of $E(\nu)$, which is easily seen with its posterior density. Also note that the 95% credible interval for λ includes the value 1; the 95% credible interval for μ includes 1/2, and that for ν contains 1/3. What does this imply about the generated holding times HTmu, HTnu, and HTlamda?

7.5.6 SI Epidemic Models

This section introduces the continuous-time version of epidemic models and is a generalization of the discrete version of stochastic epidemic model of Sections 5.6.1 and 5.6.2. Recall the dynamics of SI and SIS models where the number of susceptible people at time t is denoted by $S(t)$ and the number of infected individuals is given by $I(t)$. Infected individuals are also infectious that is there is no latent period, and the total population size $N = I(t) + S(t)$ remains constant over the period of observation. This SI model has been used to explain diseases such as the common cold and influenza, where the epidemic is best described by the system of differential equations

$$\begin{aligned} &dS(t)/dt = -(\beta/N)S(t)I(t) \\ &\text{and} \\ &dI(t)/dt = (\beta/N)S(t)I(t), \end{aligned} \tag{7.90}$$

where $S(0) + I(0) = N$ and β is the transmission rate, the number of contacts per unit time that result in an infection of a susceptible individual. Pages 302–308 of Allen[3] present a more detailed account of the SI and SIS models and are an excellent reference of the general area of stochastic epidemics.

For the stochastic version, the infinitesimal generator matrix for the number of infected individuals with state space $\{0, 1, 2, \ldots, N\}$ is given as

$$Q = \begin{pmatrix} -\beta(N-1)/N, \beta(N-1)/N, 0, 0, 0, \ldots\ldots\ldots \\ 0, -2\beta(N-2)/N, 2\beta(N-2)/N, 0, 0, 0, \ldots\ldots\ldots \\ 0, 0, -3\beta(N-3)/N, 3\beta(N-3)/N, 0, 0, \ldots\ldots\ldots \\ . \\ . \\ 0, 0, 0, 0, \ldots\ldots\ldots -\beta(N-1)/N, \beta(N-1)/N \\ 0, 0, 0, 0, \ldots\ldots\ldots\ldots\ldots\ldots 0, 0 \end{pmatrix}. \tag{7.91}$$

The only unknown parameter is the transmission rate β, the number of contacts per unit time of infected individuals with susceptible people resulting in an infection.

The holding time for state 1 (with one infected individual) is $2\beta(N-2)/N$. How should β be estimated? Remember that $\gamma = 2\beta(N-2)/N$ is the parameter for the holding time for one infected individual (until the number of infected increases to 2 infected individuals) and that holding time has an exponential distribution. Suppose that $\beta = 3$ and $N = 6$; then $\gamma = 4$. Thus, I will generate 16 holding times with parameter $\gamma = 4$ and use those to estimate γ and, consequently, $\beta = 3\gamma/4$. The main parameters of interest are β, the contact rate; γ, the parameter of the exponential distribution for the holding time of one infected individual; and E(HT1), the average holding time for one infected individual. Bayesian inferences are based on WinBUGS Code 7.10, which is executed with 35,000 observations for the simulation and a burn-in of 5,000:

WinBUGS Code 7.10

```
model;

{

# gamma is the parameter for the exponential distribution of the holding
time for one infected
# gamma has a noninformative gamma distribution
gamma~dgamma(.001,.001)
for ( i in 1:16){

HT1[i]~dexp(gamma)}
beta<-3*gamma/4
EFT1<-1/gamma
}
# the following times were generated with an exponential distribution
with parameter 4
list(
HT1 = c(
0.3804,0.03817,0.1017,0.3212,0.2032,
0.4066,0.3085,0.4867,0.2681,0.07279,
0.1096,0.04133,0.5914,0.2417,0.1273,
0.04811))
list(gamma=4)
```

The values of the parameters used to generate the holding times in the list statement of WinBUGS Code 7.11 are $N = 6$, $\gamma = 4$, and $\beta = 3$, and these values should be compared to the corresponding posterior means appearing in Table 7.11, the posterior analysis for the epidemic model. For example, the posterior mean of γ is 4.272 with a 95% credible interval of (2.452,6.621), implying informally that the generated holding times were indeed distributed as an exponential with parameter $\gamma = 4$. Note that the average holding time for one infected individual is .2496 days as estimated by the posterior mean.

TABLE 7.11

Posterior Analysis for the SI Epidemic Model

Parameter	Mean	SD	Error	2 1/2	Median	97 1/2
β	3.204	0.8015	0.004305	1.839	3.131	4.966
γ	4.272	1.069	0.00574	2.452	4.174	6.621
E(HT1)	.2496	.06625	.000362	.151	.2396	.4078

7.5.7 Stochastic SIS Epidemic Model

A generalization of the SI model is the SIS, where individuals who are infected can recover but do not develop immunity and can immediately become infected again, which is graphically represented as $S \rightarrow I \rightarrow S$. This is a birth and death process with $\mu_i = (\gamma + b)i$ and $\lambda_i = \max\{0, (\beta/N)i(N - i)\}$, where i is the size of the population.

Let $S(t)$ and $I(t)$ be the numbers of susceptible and infected individuals at time t, respectively; then the number of new susceptible individuals at the next instance equals the people that did not become infected $(1 - \beta I(t)/N)$ plus those that recovered $(\gamma I(t))$ plus newborns among the infected class $(bI(t))$.

Corresponding to the number of infected $\{I(t), t > 0\}$, the infinitesimal rate matrix is

$$Q = \begin{pmatrix} -(\beta/N)N, (\beta/N)N, 0, 0, \ldots\ldots\ldots\ldots\ldots\ldots\ldots\ldots \\ (b+\gamma), -[(\beta/N)(N-1) + (b+\gamma)], (\beta/N)(N-1), 0, 0, 0, \ldots \\ 0, 2(b+\gamma), -[2(\beta/N)(N-2) + 2(b+\gamma)], 2(\beta/N)(N-2), 0, 0, \ldots \\ . \\ . \\ . \end{pmatrix}.$$

Consider the example presented on pages 308 and 309 of Allen[3] with $\beta = 2$, $N = 100$, and $\gamma + b = 1$. Our goal is to estimate β, $(b + \gamma)$, and $R_0 = \beta - (b + \gamma)$. Note that for one infected individual, the holding time (until a new infection) is $(\beta/N)(N - 1) + (b + \gamma) = 2.98$ and the holding time for a population with two infected people is $(2\beta/N)(N - 2) + 2(b + \gamma) = 5.92$.

Using the values $\lambda_1 = 2.98$ and $\lambda_2 = 5.92$ for the parameters of the exponential distribution for the holding times for one and two infected individuals, 17 observations are generated for both and labeled as HT1 and HT2, respectively. These holding times are in the list statement of WinBUGS Code 7.11, which is the Bayesian analysis executed with 45,000 observations for the simulation and 5,000 for the burn-in:

WinBUGS Code 7.11

```
model;

{
lamda1~dgamma(.001,.001)
lamda2~dgamma(.001,.001)
for( i in 1:18){

HT1[i]~dexp(lamda1)
HT2[i]~dexp(lamda2)}
beta<-(lamda1-1)/.99

delta<-(lamda2-1.96*beta)/2
R<-beta-delta
S<-step(R)
p10<-delta/(delta+beta*.99)
}
```

```
list(
HT1 = c(
0.3496,0.9303,0.6686,0.7489,0.1625,
0.2897,0.06194,0.3293,0.1065,0.04272,
0.07717,0.2257,0.3087,0.004282,0.0615,
0.5061,0.1497,0.3919),
HT2 = c(
0.3002,0.01733,0.3756,0.02314,0.2873,
0.1072,0.2551,0.474,0.1629,0.1206,
0.4864,0.1873,0.01003,0.01907,0.3545,
0.2151,0.03929,0.007648)))

list(lamda1=2.98,lamda2=5.92)
```

The posterior analysis is reported in Table 7.12. One can see the effect of the data, the holding time on the estimation of the parameters. For example, the value of $b+\gamma$ used to generate the holding times is 1, but the posterior mean of this parameter is 0.3112 with a 95% credible interval (−1.696,2.224), which does indeed include 1. Also, the value of β was set to 2 for generating the holding times; however, its posterior mean is 2.346 with a 95% credible interval (0.0035,2.284), which does include the value 2.

Consider

$$p_{10} = (b+\gamma)/[.99\beta + (b+\gamma)], \tag{7.92}$$

the probability that the epidemic will change from one to zero infected person. It is seen that its posterior mean is .06774 and its posterior median is .1262, implying skewness in this posterior distribution.

7.6 Summary and Conclusions

The chapter begins with laying the theoretical foundation for the study of CTMCs. First, the Markov property is defined, which is the most important concept for such processes. Next to be described are the ideas of time homogeneity and the probability transition function. One of the most important concepts for CTMC is the infinitesimal transition rate matrix,

TABLE 7.12

Posterior Analysis for an SIS Epidemic

Parameter	Mean	SD	Error	2 1/2	Median	97 1/2
$R=\beta-\delta$	2.035	1.681	0.00784	−0.9977	1.935	5.607
$\Pr(R>0)$	.8947	.3069	.001435	0	1	1
β	2.346	0.789	0.003505	0.9849	2.284	4.057
$\delta=b+\gamma$	0.3112	0.9907	0.004777	−1.696	0.3269	2.224
λ_1	3.323	.7811	.00347	1.975	3.261	5.017
λ_2	5.221	1.237	.005924	3.091	5.111	7.93
p_{10}	.06774	.392	.002323	−.8665	.1262	.6561

which leads to the transition probabilities of the embedded chain. Next to be introduced is the Kolmogorov forward and backward system of differential equations involving the transition rate matrix Q and the derivatives of the transition probability function. The solution to this system is the transition probability function $P(t)$. It is demonstrated how one may use R and the exponential function $\exp(tQ)$ to compute the transition probability function. Next it is shown how to compute the stationary distribution (if it exists) by solving a system of equation involving the transition rates of the infinitesimal rate matrix Q.

In Section 7.4, a CTMC with an absorbing state is considered, and this is demonstrated with a liver disease process and R is used to determine the average time to absorption from a transient state of the chain. Also explained are the concepts of time reversibility and global stability, which depends on the stationary distribution of the chain.

Section 7.5 introduces several biological examples including three versions of DNA evolution. The first to be considered is the simplest model of DNA evolution, called the Jukes–Cantor model. One of the holding time parameters is estimated using WinBUGS Code 7.1 and the posterior reported in Table 7.1. Also presented is a test of hypothesis involving the holding time parameter and a derivation of the predictive density of a future holding time. The Kimura model is a generalization of the Jukes–Cantor model and allows one to distinguish between transversions and transcriptions of DNA substitutions. A Bayesian test of hypothesis investigates if the Kimura model reduces to the Juke–Cantor model, which is followed by an examination of the Felsenstein–Churchill model, which generalizes the Kimura model. A Bayesian analysis is conducted that estimates the alpha parameter of the generalization. WinBUGS Code 7.3 is executed to estimate alpha, and the posterior analysis is reported in Table 7.7. Lastly considered is the DNA evolution model referred to as the Hasegawa, Kishino, and Yano model, and a Bayesian test that the model reduces to the Felsenstein–Churchill model is performed.

Section 7.5.2 is regarding an important class of CTMCs, namely, birth and death processes. Its stationary distribution corresponding to the infinitesimal rate matrix Q given by Equation 7.74 is explained, and this is followed by an example with five states determined by the Q matrix in Equation 7.80. The Bayesian analysis depends on the holding times generated with an exponential distribution with known parameters. Using 45,000 observations for the simulation and a burn-in of 5,000, WInBUGS Code 7.5 is executed with the prior information defined as a noninformative gamma distribution for the holding time parameter. Posterior analysis results are reported in Table 7.5 and consist of the characteristics of the posterior distribution of the four average extinction times.

A special case of the birth and death process is the random walk with instantaneous rate matrix Q given by Equation 7.85 from which the stationary distribution is derived. The process depends on two infinitesimal rates: (1) the one move to the right (corresponding birth rate) and (2) the one move to the left (corresponding to the death rate). With known values assigned to these two rates, the corresponding holding times with exponential distributions are generated and appear in the list statement of WinBUGS Code 7.6 with the posterior analysis reported in Table 7.6.

Another special birth and death process is the Yule process, which is a birth process (no deaths are possible). This implies that all states are transient and that the stationary distribution does not exist. As with the previous examples, a particular value is assigned to the birth rate, which allows one to generate exponentially distributed holding times (for the various states), which are in turn used for the Bayesian analysis reported in Table 7.8.

A generalization of the birth and death process is the birth and death process with immigration, which allows only an increase in the population; however, births and deaths are possible. The Bayesian analysis is executed with WinBUGS Code 7.9 and the results appear in Table 7.5.

Section 7.5.7, the last part of Section 7.5, introduces stochastic epidemic processes, the so-called SI model, meaning two types of individuals are followed in the epidemic, namely, the susceptibles and those that are infected. Assigning known values to the infinitesimal rate matrix allows one to generate exponentially distributed observations for the Bayesian analysis executed with WinBUGS Code 7.10, and the results are reported in Table 7.11.

A generalization of this epidemic process to the so-called SIS model is the last example of the chapter, and the analysis is similar to that for the SI model.

Bayesian inferences for continuous-type Markov chains have been an active area of research, and the following references should be appealing to the students who want to contribute to the literature on the subject.

In the area of reliability, see Cano, Moguerza, and Insua,[9,10] while those interested in Markov-modulated Poisson processes should read Fearnhead and Sherlock.[11] Geweke, Marshall, and Zarkin[12] studied mobility indices for CTMCs, and Scott and Smyth[13] emphasized Bayesian inferences applicable to Markov-modulated Poisson chains. With regard to DNA evolutionary models, refer to Suchard, Weiss, and Sinsheimer.[14] But for a more comprehensive list of work related to Bayesian techniques with CTMCs, see pages 103 and 104 of Insua, Ruggeri, and Wiper.[2]

7.7 Exercises

1. Define the Markov property and the transition probability function for a CTMC.
2. a. Define time homogeneity for a CTMC.
   b. Give an example of a CTMC which is time homogenous.
3. Define the memoryless property of a CTMC. What is the distribution of the holding time of a CTMC?
4. What is the association between the infinitesimal transition rates and the transition probabilities of the embedded chain?
5. From the infinitesimal transition rates in Equation 7.8, derive the transition probabilities in Equation 7.9.
6. The Kolmogorov forward equations are a system of differential equations involving the derivative $P'(t)$ of the transition probability function. Show that the solution to this system is the transition probability function.
7. Using R Code 7.1, compute the probability transition function $P(t)$ as $P(t) = \exp(tQ)$, where Q is the matrix of instantaneous transition rates.
8. If it exists, define the limiting distribution of a CTMC.
9. Under what conditions does the unique stationary distribution of a CTMC exist?
10. Based on the infinitesimal rate matrix Q in Equation 7.19, find the unique stationary distribution given by Equation 7.14. R Code 7.2 should be employed to find the stationary distribution.

11. Use R Code 7.3 with the matrix Q specified by Equation 7.24 with state 3 (death) as an absorbing state and starting in state 1 (cirrhosis of the liver), compute the mean time to death.
12. Define time reversibility of a CTMC. Consider the three-state chain with probability transition matrix P of the embedded chain:
    a. What is the corresponding Q matrix?
    b. Is the process time reversible? Why? Explain.
13. Based on the Q matrix in Equation 7.32 of the Jukes–Cantor DNA evolution model, derive the corresponding probability transition function.
14. Execute WinBUGS Code 7.1 with 45,000 observations for the simulation and a burn-in of 5,000.
    a. What is the posterior mean of λ?
    b. What is the posterior mean of μ?
    c. Is this process time reversible? Why?
    d. Which posterior distributions appear as symmetric about the posterior mean?
15. Refer to the Jukes–Cantor model and test the hypothesis H: $\lambda = 1/6$ versus A: $\lambda \neq 1/6$. Show that the posterior probability of the null hypothesis is .7356.
16. Derive the predictive density in Equation 7.47 of $t(n + 1)$ for the holding time with exponential parameter λ.
17. Refer to the Felsenstein–Churchill model with the instantaneous rate matrix Q of Equation 7.61.
    a. What is its stationary distribution?
    b. What is its probability transition function matrix $P(t)$?
18. Refer to the Felsenstein–Churchill model for DNA evolution with matrix Q given by Equation 7.61. In order to estimate the parameter α of the matrix Q, execute WinBUGS Code 7.3 with 35,000 observations for the simulation and 5,000 for the burn-in. Your results should be similar to those reported in Table 7.4.
    a. What is the posterior median of α?
    b. Is the posterior distribution of α symmetric about its posterior mean?
    c. What is the 95% credible interval for α?
19. Refer to the birth and death process with infinitesimal rate matrix Q given by Equation 7.74 and derive the corresponding stationary distribution in Equation 7.78.
20. Refer to the birth and death process with matrix Q given by Equation 7.80 and with states 0, 1, 2, 3, and 4, where $\mu_1 = 2$, $\lambda_1 = 1$, $\mu_2 = 3$, $\lambda_2 = 2$, $\mu_3 = 4$, and $\lambda_4 = 3$.
    a. With WinBUGS Code 7.5, verify the posterior analysis in Table 7.6.
    b. What are the a_i, $i = 1, 2, 3$, in the posterior analysis?
    c. Assume the process is in state 3. What is the posterior mean of the average time to extinction?
    d. Is the posterior distribution of a_3 symmetric about its posterior mean? Why?

21. For the birth and death process with $\lambda = 2$ and $\mu = 5$, execute WinBUGS Code 7.7 and verify the posterior analysis in Table 7.7.
    a. What are the posterior mean and median of $P(\lambda > \mu)$.
    b. Display the posterior density of $P(\lambda > \mu)$.
    c. What does the posterior mean of $P(\lambda > \mu)$ imply about this birth and death process?
22. For the Yule process with Q matrix given by Equation 7.86, show that all states are transient and that the stationary distribution does not exist.
23. For the Yule process with the infinitesimal rate matrix of Equation 7.86, let the holding time parameter be $\lambda = 2$. The holding time observations are in the list statement of WinBUGS Code 7.9.
    a. What is the posterior mean of λ?
    b. What prior distribution is used for λ?
    c. Name the posterior distribution of λ.
    d. Does the posterior analysis suggest $\lambda = 2$? Why?
24. For the birth and death process with immigration and the Q matrix of Equation 7.89, let $\mu = 1$, $\nu = 1/3$, and $\lambda = 1$ be the assigned values for the death, immigration, and birth rates, respectively. These values are used to generate the exponentially distributed holding times appearing in the list statement of WinBUGS Code 7.9 Execute WinBUGS Code 7.9.
    a. What is the posterior mean of the immigration rate ν?
    b. What is the posterior median of the average number of immigrants per day?
    c. Do you believe $\lambda = 1$?
    d. Display the posterior density of μ.
    e. Specify the prior distribution for the three rates.
25. Refer to the SI epidemic with the Q matrix of Equation 7.91 for the number of infections. Execute WinBUGS Code 7.10.
    a. What is the posterior mean of β?
    b. For this epidemic, what is the interpretation of β?
    c. What is the average holding time for one infected individual?
    d. What prior distributions did you specify for β?

References

1. Dobrow, R. P. 2016. *Introduction to Stochastic Processes with R*. New York: John Wiley & Sons.
2. Insua, D. R., Ruggeri, F., and Wiper, M. P. 2012. *Bayesian Analysis of Stochastic Process Models*. New York: John Wiley & Sons.
3. Allen, L. J. S. 2011. *An Introduction to Stochastic Processes with Applications to Biology, Second Edition*. Boca Raton, FL: Taylor & Francis.
4. Bartolomeo, P., Trerotoli, P., and Serio, G. 2011. Progression of liver cirrhosis to HCC: An application of hidden Markov model, *BMC Medical Research Methodology* 11(380):1–8.

5. Lee, P. M. 1997. *Bayesian Statistics: An Introduction, Second Edition*. New York: John Wiley & Sons.
6. Kimura, M. 1980. A simple method for estimating evolutionary rates of base substitution of comprehensive studies of nucleotide sequences, *Journal of Molecular Evolution* 16(20):111–120.
7. Felsenstein, J., and Churchill, G. A. 1996. A hidden Markov model approach variation among sites in rate of evolution, *Molecular Biology Evolution* 13:93–101.
8. Hasegawa, M., Kishino, H., and Yano, T. 1980. Dating of human ape splitting by a molecular clock of mitochondrial DNA, *Journal of Molecular Evolution* 22(2):160–174.
9. Cano, J., Moguerza, J. M., and Insua, D. R. 2010. Bayesian reliability, availability, and maintainability analysis for hardware systems described through continuous time Markov chains. *Technometrics* 52:324–334.
10. Cano, J., Moguerza, J., and Insua, R. 2011. Bayesian analysis for semi Markov processes with application to reliability and maintenance, *Technical Report*. Madrid: Universidad Rey Juan Carlos.
11. Fearnhead, P., and Sherlock, C. 2009. An exact Gibbs sampler for the Markov-modulated Poisson process, *Journal of the Royal Statistical Society B* 68:767–784.
12. Geweke, J., Marshall, R., and Zarkin, G. 1986. Mobility indices in continuous time Markov chains. *Econometrica* 54:1407–1423.
13. Scott, S. L., and Smyth, P. 2003. The Markov modulated Poisson process cascade with application to web traffic modeling, In Bernardo, J. M., Bayarri, M. J., Berger, J. O. et al. (Eds) *Bayesian Statistics 7* (pp. 1–10). Oxford: Oxford University Press.
14. Suchard, M., Weiss, R., and Sinsheimer, J. 2001. Bayesian selection of continuous time Markov chain evolutionary models, *Molecular and Evolutionary Biology* 18:1001–1013.

8

Bayesian Inferences for Normal Processes

8.1 Introduction

Previous chapters have employed Bayesian inferences for stochastic processes with discrete time and discrete state space, and in this chapter, inferential techniques will be applied to the most general case where time and state space are continuous. Bayesian inferential methods of estimation, testing hypotheses, and forecasting will reveal interesting aspects of stochastic processes that are unique. Remember that the standard course in stochastic processes does not emphasize inference but instead focuses solely on the probabilistic properties of the model.

Subjects to be presented are the properties of the Wiener process (the Brownian motion) and the Bayesian estimation of its variance and covariance function. The Wiener process and the random walk will be generalized to a continuous state space, and Bayesian testing methods will be demonstrated with the parameters of the random walk.

The Brownian motion is a special case of normal stochastic processes where the joint distribution of the variables of model has a mean vector and variance covariance as parameters. The Bayesian approach uses the inverse normal–Wishart distribution as a prior for the mean vector and precision matrix, with the result that the marginal posterior distribution of the mean vector has a multivariate t distribution. This in turn provides easily implemented Bayesian techniques of inference.

Certain mapping or transformations of the Wiener process are of interest and have many applications. Translations, reflections, rescaling, and inversions of the Brownian motion will be presented and lead to such concepts as stopping times, first hitting times, and determining the zeros of the Wiener process. Such mapping applied to the Brownian motion produce other types of models that are amenable to Bayesian inferences and will be implemented with WinBUGS and R.

Certain variations of the Brownian motion will be studied including the Brownian motion with drift, the Brownian bridge, and the Ornstein–Uhlenbeck process. Bayesian inferences are especially interesting when applied to examples from finance such as stock options and derivatives.

The chapter is concluded with the broad subject of martingales which is a generalization of normal processes, including the Brownian motion. Martingales offer more complex stochastic processes that are a challenge to the Bayesian.

The reader is referred to Chapter 8 of Dobrow,[1] and Chapter 6 of Insua, Ruggeri, and Wiper[2] for information that is germane to this chapter. For a good introduction to WinBUGS and Bayesian inference see Ntzoufras.[3]

8.2 Wiener Process

In 1927, Robert Brown, a botanist, described a biological case where pollen particles suspended in water exhibited erratic behavior with continuous movement of tiny particles ejected from the grain in the water. On the basis of the laws of physics, Albert Einstein developed a mathematical description of the phenomena and Einstein showed that the position of the particle denoted by y at time t is described by the partial differential heat equation

$$(\partial / \partial t)f(y,t) = (1/2)(\partial^2/\partial y^2)f(y,t), \tag{8.1}$$

where $f(y, t)$ represents the number of particles per unit volume at position y at time t. It can be shown that the solution to Equation 8.1 is

$$f(y,t) = \left(1/\sqrt{2\pi t}\right)e^{-y^2/2t}; \tag{8.2}$$

that is, the solution is the density of a normal distribution with mean 0 and variance t. Thus, the process called the Brownian motion is a continuous-time continuous-state stochastic process. Wiener[4] investigated the properties of this process; thus, the model is sometimes called a Wiener process, and he showed that the sample functions of the process are continuous almost everywhere, but that the process is not differentiable at any time point. The Weiner process $\{B(t), t > 0\}$ or standard Brownian motion is defined as follows:

a. For all $t > 0$, $B(t)$ has a normal distribution with mean zero and variance t.
b. It has stationary increments, namely, letting $s, t > 0$, $B(t + s) - B(s)$ has the same distribution as $B(t)$.
c. The process has independent increments; that is, for $0 \le q < r \le s < t$, $B(t) - B(s)$ and $B(r) - B(q)$ are independent.
d. The function $B(t)$ is continuous with probability of 1.

Note that we use the notation that $B(t) \sim N(0,t)$, and recall that the Wiener process can be interpreted as the movement of a particle that diffuses randomly along a line, where at time, the particle location is normally distributed about the line with standard deviation $\sqrt{t}$. Wiener's fundamental contribution to our knowledge was to prove the existence of such a process as the Brownian motion. Bayesian inferences will be based on independent increments, because each increment $B(t) - B(s)$ has a normal distribution with mean 0 and variance $t - s$; for $s < t$, it is easy to write down the likelihood function. Independent increments make it easier to evaluate complicated probabilities involving Brownian motion. For example, consider finding the distribution of $B(s) + B(t)$ when $0 < s < t$.

Note that

$$B(s) + B(t) = 2B(s) + B(t) - B(s), \tag{8.3}$$

Since $2B(s)$ and $B(t) - B(s)$ are independent normal random variables, it follows that

$$E[B(s) + B(t)] = 0 \tag{8.4}$$

and

$$\text{Var}[B(s) + B(t)] = 3s + t. \tag{8.5}$$

Now consider the problem: Suppose the position of a particle follows a Wiener process, and then if the position of a particle is at 2 at time 3, find the probability that it is at most 4 at time 7.

In symbols,

$$\begin{aligned} P[B(7)<4|B(3)=2] &= P[B(7)-B(3)<4-B(3)|B(3)=2] \\ &= P[B(7)-B(3)<4|B(3)=2] \\ &= P[B(7)-B(3)<4] \\ &= P[B(4)<4]u \\ &= .9772499. \end{aligned} \tag{8.6}$$

The last probability is calculated with R using the code

```
> pnorm(4,0,2)
[1] 0.9772499.
```

The preceding formulation of the Brownian motion is called standard Brownian motion; however, we will be concerned with the more general form $\{B(t), t > 0\}$, where $B(t) \sim N(0, \sigma^2 t)$. The goal of Bayesian inference for such a process is to estimate σ^2, to test hypotheses about σ^2, and to predict future values of $B(t)$. R provides a way to simulate Brownian motion variables $B(t_i)$, $t_i = it/n$, where $i = 1, 2, \ldots, n$ and n is a positive integer. Now let

$$\begin{aligned} B(t_i) &= B(t_{i-1}) + [B(t_i) - B(t_{i-1})] \\ &= B(t_{i-1}) + Z(i), \end{aligned} \tag{8.7}$$

where $Z(i) \sim N(0, t_i - t_{i-1}) = N(0, t/n)$ and is independent of $B(t_{i-1})$, which leads to the recursive relation involving $X(i)$, $i = 1, 2, \ldots, n$ of independent $N(0,1)$ random variables. Thus,

$$B(t_i) = B(t_{i-1}) + \sqrt{(t/n)}X(i), \quad i = 1, 2, .., n$$

or

$$B(t_i) = \sqrt{(t/n)}\left[\sum_{i=1}^{i=n} X(i)\right]. \tag{8.8}$$

R Code 8.1 code shows how to generate standard Brownian motion variables over the interval [0,50], with adjacent observations one unit apart (Figure 8.1):

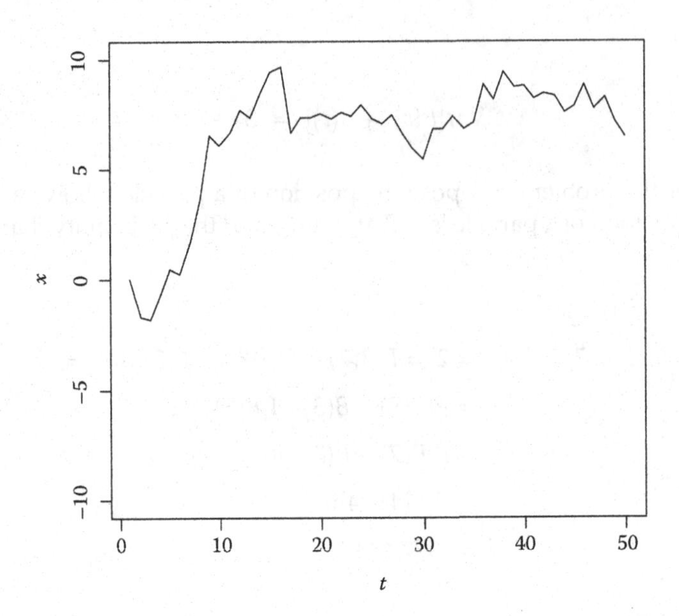

FIGURE 8.1
Fifty standard Brownian motion variables over [0,50].

R Code 8.1

```
t<-1:50
sig2<-1
x<-rnorm(n=length(t)-1,sd=sqrt(sig2))
x<-c(0,cumsum(x. t))
plot(t,x,type= "1", ylim=c(-10,10)
```

In the preceding code, sig2 is the variance of process which is one for the standard Brownian motion, and the corresponding standard deviation is listed as sd. Brownian motion-generated values are designated by the vector x, and the plot command has an abscissa range from 1 to 50 with time units of length one and with the ordinate range from −10 to 10.

R Code 8.2 generates 30 Wiener variables with $\sigma^2 = 4$.

R Code 8.2

```
t<-0:50
sig2<-4
x<-rnorm(n=length(t)-1,sd=sqrt(sig2))
x<-c(0,cumsum(x. t))
plot(t,x,type= "1", ylim=c(-8,8))
```

The x vector contains the 30 Wiener values

```
X=(0.0000000, -4.4847302, 0.4643178, -0.3437283, -0.4905968,
0.4821004, 1.8126740, -1.2513397, -1.1408787, -1.5291112,
-3.0735822, -2.9174191, -4.2614190, -6.6813461, -4.7708805,
-2.1957203, -1.3201780, 1.4937973, 1.4242956, 1.6672101,
```

```
2.7032072, 3.8358068, 4.2725358, 5.5445094, 6.5348815,
6.7983973, 5.5906146, 6.1619978, 8.8013248, 7.8247243).
```

Our goal is to use the Bayesian approach to estimate σ^2.

Let

$$B(2i) - B(2i-1) = X(i), \quad i = 1,2,..,15, \tag{8.9}$$

where the $B(i)$ are the 30 Wiener values generated with R Code 8.2; then the $X(i)$ are the independent increments and are distributed as

$$X(i) \sim B(2i) - B(2i-1) \sim B(1) \sim N(0, \sigma^2). \tag{8.10}$$

The 15 observed increments are in the list statement of WinBUGS Code 8.1. The Bayesian analysis is executed with 45,000 observations and a burn-in of 5,000. Noninformative prior distributions are assigned to μ and τ.

WinBUGS Code 8.1

```
model;
{
mu~dnorm(0,.001)
tau~dgamma(.001,.001)
for( i in 1:15){
X[i]~dnorm(mu,tau)}
sigsq<-1/tau
}

list( X=c(-4.4847,-.8083,-.14687,1.3305,.11646,-1.5493,
        -1.344,1.91076,.87553,-.06959,1.03599,.4367,
        .9903,-1.2077,2.63931))

list(tau=.24,mu=0)
```

Table 8.1 reports the analysis for the Brownian motion with 30 observations displayed in Figure 8.2.

The parameter μ is the mean of Brownian motion and should be zero. Is there sufficient evidence to declare that $\mu = 0$? The 95% credible interval for μ is (–.9842,.9512) and implies that indeed $\mu = 0$.

TABLE 8.1

Posterior Analysis for Wiener Process

Parameter	Mean	SD	Error	2 1/2	Median	97 1/2
σ^2	3.497	1.557	.00878	1.609	3.148	7.458
τ	.3333	.1245	.000727	.1341	.3177	.6215
μ	–.0204	.4877	.002738	–.9824	.01989	.9512

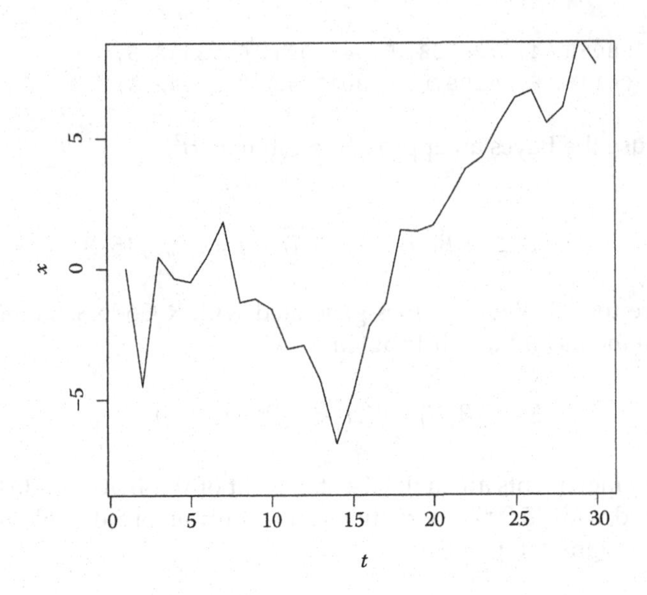

FIGURE 8.2
Brownian motion with $n = 30$.

The estimate of σ^2 is 3.497 with the posterior mean. Remember that the value $\sigma^2 = 4$ was used to generate the Brownian motion values, and this estimate appears reasonable with a 95% credible interval (1.609,7.458).

Let us test in a formal Bayesian way the null hypothesis

$$\text{H: } \mu = 0 \text{ versus the alternative A: } \mu \neq 0 \tag{8.11}$$

Recall from pages 126–128 of Lee[5] that the posterior probability of the null hypothesis is

$$p_0 = 1/(1 + r), \tag{8.12}$$

where

$$r = \pi_1 f_1(x)/\pi_0 f(x|\mu = 0) \tag{8.13}$$

with

$$f_1(x) = \int_{-\infty}^{\infty} \rho_1(\mu) f(x|\mu) d\mu, \tag{8.14}$$

$$\rho_1(\mu) = 1, \ \mu \in R,$$

and

$$f(x|\mu = 0) = \left\{\Gamma(n/2)/\left[(2\pi)^{n/2}\left((n-1)s^2\right)^{n/2}\right]\right\}\left[1 + n(\bar{x})^2/(n-1)s^2\right]^{-n/2}$$

and

$$f(x|\mu) = \left\{\Gamma(n/2)/\left[\left((2\pi)^{n/2}\left((n-1)s^2\right)^{n/2}\right)\right]\right\}\left[1 + n(\bar{x}-\mu)^2/(n-1)s^2\right]^{-n/2}$$

and

$$f_1(x) = \left\{\Gamma((n-1)/2)/\left[(2\pi)\left((n-1)s^2\right)\right]^{n/2}\right\}\sqrt{(n-1)\pi}/\sqrt{n/s^2}, \tag{8.15}$$

where $x = (x(1), x(2), \ldots, x(n))$ is the vector of observations appearing in the list statement of WinBUGS Code 8.1.

Assuming $\pi_0 = \pi_1 = 1/2$ (the prior probability of the null hypothesis is the same as that of the alternative), and computing the sample mean as $x = -.057$ and the sample variance as $s^2 = 2.9632$, it can be shown that $f_1(x) = 0$ and that $r = 0$; consequently, the posterior probability of the null hypothesis H is $p_0 = 1$. The evidence implies that $\mu = 0$ and that the Brownian motion values generated by R Code 8.2 have the desired mean of 0.

Is the Wiener process a Markov process? If so, it should satisfy the Markov property of a stochastic process $\{X(t), t \geq 0\}$, which is defined as follows: Select time points $s, t \geq 0$ and real state y, then the process is Markov if

$$P[X(t+s) \leq y|X(u), 0 \leq u \leq s] = P[X(t+s) \leq y|X(s)]. \tag{8.16}$$

In addition, the process is time homogenous if the probability in Equation 8.16 does not depend on s; that is,

$$P[X(t) \leq y|X(0)] = P[X(t+s) \leq y|X(s)]. \tag{8.17}$$

One intriguing property of the Wiener process is that its realizations are continuous but nowhere differentiable. These two properties can be demonstrated with the aid of measure theory, and pages 181–184 of Feller[6] provide a convincing account.

Given realizations from a stochastic process, can one show statistically that it is a Markov satisfying Equation 8.16? This is left as an exercise at the end of the chapter.

8.3 Random Walk and Brownian Motion

The continuous time random walk and the Brownian motion are connected as follows: Suppose $Y(1)$, $Y(2)$, ... is a sequence of independent and identically distributed (i.i.d.) binary random variables with each assuming the values ± 1 with equal probability. Now let

$$S(t) = Y(1) + Y(2) + .. + Y(t), \tag{8.18}$$

$t = 0, 1, 2...$ with $S(0) = 0$, and then the process $\{S(t), t = 0, 1, 2, \ldots\}$ is a symmetric random walk with moments $E[S(t)] = 0$ and $\text{Var}[S(t)] = t$ for $t = 0, 1, 2, \ldots$. Notice that for large t, $S(t)$ is approximately normally distributed and that the process has independent increments. To see the latter assertion, note that for $0 < q < r < s < t$,

$$S(t) - S(s) = Y(s+1) + Y(s+2) + .. + Y(t)$$

and

$$S(r) - S(q) = Y(q+1) + \ldots + Y(r),$$

and because the $Y(t)$ are independent, the two increments are also independent. In addition, the process has the time homogeneity property, namely, that the distribution of $S(t) - S(s)$ is that of $S(t - s)$, and the process has stationary increments. The process represented by Equation 8.18 is in discrete time; therefore, the pertinent question is how is it extended to continuous time. One way is to base the generalization with the function as follows:

$$\begin{aligned} S(t) &= Y(1) + Y(2) + .. + Y(t), \qquad t = 0,1,2,\ldots, \\ &= S([t]) + S([t]+1)(t - [t]), \quad t \text{ is not an integer,} \end{aligned} \tag{8.19}$$

and $[t]$ is the floor function defined as the largest integer less than or equal to t. If k is a positive integer with $k \le t \le k+1$, then $S(t)$ is the linear interpretation between the two points $(k, S(k))$ and $(k+1, S(K+1))$. Also, it follows from Equation 8.19,

$$\begin{aligned} &E[S(t)] = 0 \\ &\text{and} \\ &\mathrm{Var}\{S([t]) + X([t]+1)(t-[t])\} = \mathrm{Var}S([t]) + (t-[t])^2\mathrm{Var}(X([t]+1)) \\ &= [t] + (t-[t])^2 \approx t, \quad 0 \le t - [t] < 1 \end{aligned} \tag{8.20}$$

Now it should be stressed that the development of a symmetric random walk can be based on any i.i.d. sequence $Y(1), Y(2),\ldots$ with mean 0 and variance 1, and let

$$S(n) = \sum_{i=1}^{i=n} Y(i).$$

Then it is easily demonstrated that as $n \to \infty$, the sequence converges to the Brownian motion, that is

$$S(nt)/\sqrt{n} \to B(t)$$

and

$$\begin{aligned} \lim_{n\to\infty}\{g(S(tn/\sqrt{n}))\} &= \lim_{n\to\infty, 0<t<1} \max\{S(tn)/\sqrt{n}\} \\ &= \lim_{n\to\infty, 0<k<n} \max\{S(k)/\sqrt{n}\} = g(B(t)), \end{aligned} \tag{8.21}$$

where g is the maximum value function and $\{B(t), t > 0\}$ is the Brownian motion. The result (Equation 8.21) is the so-called invariance principle.

Of course, the function g can be any continuous function.

The following R Code computes the maximum of a symmetric random walk (Equation 8.19) using 100 replications where the generating process is the sequence of binary random variables with values of +1 and −1 with equal probability:

```
n<-50
sim<-replicate(50,
max(cumsum(sample(c(-1,1),n, replace =T))))
max(sim)
mean(sim)
sd(sim)
s<-seq(1,50,1)
plot(s,sim)
```

The 50 generated random walk values are

```
0, 1, 5, 1, 12, 4, 7, 19, 14, -1, 5, 4, -1, 4, 9, 4, 2, 2, 9, 3, 1, 5, 3, 6, 7, 5, 5, 10, 7,
5, 3, 5, 2, 2, 9, 15, 2, 0, 11, 7, 2, 7, 9, 4, 1, 5, 14, 7, 9, 2,
```

where the sample mean and standard deviation are $\bar{x} = 5.56$ and $s^2 = 4.35$, respectively, and the maximum is max = 19. Refer to Figure 8.3, which is a plot of the random walk values versus time.

The generated random walk values are only approximately distributed as the Brownian motion and should have a mean of 0 and standard deviation of 1. Do the generated values support the hypothesis that they are the realization from a Brownian motion process? The answer to this question is left as an exercise for the student.

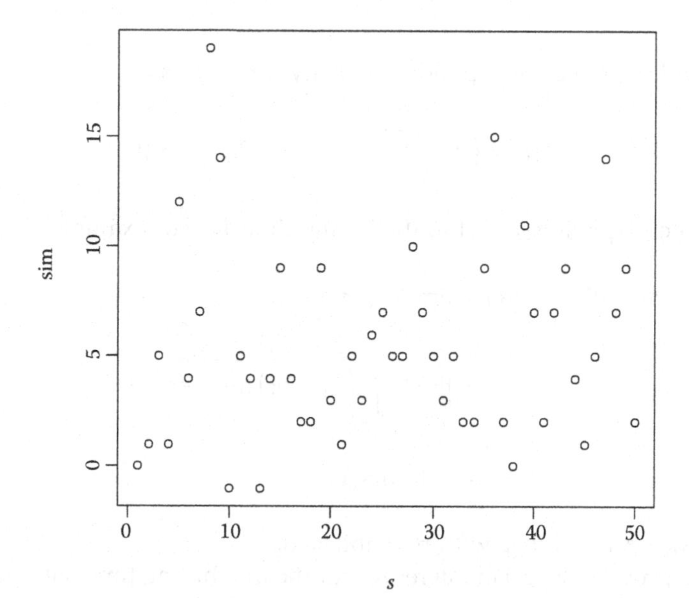

FIGURE 8.3
Symmetric random walk.

8.4 First Hitting Times

One interesting problem about the Wiener process is to determine the time until the process reaches a particular value, say, the real number v. The first hitting time is described and defined on pages 337–339 of Dobrow[1] as follows. Let T_v be the random variable starting from 0 until reaching v, then

$$T_v = \min[t : B(t) = v]. \tag{8.22}$$

If T_v is thought of as the stopping time, then the process begins with the translated process at time v.

For the standard Brownian motion at time t, the process is equally likely to be above zero as to be below zero. Assume $v > 0$, and then for the process beginning at time v, at any time $t > v$, the process is equally likely to be above the horizontal line v unit above the line $y = 0$. In symbols

$$P[B(t) > v | T_v < t] = P[B(t) > 0] = 1/2, \tag{8.23}$$

and from this, it follows that the distribution function of T_v is

$$\begin{aligned} P[T_v < t] &= 2P[B(t) > v] = 2\int_v^\infty \left(1/\sqrt{2\pi t}\right)\left[\exp - x^2/2t\right]dt \\ &= 2\int_{v/\sqrt{t}}^\infty \left(1/\sqrt{2\pi t}\right)\left[\exp - x^2/2\right]dx. \end{aligned} \tag{8.24}$$

Upon differentiation, the corresponding density function is

$$f(t) = \left(v/\sqrt{2\pi t^3}\right)\exp - v^2/2t, \quad t > 0 \tag{8.25}$$

There are some surprising aspect to the hitting time T_v. For example,

$$\begin{aligned} P(T_v < \infty) &= \lim_{t\to\infty} P(T_v < t) \\ &= \lim_{t\to\infty} 2\int_{v/\sqrt{t}}^\infty \left(1/\sqrt{2\pi}\right)\left[\exp - x^2/2\right]dx \\ &= 2(1/2) = 1; \end{aligned} \tag{8.26}$$

that is to say, the process hits v with probability one.

Our goal is to provide Bayesian inferences for the first hitting time with the more general Brownian motion process $\{B(t), t > 0\}$ with mean μ and variance σ^2.

For this general process, it is easily shown that the mean and variance of the first hitting time to target v are

$$E(T_v) = v/\mu$$

and (8.27)

$$\text{Var}(T_v) = v\sigma^2/\mu^3,$$

respectively.

Based on a realization of the general Brownian motion with unknown mean and variance, the goal of the Bayesian approach will be to estimate these two parameters as well as the preceding moments and to test hypotheses about μ and σ^2.

R Code 8.3 generates 1000 values of the Brownian motion with a mean of ½ and a standard deviation of 1 and target of 10 over the interval (0,80] with a time interval of length .016 units. The objective is to estimate the average time to hit the value 10.

R Code 8.3

```
mu<-1/2
sig<-1
v<-10
simlist<-numeric(1000)
for( i in 1:1000){
t<-80
n<-5000
bm<-c(0, cumsum(rnorm (n,0,sqrt(t/n))))
xproc<-mu*seq(0,t,t/n)+sig*bm
simlist[i]<-which(xproc >= v)*(t/n)
mean(simlist)
var(simlist)
}
```

The command mean(simlist) computes the average time to hit the target 10, while the command var(simlist) computes the sample variance of the time to hit the target with the result mean(simlist) = 19.96 and var(simlist) = 73.128.

The first 30 values of xproc of the Brownian motion values with mean ½ and variance 1 are reported in the following. Using Bayesian methods, these values will be used to estimate the mean and variance of the Brownian motion:

```
0.000000000, 0.028777582, -0.097218008, -0.269953192,
-0.227157657, -0.273306894, -0.153932689, -0.371594078,
-0.340772505, -0.315872944, -0.154263879, -0.294961244,
-0.335631249, -0.182333234, -0.157393723, -0.002417403,
0.114061076, 0.113888952, 0.140305932, 0.034385946,
0.040447019, 0.223088634, 0.313589806, 0.178910765,
0.208270820, 0.194747132, 0.235888893, 0.440015651,
0.439388151, 0.533403410.
```

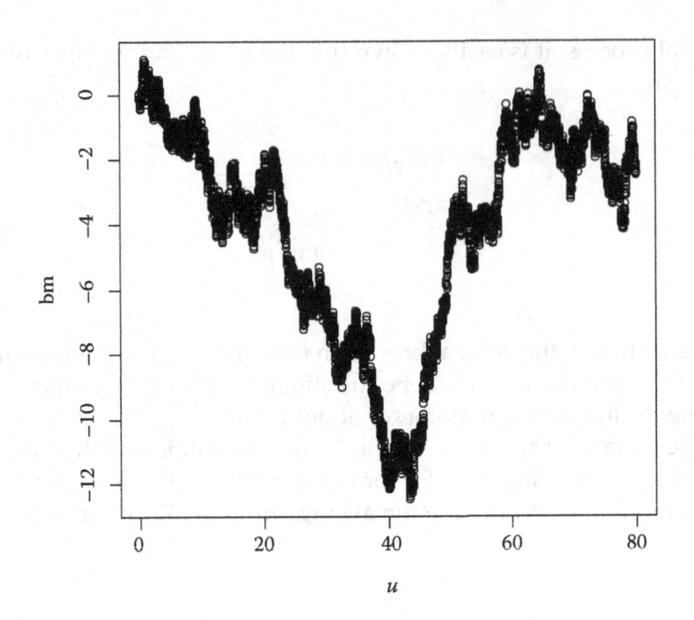

FIGURE 8.4
Simulation of 100 Brownian motions for hitting time to 10.

Based on 5000 xproc values, we have mean(xproc) = 15.884 and var(xproc) = 146.556. I generate Figure 8.4 with the following R command:

```
Plot(u,bm) where u<-seq(0,80,.016).
```

In order to estimate μ and σ^2, the Bayesian analysis will be based on the 15 independent increments corresponding to the first 30 observations of xproc generated with R Code 8.3. The increments are i.i.d normally distributed with mean 0 and variance σ^2. Based on the 15 increments, I computed the sample mean as $\bar{x} = -.02078$ and sample variance as $s^2 = .0327212$. This seems reasonable because the actual theoretical variance of the independent increments is .016.

Using noninformative prior distributions for μ and τ, namely, $\mu \sim N(0, .001)$ and $\tau \sim$ *gamma* (.001,.001), respectively, the Bayesian analysis is executed with WinBUGS Code 8.2 using 35,000 observations for the simulation and a burn-in of 5,000. The list statement of WinBUGS Code 8.2 includes the 15 increments.

WinBUGS Code 8.2

```
model;
{
mu~dnorm(0,.001)
tau~dgamma(.001,.001)
for ( i in 1:15){
X[i]~dnorm(mu, tau)}
# ET is the average time to the target 10
ET<-10/(vu)
# VT is the variance of the time to target level 10
VT<-10*sigmasq/pow(vu,3)
```

```
vu<-mu+.5
sigmasq<-1/tau

}

list(X=c(.02877,-.17274,-.14443,-.21777,.0249,-.13757,.1533,.15497,
-.000181,-.10597,.182633, -.135482,-.01353,.20413,.09401))
list(mu=0,tau=1)
```

The posterior analysis for the hitting time is reported in Table 8.2. Note that σ_o^2 is the variance of Brownian motion and σ^2 is the variance of the increments. Also, ν is the mean of Brownian motion, but μ is the mean of the corresponding increments.

The actual average time to hitting target level 10 is 20, which should be compared to the posterior mean of 20.37, implying excellent agreement with the true value. It is a different story with the variance of the hitting time which is estimated as 51.13 with the posterior median. The actual value of the variance is 80, and the difference can be attributed to the fact that the estimate is based on the first 30 Brownian motion values (or the 15 corresponding increments). The actual value of the mean is 1/2, which is estimated as .4942 with the posterior mean, which is why the posterior mean of $E(T_{10})$ is very close to the actual value of 20.

The 95% credible interval for μ is (−.08468,.0738), which implies that informally, the $\mu = 0$, but on the other hand, the 95% credible interval for σ^2 is (.0108,.05005), which does include the value $\sigma^2 = 0.016$.

I would expect that a formal Bayesian test of the hypothesis $\sigma^2 = .016$ would not be rejected in favor of the alternative $\sigma^2 \neq .016$.

If one is testing hypotheses, I would expect to not reject the hypothesis that $\nu = 1/2$ versus the alternative $\nu \neq 1/2$ (or equivalently $\mu = 0$ versus $\mu \neq 0$), and I would expect not to reject the hypothesis that $\sigma^2 = .016$ in favor of the alternative that $\sigma^2 \neq .016$.

σ_o^2 is the variance of Brownian motion which is 1, and the posterior analysis gives a 95% credible interval of (0.6753,3.129), implying that the hypothesis that the null hypothesis $\sigma_o^2 = 1$ would not be rejected.

Consider a test of the null hypothesis:

$$H: \sigma^2 = .016 \text{ versus the alternative } A: \sigma^2 \neq .016, \tag{8.28}$$

where σ^2 is the variance of each increment. Using the precision in lieu of the variance, Equation 8.28 is equivalent to testing

TABLE 8.2

Posterior Analysis for Hitting Time to Target 10

Parameter	Mean	SD	Error	2 1/2	Median	97 1/2
$E(T_{10})$	20.37	1.69	.009435	17.43	20.23	24.08
$\text{Var}(T_{10})$	127.8	76.13	.4626	51.13	108.6	315.7
μ	−.00583	.03997	.000224	−.08468	−.00579	.0738
σ^2	.02349	.01046	.000058	.0108	.02114	.05009
τ	49.63	18.68	0.1084	19.96	47.3	92.55
ν	.4942	.03997	.000224	.4153	.4942	.5738
σ_o^2	1.468	0.6529	0.00394	0.6753	1.321	3.129

$$H: \tau = 62.5 \text{ versus the alternative } A: \tau \neq 62.5 \tag{8.29}$$

Recall that the posterior probability of the null hypothesis is

$$p_0 = 1/(1+r), \tag{8.30}$$

where

$$r = \pi_1 f_1(x)/\pi_0 f(x|\tau = 62.5) \tag{8.31}$$

and

$$f(x|\tau) = \left[\tau^{(n-1)/2}/(2\pi)^{(n-1)/2}\sqrt{n}\right] \exp -\tau(n-1)s^2/2 \tag{8.32}$$

and

$$f_1(x) = \int_0^\infty f(x|\tau)\rho_1(\tau)d\tau. \tag{8.33}$$

Let the prior density of τ under the alternative be

$$\rho_1(\tau) = 1/\tau, \quad \tau > 0 \tag{8.34}$$

Suppose the prior probabilities of the null and alternative hypotheses are $\pi_0 = \pi_1 = 1/2$, then knowing that $n = 15$ and $s^2 = .0232717$, there is sufficient information to calculate $r = 0$ and the posterior probability of the null hypothesis as $p_0 = 1$. Thus, there is insufficient information to reject the null hypothesis that $\tau = 62.5$, and of course, this was implied by the 95% credible interval for τ reported in Table 8.2.

The student will be asked to derive the predictive distribution of future Brownian motion values.

8.5 Brownian Motion and Coin Tossing

The Wiener process reaches a particular level, regardless of how small or large, with certainty and reaches the level 0 infinitely often. The times the process assumes the value of 0 are called the zeros of the process. Of course, the process has infinitely many zeros occurring in the interval $(0, \varepsilon)$, regardless of how small $\varepsilon > 0$.

Our goal is to develop Bayesian inferences for the zeros of Brownian motion and the last time the origin is visited by the process. This as will be seen is related to a coin-tossing experiment.

The following theorem is given on pages 341–343 of Dobrow[1] and specifies the probability $z_{r,t}$ that the standard Brownian motion has at least one zero in the interval (r, t), namely,

$$z_{r,t} = (2/\pi)\arccos\left(\sqrt{r/t}\right), \quad 0 \le r < t. \tag{8.35}$$

Thus, for at least one zero in $(0, \varepsilon)$ is

$$z_{0,\varepsilon} = (2/\pi)\arccos(0) = 1 \tag{8.36}$$

In terms of the Brownian motion, $B(t) = 0$ for some $t \in (0, \varepsilon)$ with probability of 1. Over the interval $(0, 1)$ and based on Equation 8.35, the distribution function is

$$F(t) = (2/\pi)\arccos\sqrt{r}, \quad 0 < t < 1$$

with corresponding density

$$f(t) = (1/\pi)t^{-1/2}(1-t)^{-1/2}. \tag{8.37}$$

In this special case, the density is that of a beta (1/2,1/2) distribution.

Based on the preceding presentation, the distribution function of the time L_t to the last 0 in $(0,t)$ is

$$P(L_t \le x) = (2/\pi)\arcsin\sqrt{x/t}, \quad 0 < x < t. \tag{8.38}$$

The random variable L_t, the time to the last 0, is now shown to be related to an interesting experiment in coin tossing with two players 1 and 2. If the coins lands heads, player 1 pays \$1, and if tails occurs, player 2 pays player 1 \$1. If the coin is flipped a large number of times, when would you expect the players are even?

R Code 8.4 code generates 10,000 tosses of a fair coin, and the histogram is the last time the two players are even. Of course, this is related to the number of zeros occurring over a given time interval.

R Code 8.4

```
trials<-10000
simlist<-numeric(trials)
for ( i in 1:trials){
rw<-c(0, cumsum(sample(c(-1,1), (trials-1),
replace = T)))
simlist[i]<-tail(which(rw==0),1)
}
hist (simlist)
```

This is interesting in that it is more likely that the two players are even with a fewer (closer to 1) number of tosses and a larger number (nearer to 10,000). Is this what you would expect? Why does the histogram appear similar to the arcsin distribution (Equation 8.38)?

What type of inferences should be made for the coin-tossing experiment?

Consider the Bayesian estimation of the random walk which should approximately follow the Brownian motion process $\{B(t), t > 0\}$, where $E[B(t)] = 0$ and $\text{Var}[B(t)] = t$. Note that the increments $X(i) = B(i+1) - B(i), i = 1, 2, \ldots,$ are i.i.d. and normally distributed with mean of 0 and variance of 1.

Figure 8.5 is the graph of the random walk values versus time.

Forty random walk values are listed in the following and will be the basis for Bayesian inferences about the mean and variance of the increment process:

```
rw = (-4, -5, -6, -5, -4, -3, -2, -3, -4, -3, -4, -5, -6, -5, -4, -3, -4, -5, -4, -5,
-4, -3, -2, -1, -2, -3, -4, -3, -2, -3, -4, -3, -4, -3, -2, -1, 0, 1, 2, 3)
```

The corresponding 20 increment values are

```
X = (-1,1,1,-1,1,-1,1,1,-1,-1,1,1,-1,1,-1,1,1,1,1,1)
```

with a mean of $\bar{x} = .2$ and variance $s^2 = 1.010526$.

The Bayesian analysis is executed with 35,000 observations for the simulation and a burn-in of 5,000. Noninformative prior distributions are the set of the mean and precision.

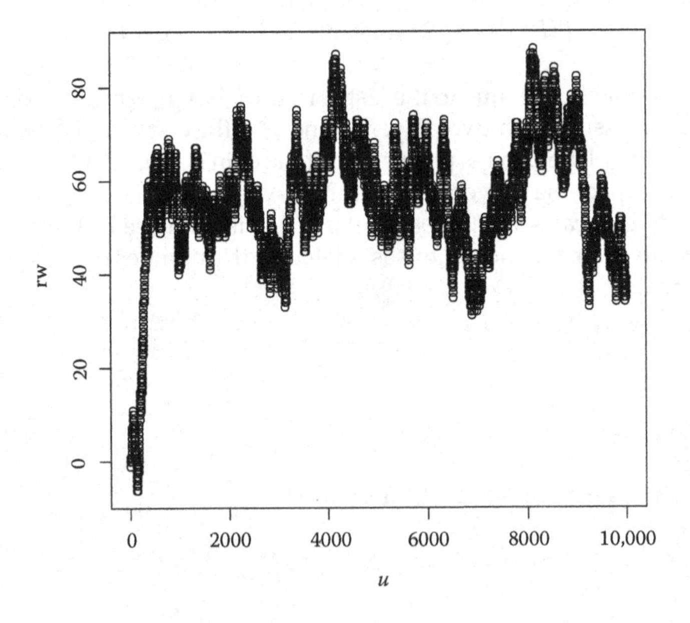

FIGURE 8.5
Random walk for coin-tossing experiment.

TABLE 8.3

Posterior Analysis for Coin-Tossing Experiment

Parameter	Mean	SD	Error	2 1/2	Median	97 1/2
μ	.299	.2334	.001394	−.1609	.2989	.7621
σ^2	1.072	0.3925	0.002341	0.5537	0.9948	2.05
τ	1.043	0.3387	0.001977	0.4877	1.005	1.806

WinBUGS Code 8.3

```
model;
{
mu~dnorm(0,.0001)
tau~dgamma(.001,.001)
for ( i in 1:20){
X[i]~dnorm(mu,tau)
}
sigmasq<-1/tau}
list(X=c(-1,1,1,-1,1,-1,1,1,-1,-1,1,1,-1,1,-1,1,1,1,1,1))
list(mu=0,tau=1)
```

See Table 8.3 for the posterior analysis.

The posterior analysis implies that the coin-tossing experiment is indeed based on the Brownian motion. Note that the mean μ of the increments has a posterior mean of .299 and a 95% credible interval of (−.1609,.7621). Since the interval includes zero, one is inclined to believe that the population mean of the increments is indeed zero. Now consider the variance σ^2, which has a posterior median of .9948 and a 95% credible interval of (0.5537,2.05). Since the interval includes the value of 1, the evidence supports the conclusion that the population variance is indeed 1.

8.6 Brownian Motion with Drift

Consider the Brownian motion with drift defined as

$$X(t) = \mu t + \sigma B(t) \tag{8.39}$$

where $\sigma > 0$, μ is any real number, and $\{B(t), t > 0\}$ is the standard Brownian motion. It can be demonstrated that Equation 8.39 is a normal process with mean μt and variance σ^2 and that the process has independent and stationary increments $[X(t + s) - X(t)]$ with mean μs and variance $\sigma^2 s$, where $s, t > 0$.

Consider the following application of the Brownian motion with a drift explained on pages 346–347 of Dobrow,[1] where the objective is to estimate the home field advantage by determining the probability that the home team leads by y points after a fraction t $(0 < t < 1)$ of the game is played. To evaluate this probability, let $X(t)$ denote the difference in scores between the home time and the visiting team after $100t$% of the game is played. This approach is presented by Stern[7] with a Brownian motion process $\{X(t), 0 < t < 1\}$ } where μ denotes the magnitude of the home field advantage.

The probability that the home team wins, conditional on the fact that they have a y point lead at time t, is

$$\begin{aligned} p(y,t) &= P[X(1) > 0|X(t) = y] \\ &= P[X(1) - X(t) = -y] \\ &= P[X(1-t) = -y] \\ &= P[\mu(1-t) + \sigma B(1-t) > y] \\ &= P\left[B(t) < \left(\sqrt{t}(y + \mu(1-t))\right)/\sigma\sqrt{1-t}\right]. \end{aligned} \tag{8.40}$$

In order to evaluate this probability, data from the 1992 National Basketball Association (NBA) season with 493 games were used, which gave a home field advantage estimated as $\mu = 4.87$ and an estimated standard deviation of $\sigma = 15.82$. Instead of substituting these two estimates into Equation 8.40, the Bayesian approach will be based on data generated (via R) with the Brownian motion model (Equation 8.39) and using the estimates $\mu = 4.87$ and $\sigma = 15.82$, then find the posterior distribution of μ and σ, and finally estimating the desired probability (Equation 8.40) with WinBUGS.

R Code 8.5 generated 100 Brownian motion values with drift value $\mu = 4.87$ and standard deviation $\sigma = 15.82$.

R Code 8.5

```
t <- seq(0,1,.01) # time
sig2 <- 15.82*15.82
## first, simulate a set of random deviates
x <- rnorm(n = length(t) - .01, sd = sqrt(sig2))
## now compute their cumulative sum
x <- (c(0, cumsum(x)))
y<-x+4.87
```

The 100 Brownian motion values with drift appear in the following.

```
4.870000 12.775059 -3.128824 19.163220 0.211200 1.740787,
14.567501, 22.314764, 34.994566, 41.803923, 20.012853,
56.819429, 63.110764, 67.496794, 57.860953, 56.782067,
44.161325, 14.235617, -5.332750, 1.740305, -3.021845, -8.936915,

-28.556834, -18.570739, -26.561301, -24.601585, -1.132670,

-1.828439, -28.955132, -30.627759, -13.233479, 5.783106,
1.731944, 25.423337, 55.475961, 102.191471, 94.618433, 104.231076,
119.971464, 114.613043, 132.181924, 147.195775, 150.642156,
104.167129, 114.355566, 114.582757, 117.562082, 114.130562,

118.511579, 115.297406, 124.145900, 114.092315, 109.761164,
116.989473, 103.600474, 93.227840, 85.069323, 106.782843,
106.149420, 115.361881, 106.849390, 110.700031, 138.594530,
127.304229, 147.243907, 136.143945, 137.707052, 128.164416,
132.109209, 129.894992, 118.604953, 122.488599, 138.562499,
121.032912, 129.626051, 112.446461, 101.233938, 91.620732,
```

```
70.385467, 79.309660, 84.624360, 75.924865, 69.799267,
102.212087, 87.112890, 91.589131, 94.869071, 80.678975,
77.385947, 105.868184, 90.561178, 73.718837, 63.782504,
68.397127, 79.890086, 89.198168, 70.810888, 76.166559, 62.198091,
71.641202, 45.719826.
```

The Bayesian estimation of μ and σ will be based on the 50 increments corresponding to the preceding 100 Wiener values. Note that the 100 values are generated over the interval [0,1] with intervals of length .01; thus, the mean of each interval is 0 and its variance is σ^2.

The Bayesian analysis is executed with WinBUGS Code 8.4 using 35,000 observations for the simulation and 5,000 for the burn-in.

Noninformative prior distributions are assigned for μ and τ, and the main goal is to determine the posterior probability (Equation 8.40) for the lead time with time t and lead y. For this example, the time $t = .5$ and lead $y = 0$.

WinBUGS Code 8.4

```
model;

{
mu~dnorm(0,.001)
tau~dgamma(.001,.001)
for ( i in 1:50){
X[i]~dnorm(mu,tau)
}
# mu is the mean of the independent increments
#sigmasq is the variance of the increments
sigmasq<-1/tau
y<-0
t<-.5
d<-sqrt(sigmasq)*sqrt(1-t)
a<-n/d
n<-sqrt(t)*(y+vu*(1-t))
#posterior probability of the lead time at time t and lead y.
probability<-step(y-m)
# vu is the mean of the original Brownian motion values
vu<-4.87+mu
m~dnorm(0,t)
}
list(X=c(7.905,16.035,1.5288,7.7472,6.8094,36.8066,
.386,-1.0249,-29.9257,7.047,-6.6081,9.9861,1.9598,
.6958,-1.6726,19.0165,23.6914,46.7164,9.6126,
-5.3584,2.9793,-3.432,-3.218,-10.0563,-7.2254,
-10.3726,21.7138,9.2199,3.8507,11.2905,-11.1,
-9.5426,-2.215,3.884,-17.5295,-17.1796,-9.6132,
-4.6995,-8.7003,32.4128,4.4762,-14.19,28.4822,
-16.8423,4.6145,9.3081,5.3556,9.5262,-8.3677,
5.2556))
list(mu=4.86,tau=.00428)
```

TABLE 8.4

Posterior Analysis Home Team Advantage: $y = 0$, $t = .5$

Parameter	Mean	SD	Error	2 1/2	Median	97 1/2
μ	2.964	2.091	0.009762	−1.199	2.977	7.077
Prob	.5025	.5	.001791	0	1	1
σ^2	220.7	46.54	0.2216	147.8	214.5	328.4
τ	.004725	.000955	.000045	.003045	.004661	.006766
ν	7.834	2.091	0.009762	3.671	7.847	11.95

TABLE 8.5

Probabilities the Home Team Wins

	Lead y		
Time t	**0**	**2**	**10**
0	.62		
.25	.61	.66	.84
.5	.59	.65	.87
.75	.56	.66	.92
.9	.54	.69	.98
1.0		1.0	1.0

The Bayesian analysis for the parameters of the home team advantage example appears in Table 8.4. Note that the time is .5 (half way through the game) and the lead at this time is 0.

The parameter ν is the mean of the original Brownian motion values generated by R Code 8.5, and σ^2 is the variance of the increments appearing in the list statement of WinBUGS Code 8.4.

The posterior mean of the probability of winning the game half way through the game when the teams are tied is .5025, with a standard deviation of .5 for the posterior distribution of the probability.

The probabilities of the home team winning for a given time t and lead y is computed directly using Equation 8.40 and is reported in Table 8.5. When $y = 0$ and time = .5, the value computed is .59 compared to .5025 determined with Bayesian analysis executed with WinBUGS Code 8.4. Why the difference in these two values? The reader will be asked to answer this question in the exercises at the end of the chapter.

Notice that for a given time, the probability that the home time wins increases as the lead increases, as it should.

8.7 More Extensions of the Wiener Process

In this section, two variations of Brownian motion are studied: (1) the geometric Brownian motion and (2) the Brownian bridge motion along with applications that demonstrate the versatility of the process. For example, with the geometric Brownian motion where the example presented is from finance.

8.7.1 Geometric Brownian Motion

These processes are used to model exponential growth and decay and are often used in finance to model stock prices and to represent the return from stock options.

The process is defined as

$$G(t) = G(0)\exp[X(t)], \quad t \geq 0, G(0) > 0, \tag{8.41}$$

where $\{X(t), t > 0\}$ is Brownian motion with drift μ and variance σ^2.

When Equation 8.41 is represented in logarithmic form

$$\ln G(t) = \ln G(0) + X(t) \tag{8.42}$$

and moments

$$E(\ln G(t)) = \ln G(0) + \mu t, \tag{8.43}$$

$$\mathrm{Var}(\ln(G(t)) = \sigma^2 t \tag{8.44}$$

Thus, the log is a normal process with a log normal distribution.

It can be shown that the first two moments for $G(t)$ are

$$E(G(t)) = G(0)\exp\left(t\left(\mu + \sigma^2/2\right)\right)$$
and
$$\mathrm{Var}(G(t)) = \left(e^{t\sigma^2} - 1\right)G^2(0)\exp 2t\left(\mu + \sigma^2/2\right). \tag{8.45}$$

It is obvious that the mean of the process exhibits exponential growth with growth rate of $(\mu + \sigma^2/2)$.

The goal is to develop Bayesian inferences for the parameters of the geometric Brownian motion using an example about stock prices.

Realizations for the geometric version are easily generated by transforming standard Brownian motion.

It is often the case that the geometric process can be expressed as the product of random multipliers. Let $s, t \geq 0$ and consider the ratio

$$G(t+s)/G(s) = \exp(\mu s + \sigma(X(t+s) - X(s))). \tag{8.46}$$

Because the underlying Wiener process $\{X(t), t > 0\}$ has stationary independent increments, the ratios $G(t)/G(0)$ and

$$G(r)/G(q) = \exp(\mu(r-q) + \sigma(X(r) - X(q))). \tag{8.47}$$

are independent random variables for $0 \leq q < r < s < t$.

Now let

$$Y(k) = G(k)/G(k-1), \quad k = 1, 2, \ldots$$

Then this sequence consists of i.i.d. random variables, and an alternative representation of the geometric process can be expressed as

$$G(n) = G(0)Y(1)Y(2)..Y(n). \tag{8.48}$$

The representation of the process as ratios or products (Equation 8.48) is often employed when analyzing stock prices and options. See pages 352–356 of Dobrow[1] for additional insights into the geometric process.

Consider the following example of geometric Brownian motion with parameters $\mu = 1$ and $\sigma^2 = .5$, then the corresponding Brownian motion with drift has mean μt and variance $\sigma^2 t$, where $0 \le t \le 1$. R Code 8.6 generates realizations from a geometric Brownian motion process.

R Code 8.6

```
{
t <- seq(0,365,1) # time
sig2 <- 0.5
## first, simulate a set of random deviates
x <- rnorm(n = length(t) - 1, sd = sqrt(sig2))
## now compute their cumulative sum
x <- c(0, cumsum(x))
y<-x+.1   g<-.01*exp(y)}
```

The following y values are the Brownian motion values with drift parameter $\mu = 1$ and $\sigma = .5$.

```
X = (0.00000000, -0.05001197, -0.32318259, 0.16780378, -0.42227482,
-0.05281918, 0.21517819, -0.43398929, -0.56569527, -1.41766731,
-1.04528536, -1.47824415, -2.26287486, -2.63555567, -2.16807106,
-2.68234271, -2.74855017, -2.52898434, -1.75166436, -2.38117730,
-3.07062934, -3.28345939, -2.14394150, -2.69041558, -3.10910347,
-3.46575120, -3.28469407, -3.29709531, -3.56759687, -3.45335676)
```

Using the 30 values, the corresponding 15 increments will be the foundation for the Bayesian analysis.

Recall that the initial value of the geometric Brownian motion process is .001, which is portrayed in Figure 8.6, and notice the explosive growth beginning at day 275. The geometric Brownian motion values in the figure mimic the Brownian motion value with drift depicted in Figure 8.7.

Using these 30 values, the corresponding 15 increments will allow one to estimate these two parameters and to estimate the probability

$$P[G(281) < 6] = P[\exp(\mu t + \sigma X(281)) < 6], \tag{8.49}$$

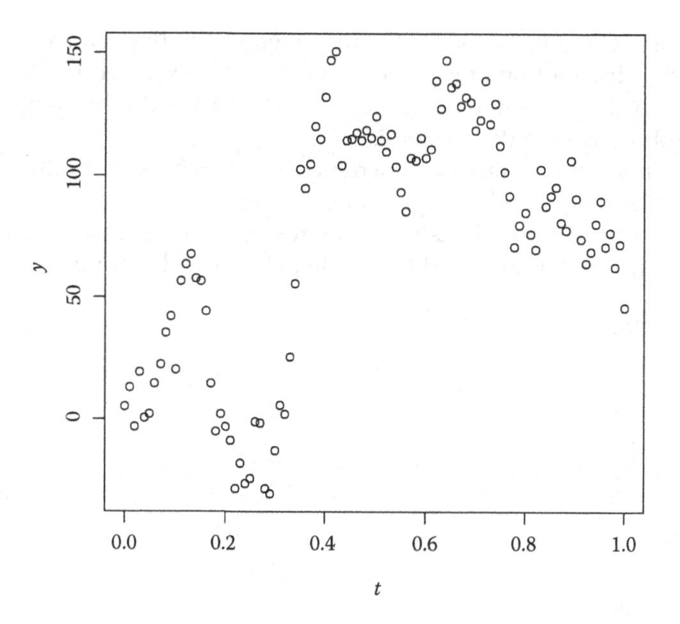

FIGURE 8.6
Three geometric Brownian motion processes.

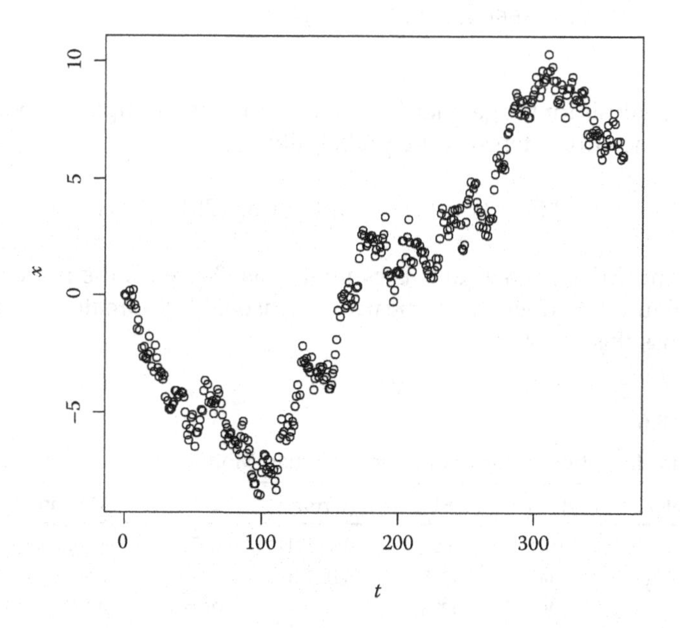

FIGURE 8.7
Geometric Brownian motion.

where μ and σ are estimated based on the increments. The Bayesian approach is to determine the posterior distribution of Equation 8.49, and it is executed by WinBUGS Code 8.4 with 35,000 observations for the simulation and 5,000 for the burn-in, where the main parameter is prob in the WinBUGS Code.

The parameter mu is the mean of the increments, which hypothetically is 0. The variance of each increment is designated by sigma in the code.

Note the noninformative prior distributions for μ, which is assigned a normal (0,.001) distribution, and τ, which is assigned a gamma (.001,.001) distribution.

WinBUGS Code 8.5

```
model;
{
mu~dnorm(0,.001)
tau~dgamma(.001,.001)

for ( i in 1:15){

X[i]~dnorm(mu,tau)}
sigma<-1/tau
a<-exp(mu*281+sigma*6.92001681)
# prob is the probability (8.49)
prob<-step(6-a)

}

list(X=c(-.05001,.49098,.36946,-.64969,.14803,
     -.43292,-.37262,-.51434,.21965,-.62951,
     -.21283,-.54647,-.35664,-.01231,.11429))
list( mu=.1,tau=2)
```

The Bayesian analysis for the geometric Brownian motion example is reported in Table 8.6. The main parameter of interest is the probability

$$P[G(281) < 6] = P[\exp(\mu t + \sigma X(281)) < 6],$$

where $t = 281$ and $X(281) = 6.92$ and is estimated as .944 with the posterior mean. Upon inspection of Figure 8.6, does this appear as a reasonable estimate? It will be left to the student to discuss this question.

TABLE 8.6

Posterior Distribution for Geometric Brownian Motion

Parameter	Mean	SD	Error	2 1/2	Median	97 1/2
μ	–.1628	.1053	.000591	–.3704	–.1627	.04698
Probability	.9444	.2295	.001371	0	1	1
σ^2	.1629	.07254	.000409	.07486	.1456	.3475
τ	7.154	2.693	0.01562	2.878	6.818	13.34

Recall that the geometric process is defined as

$$G(t) = G(0)\exp[X(t)], \quad t \geq 0, G(0) > 0 \tag{8.41}$$

and is nothing but an exponential transformation of the underlying process $\{X(t), t \geq 0\}$ with drift parameter μ and variance σ^2. Also note that $X(t) = \mu t + \sigma B(t)$, where $B(t)$ is standard Brownian motion.

Consider the posterior predictive distribution of m future observations $y = (y_1, y_2, ..., y_m)$ given the past n observations $x = (x_1, x_2, ..., x_n)$ of the incremental process of the preceding geometrical Brownian motion.

The Bayesian predictive density of y given x is

$$f(y|x) = \int_0^{\infty}\int_{-\infty}^{\infty} f(y|\mu,\tau)f(x|\mu,\tau)\rho(\mu,\tau)d\mu\, d\tau, \tag{8.50}$$

where $\rho(\mu, \tau)$ is the prior density of μ and τ.

$$f(x|\mu,\tau) = \left[\tau^{n/2}/2\pi^{n/2}\right]e^{-(\tau/2)\left[n(\mu-\bar{x})^2+(n-1)s_x^2\right]}, \tag{8.51}$$

$$f(y|\mu,\tau) = \left[\tau^{m/2}/2\pi^{m/2}\right]e^{-(\tau/2)\left[m(\mu-\bar{y})^2+(m-1)s_y^2\right]}, \tag{8.52}$$

and let the prior distribution be

$$\rho(\mu,\tau) = 1/\tau, \quad \tau > 0, \mu \in R.$$

Under these assumptions, it can be shown that the predictive density of y given x is

$$f(y|x) = f_1(y|x)f_2(y|x), \tag{8.53}$$

where

$$f_1(y|x) = \{\Gamma((m+n-1)/2)\}/\left\{(2\pi)^{(m+n-1)/2}\sqrt{m+n}\left[(m-1)s_y^2 + (n-1)s_x^2\right]^{(m+n-1/2)}\right\}$$

and

$$f_2(y|x) = \left\{1/(1+mn(\bar{y}-\bar{x})^2)/(m+n)\left[(m-1)s_y^2 + (n-1)s_x^2\right]\right\}^{(m+n-1)/2}.$$

Also, note that the sample means for the future and past observations are $\bar{y}$ and $\bar{x}$, respectively, and the corresponding sample variances are s_y^2 and s_x^2, respectively.

The density of the future sample mean $\bar{y}$ is recognized as a t density with parameters mean $\mu = \bar{x}$, degrees of freedom $\nu = (m+n-3)/2$, and precision $\tau = mn(m+n-3)/2(m+n)[(m-1)s_y^2 + (n-1)s_x^2]$. See page 279 of Insua, Ruggeri, and Wiper[2] for the form of the t density.

8.7.2 Brownian Bridge

The Brownian bridge is a Wiener process conditioned so that the process ends at the same level that it began.

From standard Brownian motion, the conditional distribution, the conditional process $\{B(t),\ 0 \le t \le 1\}$ conditional on $B(1) = 0$ is referred to as a Brownian bridge, and the process is anchored to 0 at the end points of [0,1]. Suppose $\{X(t), t \ge 0\}$ is a Brownian bridge, then if $0 \le t \le 1$, the distribution of $X(t)$ is the same as the conditional distribution of $B(t)$ given $B(1) = 0$, and it follows that a Brownian bridge process is also a Gaussian bridge; consequently, the Brownian bridge has stationary and independent increments. Let us find the mean and variance of Brownian bridge.

From the properties of standard Brownian motion, one can show that

$$E[X(t)] = 0 \tag{8.54}$$

Also, it can be shown that

$$\begin{aligned} \text{Cov}[X(s), X(t)] &= E[X(s)X(t)] \\ &= s - st, \end{aligned}$$

and by symmetry for $t < s$ that $E[X(s)X(t)] = t - st$; hence, in general,

$$\text{Cov}[X(s), X(t)] = \min(s, t). \tag{8.55}$$

As a special case, the variance of the Brownian bridge is

$$\text{Var}[X(t)] = t. \tag{8.56}$$

In conclusion, a Brownian bridge process acts like a standard Brownian motion, which is anchored at 0 at the end points 0 and 1.

The process is a special case of the process defined by the normal process $\{B(t), t \ge 0\}$, where $B(t_1) = a$ and $B(t_1) = b$, and $B(t)$ is standard Brownian motion with mean and covariance given by

$$E[B(t)] = a + [(t - t_1)/(t_2 - t_1)](b - a) \tag{8.57}$$

and

$$\text{Cov}[B(s), B(t)] = (t_2 - t)(s - t_1)/(t_2 - t_1), \tag{8.58}$$

where $t_1 < s < t < t_2$, respectively.

Of course, this implies that

$$\text{Var}[B(t)] = (t_2 - t)(t - t_1)/(t_2 - t_1). \tag{8.59}$$

Refer to pages 348 and 349 of Dobrow[1] for the R Code that simulates the Brownian bridge process which appears here as R Code 8.7:

R Code 8.7

```
mu<-0
sig<-1
n<-1000
t<-seq(0,1, length =n)
bm<-c(0,cumsum(rnorm(n-1,0,1)))/sqrt(n)
x<-mu*t+sig*bm
bb<-x-t*x[n]
```

The first 30 values are:

```
bb = (0.000000000, -0.013604458, -0.062329936, -0.075917470,
-0.094821232, -0.037913795, -0.044450710, -0.034984275,
-0.065209187, -0.004308567, 0.019417133, 0.022046633,
0.058177422, -0.042812927, -0.056317595, -0.031195645,
-0.044878532, -0.042353499, -0.055827645, -0.075289453,
-0.107196523, -0.122376326, -0.163769974, -0.189494716,
-0.160949115, -0.128565969, -0.135735785, -0.149629010,
-0.214404228, -0.215074953).
```

The Brownian bridge values are plotted in Figure 8.8. Note how the realization is anchored at the end points of [0,1]. The time interval length is .001 over [0,1] for a total of 1000 values.

The Bayesian analysis will be based on the 15 increments corresponding to the first 30 Brownian bridge values generated by R Code 8.7.

The Bayesian analysis is executed with WinBUGS Code 8.6 with 35,000 observations for the simulation and 5,000 for the burn-in. Noninformative prior distributions were assigned to τ and μ.

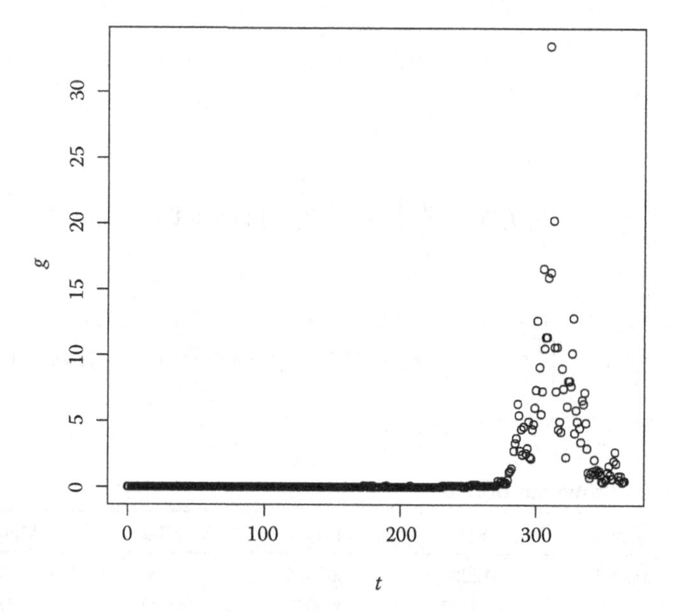

FIGURE 8.8
Plot of 1000 Brownian bridge values over [0,1].

WinBUGS Code 8.6

```
model;
{
mu~dnorm(0,.001)
tau~dgamma(.001,.001)
sigma<-1/tau
for ( i in 1:14){
X[i]~dnorm(mu,tau)}
}
list(X=c(-.0136,-.01358,.056921,.00947,.060901,
        .002629,-.100987,.025115,.002528,-.01945,
        -.015174,-.02573,.03238,-.01309,-.000674))
list(mu=1, tau=1)
```

The Bayesian analysis for the Brownian bridge example appears in Table 8.7.

The posterior distribution of τ appears to be nonsymmetric as does that for σ^2. The estimates for μ and σ^2 are very reasonable. For example, the hypothetical value of σ^2 should be .001 (the interval length between observations of the Brownian bridge values) and the posterior median is .001847 with a 95% credible interval of (.00092,.00454). Also, the hypothetical value of μ is zero, and its posterior median is –.00089 with a 95% credible interval of (–.02504, .023590).

It would be interesting to test the hypothesis H: $\mu = 0$ and $\tau = 1000$ versus the alternative A that H is not true.

Recall that the posterior probability of the null hypothesis is

$$p_0 = 1/(1+r), \tag{8.60}$$

where

$$r = \pi_1 f_1(x)/\pi_0 f(x|\mu = 0, \tau = 1000), \tag{8.61}$$

where

$$f_1(x) = \int_0^{\infty}\int_{-\infty}^{\infty} f(x|\mu,\tau)\rho_1(\mu,\tau)d\mu d\tau,$$

$$f(x|\mu,\tau) = \left\{\left[\tau^{n/2}/2\pi\right]\exp-(\tau/2)\left[n(\mu-\bar{x})^2+(n-1)s_x^2\right]\right\}d\mu\, d\tau,$$

TABLE 8.7

Posterior Analysis for Brownian Bridge

Parameter	Mean	SD	Error	2 1/2	Median	97 1/2
μ	–.00089	.01228	.000074	–.02504	–.00089	.02359
σ^2	.002071	.000972	.000005	.00092	.001847	.00454
τ	570.1	222.9	1.311	220	541.4	1086

where $x = -0.00082$ and $s_x^2 = 0.001481$, and $\pi_0 = \pi_1 = 1/2$ is the prior probability of the null and alternative hypotheses. Also, the improper prior is assigned as

$$\rho_1(\mu, \tau) = 1/\tau, \quad \tau > 0$$

In addition,

$$f(x|\mu = 0, \tau = 1000) = \left[(1000)^{15/2}/(2\pi)^{15/2}\right]\exp - 500(10.367) \sim 10^{12} \tag{8.62}$$

and $f_1(x) = .000419561$; thus, the posterior probability of the null hypothesis in Equation 8.60 is $p_0 \sim 1$, and there is not enough evidence to reject the null hypothesis.

For additional information about Brownian bridge processes, see the book by Revuz and Marc.[9]

8.8 Martingales

The martingale class of stochastic processes conveys the idea of a fair game in the sense that after n plays of gambling, if your winnings are z, your future earning on average will be z, regardless of the past history of the gamble. To be more precise, a stochastic process $\{Y(t),\ t \geq 0\}$ is a martingale if for all $t \geq 0$,

1. $[Y(t), Y_r, 0 \leq r \leq s] = Y(s), \quad 0 \leq s \leq t$ (8.63)
2. $E[|Y(t)|] < \infty$ (8.64)

An interesting property of martingales is that the expectation for all times is the same; that is,

$$E[Y((t))] = E\{E(Y(t)|0 \leq r \leq s)\} = E[Y(s)] \tag{8.65}$$

for all s and t, such that $0 \leq r \leq s$.

Is standard Brownian motion $\{B(t),\ t \geq 0\}$ a martingale? Consider

$$\begin{aligned} E[B(t)|B(r), 0 \leq r \leq s] &= E[B(t) - B(s) + B(s)|B(r), 0 \leq r \leq s] \\ &= E[B(t) - B(s)|B(r), 0 \leq r \leq s] + E[B(s)|B(r), 0 \leq r \leq s] \\ &= E[B(t) - B(s)] + B(s) = B(t) \end{aligned} \tag{8.66}$$

It is often useful to know that the following extension involving the idea of a stochastic process is a martingale with respect to another process; hence the process $\{Y(t),\ t > 0\}$ is a martingale with respect to $\{X(t), t > 0\}$ if for all $t > 0$,

1. $E[Y(t)|X(r), 0 \leq r \leq s] = Y(s), \quad 0 \leq s \leq t$
2. $E[\,|\,Y(t)\,|\,] < \infty$

(8.67)

Examples of such a situation are demonstrated when $Y(t)$ is a function g of $X(t)$, say, $Y(t) = g(X(t))$. Consider the quadratic martingale

$$Y(t) = B^2(t) - t, t \geq 0 \tag{8.68}$$

Is the process a martingale with respect to Brownian motion? If it is, one needs to show that

$$\begin{aligned} E[Y(t)|B(r), 0 \leq r \leq s] &= Y(s) \\ &= B^2(s) - s \end{aligned}$$

and that

$$E[|Y(t)|] < \infty.$$

This will be left as an exercise for the reader. See the problems at the end of the chapter.

The next example returns to the geometric Brownian motion and an application to pricing options in finance. Suppose

$$G(t) = G(0)\exp X(t)$$

is a geometric Brownian motion where $\{X(t), t \geq 0\}$ is Brownian motion with drift μ and variance σ^2 and suppose $r = \mu + \sigma^2/2$, then $e^{-rt}G(t)$ is a martingale with respect to standard Brownian motion.

The following proof can be found on page 359 of Dobrow.[1]

Let $0 \leq s < t$, then consider

$$\begin{aligned} E\left[e^{-rt}G(t)|B(r), 0 \leq r \leq s\right] &= e^{-rt}E\left[G(0)e^{\mu t + \sigma B(t)}|B(r), 0 \leq r \leq s\right] \\ &= e^{-rt}e^{\mu s + \sigma B(s)}E\left[G(t-s)\right] \\ &= e^{-s\left(\mu + \sigma^2/2\right)}G(0)e^{\mu s + \sigma B(s)} \\ &= e^{-rs}G(s), \end{aligned} \tag{8.69}$$

which shows that the process $\{e^{-rs}G(s), s \geq 0\}$ is a martingale with respect to Brownian motion.

The geometric Brownian motion is often used to model stock and option prices and can be applied to the Black–Scholes[10] model for pricing options. It provides a way to price options and other financial instruments such as derivatives, which soon developed into a giant global market for trading even more complicated financial options. There are some objections to this model: (1) the prices follow the geometric Brownian motion and (2) the expected rate of return should be risk free such as, for example, government bonds.

Suppose r is the risk-free interest rate, and P is the initial investment, then in t years, one would expect to accumulate $P(1 + r)^t$; thus, after a long period under continuous compounding, one could accumulate a future value of $F = Pe^{rt}$ dollars. On the other hand, and under the same circumstances, suppose in the future, one would accumulate F dollars; then to know its present value by discounting the future amount by e^{-rt} gives the present value (the price of the instrument) as $P = e^{-rt}F$. This is the basis for using the geometric Brownian motion for the price of the option.

Let $G(t)$ denote the price of the instrument t years from now, then the present price is $e^{-rt}G(t)$. The model is risk free, which implies that the discounted instrument is a fair game, and the return is a martingale; that for times $0 < s < t$, the average present value at time t should be the same as it was up to the past time s. In symbols, this is represented as

$$E\left[e^{-rt}G(t) \middle| G(y), 0 < y < s\right] = e^{-rs}G(s). \tag{8.70}$$

If Equation 8.70 holds, then it can be shown that $r = \mu + \sigma^2/2$ or $\mu = r - \sigma^2/2$, then the Black–Scholes formula for the price of the instrument is

$$\text{Price} = E\left[e^{-rt}\max\{G(t) - k, 0\}\right] = G(0)P\left[Z > (\alpha - \sigma t)/\sqrt{t}\right] - e^{-rt}KP\left[Z > \alpha/\sqrt{t}\right], \tag{8.71}$$

where

$$\alpha = \left[\ln(K/G(0)) - (r - \sigma^2/2)t\right]/\sigma. \tag{8.72}$$

Consider the following example, where $G(0) = 80$, $K = 100$, $\sigma^2 = .25$, $t = 90/365$, and interest rate $r = .02.$, which determines $\alpha = .498068$. (Note that $\mu = r + \sigma^2/2 = .145$.) Based on the Black–Scholes price (Equation 8.71), the price is computed as \$2.426.

Bayesian inferential techniques will be based on 30 observations generated from a geometric Brownian motion process using R and parameter values $\sigma^2 = .25$ and $\mu = .145$, and then based on the 15 corresponding increments, determine the posterior distribution of μ and σ^2, and the main objective of the posterior distribution of the price of the option, given by Equation 8.72. This implies that the posterior distribution of P will have a posterior standard deviation and credible interval that project the uncertainty in making inferences about the price. The value of \$2.426 was computed on page 361 of Dobrow[1] by direct substitution into the formula for the price (Equation 8.72); however, the Bayesian approach is much more realistic because it is based on data which reflect the uncertainty in investing in financial instruments.

R Code 8.8

```
t <- 1:365 # time
sig2 <- 0.25
## first, simulate a set of random deviates
x <- rnorm(n = length(t) - 1, sd = sqrt(sig2))
## now compute their cumulative sum
x <- c(0, cumsum(x))
y<-x+.145
# g values are Geometric Brownian motion values
g<-exp(y)

1.156040e+00, 5.106385e-01, 3.938965e-01, 2.178212e-01,
2.096127e-01, 1.042481e-01, 1.072979e-01, 8.753381e-02,
1.745498e-01, 2.404782e-01, 3.024356e-01, 3.847971e-01,
1.894463e-01, 2.284997e-01, 1.089254e-01, 1.760200e-01,
```

```
3.511947e-01, 3.764326e-01, 3.143635e-01, 7.524380e-01,
5.496282e-01, 1.184883e+00, 8.288976e-01, 5.464548e-01,
9.698991e-01, 1.722712e+00,, 2.544808e+00, 5.034664e+00,
6.609783e+00, 6.430235e+00.
```

Using noninformative prior distributions for μ and σ^2, and based on 15 increment values appearing in the list statement, the Bayesian analysis is executed with WinBUGS Code 8.7 using 35,000 observations for the simulation and 5,000 for the burn-in.

WinBUGS Code 8.7

```
model;
{
mu~dnorm(0,.001)
tau~dgamma(.001,.001)
for ( i in 1:15){
X[i]~dnorm(mu,tau)}
K<-100
sigma<-1/tau
r<-.02
Price<-P1-P2
t<-90/365
alpha<-.05756-mu*t/sqrt(sigma)
Z~dnorm(0,1)
P1<-80*step(Z-(alpha-(sqrt(sigma)*t))/sqrt(t))
P2<-K*step(Z-alpha/sqrt(t))*exp(-r*t)
}

list(X=c(1.6666,-.17607,-.28964,-.12208,.06593,
        .08236,.03905,.0671,.02531,.43813,.63526,
        -.28224,.75282,2.48986,-.17967))
list(mu=0, tau=4)
```

The Bayesian analysis for estimating the stock price appears in Table 8.8. Note that the main parameter of interest is the stock price which has a posterior mean of \$1.924 compared to \$2.42 dollars directly by substitution in Equation 8.71.

TABLE 8.8

Posterior Analysis for Price of Stock

Parameter	Mean	SD	Error	2 1/2	Median	97 1/2
Price	1.925	34.54	.1011	−19.51	−10.51	80
α	−.04975	.06678	.000204	−.1806	−.04954	.08162
μ	.3473	.2184	.000681	−.08876	.3474	.7798
σ^2	0.7167	0.3183	0.00104	0.3293	0.6488	1.525
τ	1.627	0.6152	0.00199	0.6559	1.551	3.037

The data generated for the analysis induce a mean for the increments as zero, but the Bayesian analysis for μ has a posterior mean of .3473 and a 95% credible interval of (−.08876,.7798). The interval barely includes the number 0, implying weakly that the mean is indeed 0. Also, the value for σ^2 used to generate the data was .25 compared to its posterior median .6488, and its 95% credible interval is (0.3293,1.525), which does not include the value of .25. Of course, one should not be surprised that the Bayesian estimates do not agree with the values used to generate the data. If another set of 30 generated values had been used for the Bayesian analysis, the posterior results would differ from those reported in Table 8.9. The sampling variability in choosing the data set induces variability in the posterior estimates of the parameters reflected by the posterior standard deviations and credible intervals.

8.9 Bayesian Analysis for Stock Options

An option is a contract that gives its owner the right to buy shares of a stock sometime in the future at a fixed price. The following approach is not based on the Black–Scholes model of Section 8.8. Assume that the stock is selling for \$90 per share, and under the terms of contract, in 80 days, you may buy a share of stock for \$110. Once having bought the option, several alternatives need to be considered. Assume that in 80 days, the prices of the stock exceed \$110, so if you exercise the option and buy the stock for \$110 and sell it at the current price, your payoff would be $G(80/365) - 110$, where $G(80/365)$ is the price of the stock in 80 days.

However, the other alternative has to be considered, namely, that in 80 days, the stock price is less than \$110 and the option would not be exercised; consequently, you receive nothing. In general, the two alternatives imply that the payoff is $\max\{G(80/365) - 110, 0\}$. What is the profit for such a situation? It is the payoff minus the cost of the option of \$15. Assuming that the stock price follows the geometric Brownian motion, pages 355 and 356 of Dobrow explain how to find the future average payoff, and this description is repeated here.

Let $G(0)$ denote the current stock price and t, the future time the option is exercised. Suppose k denotes the strike price, the price you can buy the stock if you exercise the option. Note that for this illustration, $G(0) = 90$, $t = 80/365$, and $k = \$110$. Our goal is to determine the average profit of the option, namely,

$$\begin{aligned} E[\max\{G(t) - k, 0\}] &= E[\max\{G(0)\exp(\mu t + \sigma B(t) - k, 0)\}] = \\ &G(0)\exp t(\mu + \sigma^2/2)\Pr[Z > (\beta - \sigma t)/\sqrt{t}] - k\Pr[Z > \beta/\sqrt{t}], \end{aligned} \tag{8.73}$$

where

$$\beta = [\ln(k/G(0) - t\mu)]/\sigma.$$

Notice the similarity in the expected payoff of an option given by Equation 8.73 and the price of stock given by the Black–Scholes approach given by Equation 8.71. The two parameters μ and σ^2 are the drift and variance of the geometric Brownian motion process, respectively. R Code 8.9 generates 365 values for the stock process with parameters $\sigma^2 = .25$ and $\mu = .1$ for the underlying Brownian motion process.

R Code 8.9

```
t <- 1:365 # time
sig2 <- 0.25
## first, simulate a set of random deviates
x <- rnorm(n = length(t) - 1, sd = sqrt(sig2))
## now compute their cumulative sum
x <- c(0, cumsum(x))
y<-x+.1
g<-exp(y)
```

Forty values of the underlying Brownian motion process are given value, and they include values 256–304:

```
Y = (0.96676506, 0.61594285, 0.66534596, 0.36374412, 0.48842212,
0.41844021, 0.06161046, 0.18435588, 0.69238297, 0.13834786,
0.40994016, -0.13303306, -0.48741286, -0.79124510, -1.46240081,
-1.18034531, -1.01237042, -1.28732419, -1.48092808, -0.45649775,
-0.93371867, -0.79224386, -0.68857902, -1.13706716, -1.24114566,
-0.34094420, -0.82473306, -1.35704727, -1.67109217, -1.52524371,
-1.16200053, -0.86781197, -1.44895509, -1.27968528, -1.72116867,
-1.42271242, -2.24197601, -2.46238107, -1.47948716, -1.03257688).
```

The Bayesian posterior analysis is executed with WinBUGS Code 8.8 with 35,000 observations for the simulation and burn-in of 5,000. Information for the analysis consists of the 20 increment values corresponding to the 40 geometric Brownian values generated with R Code 8.9. Such values were generated assuming $\sigma^2 = .25$ and $\mu = .1$. Note the non-informative prior distributions, namely, a normal (0,.001) for μ and a gamma (.001,.001) for τ.

WinBUGS Code 8.8

```
model;
{

mu~dnorm(0,.001)
tau~dgamma(.001,.001)
for ( i in 1:20){
X[i]~dnorm(mu,tau)}

sigma<-1/tau
# current value of stock
g0<-90

# striking value
k<-110
t<-80/365
# p is the payoff
p<-p1-p2
```

```
p1<-g0*(exp(t*(mu+sigma/2))*step(Z-(beta-sqrt(sigma)*t))/sqrt(t))
p2<-k*step(Z-beta/sqrt(t))
Z~dnorm(0,1)
beta<-(log(k/g0)-t*mu)/sqrt(sigma)

}
# the 20 increment values are given below in the list

list(X=c(-.3507,-.3016,-.06998,.12274,-.55404,
         -.54297,1.27865,.28206,-.27495,1.02443,
         .14147,-.44845,.900205,-.532307,.14566,
         .294189,-.13073,.29245,-.2204,.44691))
list(mu=0,tau=4)
```

The main parameter is the payoff with a posterior mean of $57.34 for an option that expires in 90 days, with a current value of $80 and a striking value of $110 (Table 8.9).

The simulation of the geometric Brownian motion values appears reasonable, and the increments have hypothetical mean of 0 and variance of .25. Also, 95% credible intervals for μ and σ^2 include 0 and .25, respectively, indicating that the posterior inferences for these parameters are not misleading.

Lastly, it is worth mentioning the subject of fractional Brownian motion, a subject that will not be developed here, but will be left to the student in the exercises at the end of Chapter 8. Such processes are examples of long-term processes, and Beran[11] should be read for additional information.

The fractional Brownian motion $\{B_H(t), t \in [0,T]\}$ is a Gaussian process with mean value of 0 and covariance function

$$E[B_H(t), B_H(s)] = (1/2)[|t|^{2H} + |s|^{2H} - |t-s|^{2H}], \tag{8.74}$$

where H is the Hurst index. When $H = 1/2$, the process is a standard Brownian motion, while when $H > 1/2$, one may show that the correlation between observations is positive, but on the other hand, when $H < 1/2$, the correlation is negative.

TABLE 8.9

Posterior Analysis for Option Price

Parameter	Mean	SD	Error	2 1/2	Median	97 1/2
β	.3428	.07432	.000469	.1998	.3411	.4926
μ	.0746	.1255	.000724	−.1755	.0746	.3233
Payoff	57.34	76.95	0.4676	0	0	208
P_1	88.24	99.81	0.5969	0	0	212
P_2	25.9	46.67	0.2622	0	0	110
σ^2	.3146	.1145	.000694	.1622	.2916	.5992

8.10 Comments and Conclusions

This last section reviews the chapter and makes comments on the content presented for making Bayesian inferences about stochastic processes that are continuous in state and time. The majority are Wiener processes and various generalizations. The chapter begins with the Brownian motion process, which is a normal process first used to describe the motion of tiny particles in a solution. Einstein provided a sound theoretical basis by showing that the Brownian motion $B(t)$ is a solution to a partial differential equation (Equation 8.1), namely, that $B(t) \sim$ normal with mean of 0 and variance of t. The Wiener process (Brownian motion) possesses stationary independent increments. Section 8.1 contains the R Code 8.1 that generates standard Brownian motion, and the 30 values generated are displayed in Figure 8.1. In addition, R Code 8.2 generates values from a slight generalization of the Brownian motion, namely, where $B(t) \sim N(0, \sigma^2 t)$, and the results are plotted in Figure 8.2. Using 30 Brownian motion values generated by R Code 8.2, Bayesian inferences for σ^2 are made, and the posterior analysis is reported in Table 8.1. Bayesian inferences are computed via WinBUGS Code 8.1 and consist of determining the posterior distribution of σ^2 and, in addition, a test of the hypothesis that the mean of the incremental process is zero, and it was found that the posterior probability of this process is almost 1.

Section 8.3 is a description of how a symmetric random walk is approximately a Brownian motion process, and it was shown how for large t, the sum of independent discrete binary variables has mean of 0 and variance of t. In addition, it can be shown that the process had independent and stationary increments. Based on the symmetric random walk, R Code 8.2 generates approximately Brownian motion values, and Figure 8.3 displays 50 values. A pertinent question is do the generated values support the ideas that this is indeed a Brownian motion? Based on the 25 increments, a Bayesian analysis can be performed, but is left as an exercise for the student.

Next, in Section 8.4, the first time T that Brownian motion $B(t)$ "hits" a particular value v, called the target, is defined and described. The mean and variance of the first hitting time T are derived and displayed by Equation 8.27. R Code 8.3 generated Wiener process values with $\mu = 0$ and $\sigma^2 = 1$ and tracks the time that the target $v = 10$ is hit. It was found that the average time to the first hit is 19.86 and the variance is 73.12. Bayesian inferences about the underlying Brownian process $B(t) \sim N(1/2, 1)$ are conducted by using the values of the increments, which hypothetically have a mean of 0 and a variance of .016, and then, using noninformative prior distributions for μ and σ^2, the posterior analysis is reported in Table 8.2, and it was concluded that the increment process supported the parameter values $\mu = 0$ and $\sigma^2 = .016$. In addition, a formal Bayesian test of the null hypothesis H: $\sigma^2 = .016$ is conducted using noninformative priors, and the posterior probability of the null hypothesis is .1.

An interesting example of Brownian motion of coin tossing is presented in Section 8.5. The main interest is the number of zeros encountered by the Brownian motion and the probability of at least one zero in the interval (r,t).

This is related to the problem that two players playing a fair game are even, and the game is simulated 10,000 times, and a corresponding histogram of the times the two are even displayed in Figure 8.4.

Section 8.6 presents Bayesian inferences for the parameters of Brownian motion with drift parameters μ and σ^2. An interesting example is home field advantage of NBA teams where the parameter of interest is the probability that the home team wins, given the home team leads by y points at time t through the game, where $t \in [0,1]$. Data from 493 games of the 1992 season provided estimates $\mu = 4.87$ and $\sigma = 15.82$ for the underlying Brownian motion with drift and are used to evaluate the probability that the home team will win, given that the home team score was y ahead of the other team at a given fraction time t of the game. R Code 8.5 generated 100 Brownian motion values with drift parameters $\mu = 4.87$ and $\sigma = 15.82$, and then based on the 50 corresponding increment values, with WinBUGS Code 8.4, a Bayesian analysis estimates the probability (Equation 8.40) that the home team wins, and the posterior analysis is reported in Table 8.5 for $y = 0$ (the teams are tied) and $t = 1/2$ (half time). Some exercises at the end of the chapter expand on the Bayesian analysis for home team advantage.

Next to be considered in Section 8.7 is an extension of Brownian motion referred to as the geometric Brownian motion, where the process is defined (Equation 8.41), and several examples are given. Recall that the geometric version is defined as the exponential function with exponent of a regular Brownian motion process $X(t) \sim N(\mu t, \sigma^2 t)$, with moments expressed with formulas in Equation 8.45.

The main parameter of interest is $P[G(281) < 6]$ (Equation 8.49), and the Bayesian analysis is executed with WinBUGS Code 8.5. Note that the sample information is the 15 increment values corresponding to the 30 values of the geometric process. Noninformative prior distributions are assigned to the parameters, and the posterior results are reported in Table 8.6. The next example involves several geometric processes plus the underlying exponent for a regular Brownian motion. See Figure 8.6 for a plot of these three geometric processes. R Code 8.8 generates the geometric values with various values of μ and σ^2 of the underlying Brownian motion. A Bayesian analysis executed with WinBUGS Code 8.7 based on the 15 increments of the underlying Brownian motion is completed. Table 8.7 reports the posterior analysis for estimating μ and σ^2, and separately, the predictive distribution for a future value from a Geometric Brownian motion process is derived and appears in Equation 8.53.

Still another generalization is described in Section 8.7.2, the so-called Brownian bridge, which is a Brownian motion over [0,1], where $B(1) = 0$. Moments of the process are expressed by Equations 8.54 and 8.55, and the Brownian bridge is generalized as a normal process anchored at arbitrary time points s and t with $s < t$ and $B(s) = a$ and $B(t) = b$. As in the previous section, R Code 8.0 generates values from the Brownian bridge with parameters $\mu = 0$ and $\sigma^2 = 1$ and the increments used as data for the Bayesian analysis. Remember that the increment process has mean of 0 and variance of .001.

Lastly, a formal Bayesian test of H: $\mu = 0$ and $\sigma^2 = .001$ is performed using the 15 increment values, and the posterior probability of the null hypothesis calculated as $p_0 \approx 1$.

A generalization of Brownian motion is the stochastic process called a martingale defined by Equation 8.63 that has the property that its expectation is the same for all time. Many functions of the Brownian motion are martingales, for example, the process $Y(t) = B^2(t) - t$ and the process $X(t) = e^{-rt}G(t)$, where $G(t)$ is a geometric Brownian process is a martingale. Also presented is the Black–Scholes model of pricing a financial instrument such as a stock, where observations for this process are generated by R Code 8.10, and using the observations for data, a Bayesian analysis is conducted and reported in Table 8.9, with the main parameter being the price of financial instrument.

A similar example is provided by Bayesian inferences for the payoff of buying a stock option. In order for the option to be exercised, one must know the current price of the option, the price of the option, the length of the option, and the striking value of the option. The value of the underlying stock follows a geometric Brownian motion whose 40 values are generated via R Code 8.11, and the Bayesian analysis is executed with R Code 8.8. The average payoff of the option is the main parameter of interest with its posterior distribution reported in Table 8.9.

8.11 Exercises

1. Write a short essay on the content of Chapter 8.
2. a. Show that the density (Equation 8.2) is that of the position of a particle at point y at time t. This is a $N(0,t)$ distribution.
   b. Demonstrate that the density is a solution to the partial differential equation (Equation 8.1).
3. Define the standard Brownian motion process $\{B(t), t \geq 0\}$.
4. Show that standard Brownian motion has (1) independent increments and (2) stationary independent increments.
5. Derive the mean and variance of $B(t)$.
6. a. Derive the recursive relation (Equation 8.8) for the Brownian motion.
   b. Show how the recursive relation can be used to generate the Brownian motion.
7. a. Based on R Code 8.1, generate 50 standard Brownian motion values.
   b. Graph the 50 values versus time $t = 1, 2, \ldots, 50$. Your graph should be similar to Figure 8.1.
8. a. With R Code 8.2, generate 30 Brownian motion values with $\sigma^2 = 1/2$ and time $t = 1, 2, \ldots, 30$.
   b. Plot these 50 values versus time. Is your plot similar to Figure 8.2?
9. a. Based on the 15 increment values corresponding to the 30 values generated by R Code 8.2, perform a Bayesian analysis using WinBUGS Code 8.1.
   b. What prior distributions did you use for μ and τ.
   c. Verify the posterior analysis reported in Table 8.1.
   d. What are the posterior mean and variance of σ^2?
   e. What is the 95% credible interval for σ^2?
   f. Are the posterior distributions reported in Table 8.1 symmetric about their posterior means? Explain!
10. The theoretical mean of the increments is 0.
    a. What is the posterior distribution of μ as reported in Table 8.1?
    b. Based on the approach on page 126 of Lee,[5] test the hypothesis H: $\mu = 0$ versus the alternative A: $\mu \neq 0$. Use $\pi_0 = \pi_1 = 1/2$ and follow the testing procedure given by Equations 8.12 through 8.14, and using the 15 increment values in the list statement of WinBUGS Code 8.1, compute the posterior probability of the null hypothesis.
11. Refer to Equations 8.16 through 8.17 and demonstrate that Brownian motion satisfies the Markov property.
12. Explain how the consecutive sums of a symmetric random walk, consisting of an independent binary variable and taking values of −1 and +1 with equal probability, is approximately a Brownian motion. See Equation 8.18.

13. a. Using R Code 8.2, generate 50 Brownian motion values.
    b. Show that the sample mean and variance of the 50 values are 5.56 and 4.3, respectively.
    c. Plot the 50 values versus time $t = 1, 2, \ldots, 50$. Is your graph similar to Figure 8.2?
14. a. For Brownian motion, define the first hitting time random variable T_v with target v.
    b. Show that the mean and variance of T_v are $E(T_v) = v/\mu$ and $\text{Var}(T_v) = v\sigma^2/\mu^3$, respectively, where μ and σ^2 are the mean and variance of the Brownian motion process, respectively.
    c. In order to estimate σ^2, $E(T_{10})$, and μ, and using WinBUGS Code 8.2, perform a Bayesian analysis.
    d. Verify the posterior analysis reported in Table 8.2.
    e. What are the posterior mean and 95% credible interval for $E(T_{10})$?
15. Define the Brownian motion with drift and parameters μ and σ^2.
16. a. Derive the probability $p(y,t)$ that the home team wins, given that the score is y at time t through the game, where $t \in [0, 1]$. See Equation 8.40.
    b. R Code 8.5 generates 100 Brownian motion values with parameters $\mu = 4.87$ and $\sigma = 15.82$. The mean and variance values are estimated from 493 games of the 1992 NBA season, and the 100 values are generated over [0,1] with increments of .01.
    c. Graph the 100 values over (0.1] with width of .01.
    d. The goal of the Bayesian analysis is to determine the posterior distribution of $p(0,.5)$, the probability of the home team winning when the score at half time is tied.
    e. What is the posterior mean of $p(0,.5)$?
    f. Is this posterior mean a reasonable estimate?
17. a. Define the geometric Brownian motion.
    b. Describe the stochastic process $\{X(t), t \geq 0\}$, that is, the exponent of geometric Brownian motion.
    c. Use R Code 8.6 to generate 365 geometric Brownian motion values over the times $t = 1, 2, \ldots, 365$.
    d. Plot the geometric Brownian motion values over the times $t = 1, 2, \ldots, 365$ and describe the pattern that the graph makes.
    e. The goal of the Bayesian analysis is to estimate the following probability: $P[G(281) < 6]$.
    f. WinBUGS Code 8.5 is for estimating this probability with noninformative prior for μ and σ^2. The data are the 15 increment values in the list statement of WinBUGS Code 8.5. The posterior analysis is reported in Table 8.6.
    g. What is the posterior mean of $P[G(281) < 6]$?
    h. Explain why this is a reasonable estimate.

18. a. Refer to Section 8.7.2 and define the Brownian bridge.
    b. What are the mean and variance of a Brownian bridge process? See Equations 8.54 and 8.55.
    c. R Code 8.7 generates 1000 Brownian bridge values over [0,1] with increments of width .001, where the underlying Brownian bridge values have parameters $\mu = 0$ and $\sigma^2 = 1$.
    d. Verify Figure 8.8, the graph of the 1000 generated Brownian bridge values over [0,1] with increment width of .001.
    e. Execute the Bayesian analysis with WinBUGS Code 8.6 for estimating μ and σ^2.
    f. The posterior analysis is reported in Table 8.8. Does the posterior analysis support the idea that $\mu = 0$ and $\sigma^2 = 1$ used to generate the Brownian motion values in the list statement of WinBUGS Code 8.6?
19. a. Describe the Black–Scholes model for determining the price of a stock.
    b. The price of the stock is given by Equation 8.71.
    c. Explain the role that the geometric Brownian motion plays in the Black–Scholes model.
    d. The goal of the Bayesian analysis is to find the posterior distribution of the price of a stock given by Equation 8.71.
    e. WinBUGS Code 8.8 generates 365 daily values from the relevant geometric Brownian motion where the underlying Brownian motion values have parameters $\mu = 0.145$ and $\sigma^2 = 0.25$. Thirty values of this process are in WinBUGS Code 8.8.
    f. Execute the Bayesian analysis with WinBUGS Code 8.8 using noninformative prior distributions for the parameters and 15 increment values in the list statement of WinBUGS Code 8.7.

20. Suppose $\{X(t), t \geq 0\}$ is Brownian motion with drift of $\mu \neq 0$ and variance of σ^2. The objective is to find the posterior distribution of p, the probability of hitting a before hitting the value $-b$, where a and $b > 0$ and

$$p = \left[1 - \exp(2\mu b/\sigma^2)\right] / \left[\exp(-2\mu a/\sigma^2) - \exp(2\mu b/\sigma^2)\right]. \tag{8.75}$$

Assume that $\mu = 2$ and $\sigma^2 = 4$, a= b=3, then R Code 8.10 generates the following 30 observations from Brownian motion with $\mu = 2$ and $\sigma^2 = 4$.

R Code 8.10

```
t <- 1:30 # time
sig2 <- 4
mu<-2
## first, simulate a set of random deviates
x <- rnorm(n = length(t) - 1, sd = sqrt(sig2))
## now compute their cumulative sum
x <- c(10, cumsum(x))

X = (10.0000000, -3.2683732, -4.3066682, -6.6763235,
-6.8299745, -9.6239269, -9.5085864, -10.3229206, -7.5621807,
-6.2805042, -5.3635471, -4.4001567, -7.2345987, -6.4848822,
-9.4483685, -7.5286309, -4.7656582, -4.4880655, -5.2088219,
-1.7177467, -2.9740531, 0.0985764, -1.3306347, -2.9972144,
-0.7022531, 1.5955997, 3.1562212, 5.8853872, 6.9742033,
6.8640443).
```

With the objective of estimating p, and based on the corresponding 15 increments (included in the list statement of WinBUGS Code 8.9) corresponding to the preceding 30 values, WinBUGS Code 8.9 will execute the Bayesian analysis. The data for the Bayesian analysis are the 15 increment values, which hypothetically have a mean of 0 and variance $\sigma^2 = 4$ and noninformative prior distributions for μ and σ^2. Note that in R Code 8.10, the number 4 has been added to the parameter mu.

The Bayesian analysis is executed with WinBUGS Code 8.10.

WinBUGS Code 8.9

```
model;

{

mu~dnorm(0,.001)
tau~dgamma(.001,.001)

for ( i in 1:15){

y[i]~dnorm(mu,tau)}
sigma<-1/tau
p<-p1/(p2-p3)
a<-3
b<-3
p1<-1-exp(2*newmu*b/sigma)
p2<-exp(-2*newmu*a/sigma)
p3<-exp(2*newmu*b/sigma)

newmu<-mu+2
}

list(y=c(-13.2683,-2.3697,-2.794,-.8144,1.2816, .9637,.7493,
          1.9197,.2776,3.4823,3.0725, -1.666,1.798,2.7291,-.1102))

list( mu=0, tau=.25)
```

The posterior analysis is reported in Table 8.10.

Referring to R Code 10, WinBUGS Code 8.9, and Table 8.10, execute the Bayesian analysis reported in Table 8.10.

a. What data are used for the Bayesian analysis?
b. The posterior mean of the desired probability is .6415. Is this a reasonable estimate?
c. Referring to Table 8.10, does the posterior distribution for μ imply that it is zero?
d. Does the posterior distribution of σ^2 imply that it is 4?

TABLE 8.10

Posterior Analysis for Hitting Times

Parameter	Mean	SD	Error	2 1/2	Median	97 1/2
μ	−0.3205	1.142	0.005827	−2.568	−0.3198	1.945
Newmu	1.68	1.142	0.005827	−0.5682	1.68	3.945
p	.6415	.09599	.000493	.4626	.639	.8327
σ^2	19.25	8.616	0.04959	8.853	17.33	40.95
τ	.06056	.02283	.000129	.02442	.0577	.113

References

1. Dobrow, R. P. 2016. *Introduction to Stochastic Processes with R*. New York: John Wiley & Sons.
2. Insua, D. R., Ruggeri, F., and Wiper, M. P. 2012. *Bayesian Analysis of Stochastic Process Models*. New York: John Wiley & Sons.
3. Ntzoufras, I. 2009. *Bayesian Modeling*. New York: John Wiley & Sons.
4. Wiener, N. 1923. Differential space, *Journal of Mathematics and Physics/Massachusetts Institute of Technology* 2:131–174.
5. Lee, P. M. 1997. *Bayesian Statistics, An Introduction, Second Edition*. London: Arnold.
6. Feller, W. 1950. *An Introduction to Probability Theory, Volume 1*. London: John Wiley & Sons.
7. Stern, H. S. 1994. A Brownian motion model for the progress of sports scores, *Journal of the American Statistical Association* 89(427):1128–1134.
8. Revuz, D., and Yor, M. 1999. *Continuous Martingales and Brownian Motion, 12th Edition*. New York: Springer-Verlag.
9. Black, F., and Scholes, M. 1973. The pricing of options and corporate liabilities, *Journal of Political Economy* 81(3):637–654.
10. Beran, J. 1994. *Statistics for Long Term Processes*. Boca Raton, FL: Chapman & Hall.

9

Queues and Time Series

9.1 Introduction

This chapter presents Bayesian inferences for two types of stochastic processes, queues and time series. For example, the first process to be studied is the M/M/1 queue, and then the chapter proceeds to more complex queues. Next to be presented is an introduction to time series models, including the autoregressive, the moving average, the autoregressive moving average processes, and the regression models with residuals that are correlated time series.

As in previous chapters, in order to simulate the various stochastic processes where the parameters of the processes are known, R is implemented. Then using those observations generated by R as the sample information and assuming that the parameters are now not known, a Bayesian analysis will be executed with WinBUGS. This allows one to make Bayesian inferences about those unknown parameters. Bayesian inferences consist of three phases: estimation, testing hypotheses, and prediction of future observations. It is important to remember that Bayesian inferences depend on prior information about the unknown parameters, the sample information expressed by the likelihood function, and the resulting posterior distribution about those parameters. Of course, all Bayesian inferences are based on the posterior distribution.

9.2 Queuing Analysis

9.2.1 Introduction

The fundamental properties of all queues are described, and then this section will define the various models for the queuing process including the M/M/1, then progressing to G/M/1, and finally the M/G/1 model. For each model, using the WinBUGS package and sometimes R, data will be generated for the arrival times and the service times, where the parameters are known, and then based on those observations, Bayesian inferences are described and implemented using WinBUGS.

There are many situations where the model for a queue can reveal interesting behavior. For example, customers at a shopping mall or people who need a heart transplant or other organs all have the same concern: the time to wait while in line to be served and, once being served, the time it takes to complete being served. This section will employ R to generate observations for the queuing model, such as the arrival time of the customers entering the queue and the service time of those doing their transactions. Based on those observations generated by R with known parameters for the queuing model, Bayesian inferences for those parameters (now assumed unknown) will be implemented with WinBUGS. This section on queuing will closely follow Chapter 7 of Insua, Ruggeri, and Wiper.[1]

In what is to follow, the fundamental properties of a general queuing model is described, and then focus is centered on the special case of the M/M/1 system, followed by explanations of non-Markov processes.

Generally speaking, a queuing system is a family of several stochastic processes describing the waiting and service times of the people in the queue. They arrive according to some process (which can be represented by deterministic or stochastic mechanisms), and then they have to wait, if required, before being attended to by one or more servers.

9.2.2 Fundamental Properties of Queues

The customer will be serviced if there is at least one free server, but if there are no servers available, the customer will leave at once or wait a specified time period until they can no longer wait until a server is available. Think of going to the grocery store, where there is seemingly a multitude of alternatives about checking out. There are six characteristics that describe a queuing model denoted by A/S/c/K/M/R, where A is the arrival time process, S is the service time distribution, and c is the number of servers. Also, K is the finite or infinite capacity of the queue, M denotes the size of the customer population, and R is the service discipline. Thus, M/D/2/10/∞ /FIFO denotes a queue where the arrival process is Markovian, so that the interarrival times between customer arrivals are independent and identically distributed (i.i.d.) exponential, the service times are fixed or nonstochastic, there are two servers, the process can hold a maximum of 10, the customer population is infinite, and, finally, the service is described as FIFO = first in first out. We will consider queues, where the customer population size is infinite as is the capacity, and the service is FIFO. Such systems will be represented by M/G/c, with a Markovian arrival process and a general service time mechanism. Of primary concern to the client is the waiting time, which depends not only on the number of servers but also on the number of customers who arrived earlier. The queue is described by the following measures:

1. $N_q(t)$ is number of customers waiting at time t.
2. $N_b(t)$ is the number of busy servers at time t.
3. $N(t) = N_q(t) + N_b(t)$ is the number in the system at time t (9.1)
4. $W_q(t)$ is the time in queue by a person arriving at time t.
5. $W(t) = W_q(t) + S$ is the time spent in the system by a customer arriving at t.

S is the service time. For our purposes, the preceding variables will be considered random and with distributions that, in reality, are difficult to know. A key concept in the study of queues is that of the stability of the system (what happens in the long run), which is defined for a G/G/c queue as follows.

Consider a G/G/c process with general interarrival time distribution, as well as a general service time distribution and c servers, infinite capacity and customer population, and FIFO service discipline, and then the traffic intensity is

$$\rho = \lambda E(S)/c, \tag{9.2}$$

where λ is the average interarrival time, and $E(S)$ is the average service time. Note that when $\rho > 1$, it appears that the interarrival time average is greater than the mean service time, and consequently, the size of the queue will increase over time, and the stability of the system is in doubt. On the other hand, when $\rho < 1$, that stability occurs in the sense that the distributions of $N(t)$, the total number in the system at time t, and $W(t)$ approach stability defined as

$$\begin{gathered} \lim_{n \to \infty} P[N(t) = n] = P[N = n] \\ \text{and} \\ \lim_{t \to \infty} P[W(t) < w] = P[W < w]. \end{gathered} \tag{9.3}$$

Similar expressions representing stability are evident for N_q, N_b, and W_q.

Also, as pointed out on page 165 of Insau, Ruggeri, and Wiper, Little[2] showed that under stability,

$$\begin{gathered} E(N) = \lambda E(W) \\ \text{and} \\ E(N_b) = \lambda E(W_q). \end{gathered} \tag{9.4}$$

9.2.3 Interarrival and Service Times

Analytical results for the queueing model are often difficult to determine; thus, interest will focus on a case where it is possible, namely, with the M/M/1 process, where the arrival process is Poisson with parameter λ, hence, with exponentially distributed independent interarrival times with mean $1/\lambda$ and independent exponentially distributed service times with mean $1/\mu$. The system is denoted by $M(\lambda)/M(\mu)/1$. Notice the similarity of the queue process to a birth and death process, where the arrival of a customer is interpreted as a birth and the completion of service as a death. Obviously, the traffic intensity is

$$\rho = \lambda/\mu; \tag{9.5}$$

Hence, the system is stable if the arrival rate is less than that of the service rate. From previous considerations, the equilibrium distributions exist, and according to Gross et al.,[3] the limiting distribution for the number of people in the system is geometric:

$$\rho \sim Ge(1-p) \tag{9.6}$$

with mean $E[N] = \rho/(1-\rho)$, and that for the number of clients in the queue waiting for service has mass function of

$$\begin{aligned} P\left[N_q = n\right] &= P[N=0] + P[N=1], \quad n = 0 \\ &= P[N = n+1], \quad n \geq 1. \end{aligned} \tag{9.7}$$

In addition, the limiting distribution for the time W spent by an arriving customer in the system is

$$W \sim \exp(\mu - \lambda) \tag{9.8}$$

and mean of $E(W) = 1/(\mu - \lambda) = 1/(\mu(1-\rho))$.

Lastly, the limiting distribution of the time W_q has cumulative distribution:

$$P\left[W_q \leq t\right] = 1 - \rho e^{-(\mu-\lambda)t}, \quad t \geq 0, \tag{9.9}$$

and that for the idle period time J of a server has density of

$$f_J(t) = \lambda e^{-\lambda t}, \quad t > 0. \tag{9.10}$$

The purpose of a Bayesian analysis will be to provide inferences for the unknown parameters μ and λ. Remember that in practice, one would have data organized as follows: For each customer, the time of arrival to the queue is recorded, the waiting time is also recorded, and the service time for that client would be noted. For the statistician, a distribution needs to be assigned to the waiting times and service times. These are determined empirically with goodness-of-fit tests etc., and then classical inferences such as maximum likelihood made for parameters μ and λ.

For the M/M/1 process, one is assuming that the interarrival times and service times are exponential, but one must remember that this assumption needs to be justified. For this case, Bayesian inferences are quite simple. Suppose one has the following information: the total time t_a taken for the first n_a arrivals and the total time t_s taken to service the first n_s. It is obvious that the likelihood function is

$$L((\mu, \lambda)|\text{data}) \propto \lambda^{n_a} e^{-\lambda t_a} \mu^{n_s} e^{-\mu t_s}, \quad \lambda, \mu > 0. \tag{9.11}$$

Prior distributions must be assigned to λ and μ; thus, consider the improper prior

$$\xi(\mu, \lambda) = 1/\lambda\mu, \quad \lambda, \mu > 0, \tag{9.12}$$

and then the posterior distribution of λ is gamma (n_a, t_a) and that of μ is gamma (n_s, t_s), and λ and μ are independent. Thus, Bayesian inferences are somewhat straightforward if one knows the sufficient statistics n_a, t_a, n_s, and t_s; however, it should be remembered that these are computed from the individual waiting and service times.

Suppose it is assumed that the arrival rate to the queue is Poisson with $\lambda = 2$, and the service rate is Poisson with $\mu = 4$; then WinBUGS Code 9.1 generates 32 interarrival times with parameter $\lambda = 2$ and 24 service times with $\mu = 4$, where the data so generated are in the list statement of WinBUGS Code 9.1. The vector y comprises the 32 interarrival times, while the vector x denotes the 24 service times.

WinBUGS Code 9.1

```
model;
{
for( i in 1:32){

y[i]~dexp(2)
}

for( j in 1:24){
x[j]~dexp(4)}

}
list(
x=c(
0.3089,0.05381,0.2722,0.3297,0.1448,
0.2653,0.4153,0.1016,1.147,0.2276,
0.2444,0.1064,0.01524,0.2323,0.2039,
0.529,0.1551,0.1974,0.8159,0.2866,
0.04165,0.1458,0.009616,0.4576),
y=c(
0.9452,1.375,0.1224,0.265,0.4189,
0.3082,0.0874,0.02587,0.002924,0.7175,
0.1315,0.5326,0.1267,0.6279,0.1355,
2.008,0.1355,0.3271,0.5271,0.1954,
0.1059,1.266,0.5406,0.1794,0.7701,
0.4849,0.3033,0.2122,1.159,0.4769,
1.773,0.01511))
```

Using the observations generated by WinBUGS Code 9.1, and tacitly assuming that the two parameters are unknown, the posterior distribution for λ, μ, and $\rho = \lambda/\mu$ is executed with 35,000 observations for the simulation and 5,000 for the burn-in. Noninformative gamma priors are assigned to the two parameters.

WinBUGS Code 9.2

```
model;
{
lamda~dgamma(.001,.001)
mu~dgamma(.001,.001)
rho<-lamda/mu

for( i in 1:32){
y[i]~dexp(lamda)
}
for( j in 1:24){
x[j]~dexp(mu)}
# the following is the posterior probability of the null hypothesis
prob<-step(1-rho)
```

```
}
list(
x=c(
0.3089,0.05381,0.2722,0.3297,0.1448,
0.2653,0.4153,0.1016,1.147,0.2276,
0.2444,0.1064,0.01524,0.2323,0.2039,
0.529,0.1551,0.1974,0.8159,0.2866,
0.04165,0.1458,0.009616,0.4576),
y=c(
0.9452,1.375,0.1224,0.265,0.4189,
0.3082,0.0874,0.02587,0.002924,0.7175,
0.1315,0.5326,0.1267,0.6279,0.1355,
2.008,0.1355,0.3271,0.5271,0.1954,
0.1059,1.266,0.5406,0.1794,0.7701,
0.4849,0.3033,0.2122,1.159,0.4769,
1.773,0.01511))
list(lamda=2,mu=4)
```

The Bayesian analysis for this queue is reported in Table 9.1.

The posterior mean of λ is 2.104 with a 95% credible interval of (1.376,2.986), which includes the value $\lambda = 2$ used to generate the interarrival times.

On the other hand, the posterior median of μ is 3.24 with a 95% credible interval of (2.012,4.895), which includes the value $\mu = 4$ used to generate the service times. Also, the credible interval for ρ is (.3246,.9454), implying that the process is stable. A more formal Bayesian test of the hypothesis H: $\rho < 1$ versus A: $\rho \geq 1$ is conducted as follows. Note that the posterior probability of the null hypothesis is computed directly as reported in the last row of Table 9.1.

The WinBUGS Code WinBUGS Code 9.2 command for the probability is given as a step function. One sees that there is overwhelming evidence that the process is stable.

From the list statement of WinBUGS Code 9.2, the usual estimate of λ is $\hat{\lambda} = n_a/t_a = 32/16.302 = 1.962$, and for μ, the estimate is $\hat{\mu} = n_s/t_s = 24/6.7071 = 3.578$. This compares to the posterior means for λ and μ as 2.104 and 3.3, respectively. Why the difference?

Now consider a stable system and the limiting distribution for the number of clients in the system, given by Equation 9.6, namely, the geometric distribution $\rho \sim Ge(1-p)$, which has a mean of $E[N] = \rho/(1-\rho)$. Its posterior median is given as 1.2. The posterior median

TABLE 9.1

Posterior Analysis for M/M/1 Queue

Parameter	Mean	SD	Error	2 1/2	Median	97 1/2
λ	2.104	0.4138	0.002	1.376	2.074	2.986
μ	3.3	0.7381	0.003639	2.012	3.243	4.895
ρ	.5721	.1591	.000854	.3246	.5517	.9454
$P(\rho < 1)$	.9838	.1264	.000638	1	1	1
$E(N)$	1.906	48.31	0.2823	.4125	1.2	7.586
$E(W)$	0.9079	18.96	0.1112	0.2701	0.6221	3.31

should be used as an estimate of $E(N)$ because of the asymmetry in its posterior distribution. In a similar fashion, the posterior distribution of the average time spent by an arriving customer has an exponential distribution with mean of $E(W) = 1/(\mu - \lambda)$, which has a posterior median of 1.6221. Again because of the asymmetry in its posterior distribution, I recommend the posterior median as an estimator of $E(W)$.

R Code 9.1 to execute a simulation of an M/M/1 queue was downloaded from https://www.r-bloggers.com/simulating-a-queue-in-r/.

This example assumes that the arrival rate is Poisson with $\lambda = 2$, and with the service time, exponentially distributed with parameter $\mu = 4$, which are the values used to generate the data in R Code 9.1.

R Code 9.1

```
t.end  <- 10^3 # duration of sim
t.clock <- 0   # sim time
Ta <- 2  # interarrival period (λ = 2)
Ts <- 4  # service period (μ = 4)
t1 <- 0      # time for next arrival
t2 <- t.end    # time for next departure
tn <- t.clock  # tmp var for last event time
tb <- 0      # tmp var for last busy-time start
n <- 0       # number in system
s <- 0       # cumulative number-time product
b <- 0       # total busy time
c <- 0       # total completions
tc <- 0      # plot time delta
plotSamples <- 100
set.seed(1)
while (t.clock < t.end) {
  if (t1 < t2) {    # arrival event
    t.clock <- t1
    s <- s + n * (t.clock - tn) # delta time-weighted number in queue
    n <- n + 1
    if (t.clock < plotSamples) {
      qc <- append(qc,n)
      tc <- append(tc,t.clock)
    }
    tn <- t.clock
    t1 <- t.clock + rexp(1, 1/Ta)
    if(n == 1) {
      tb <- t.clock
      t2 <- t.clock + rexp(1, 1/Ts) # exponential  interarrival period
    }
  } else {        # departure event
    t.clock <- t2
    s <- s + n * (t.clock - tn) # delta time-weighted number in queue
    n <- n - 1
```

```
      if (t.clock < plotSamples) {
        qc <- append(qc,n)
        tc <- append(tc,t.clock)
      }
      tn <- t.clock
      c <- c + 1
      if (n > 0) {
        t2 <- t.clock + rexp(1, 1/Ts) # exponential service period
      }
      else {
        t2 <- t.end
        b <- b + t.clock - tb
      }
    }
  }
u <- b/t.clock      # utilization B/T
N <- s/t.clock      # mean queue length (see the Load Average notes)
x <- c/t.clock      # mean throughput C/T
r <- N/x          # mean residence time (from Little's law: Q = XR)
q <- sum(qc)/max(tc) # estimated queue length for plot
```

I computed the following measures of the simulation performance:
$U = 0$, $N = 12287.58$, $x = .2501833$, $r = 49114$, and $q = 6.633136$. Figure 9.1 is a plot of the queue size versus time for the first 71 time units of the simulation.

Compared to WinBUGS Code 9.2, R Code 9.1 gives much more information about the simulation performance, but such measures are not given a Bayesian interpretation.

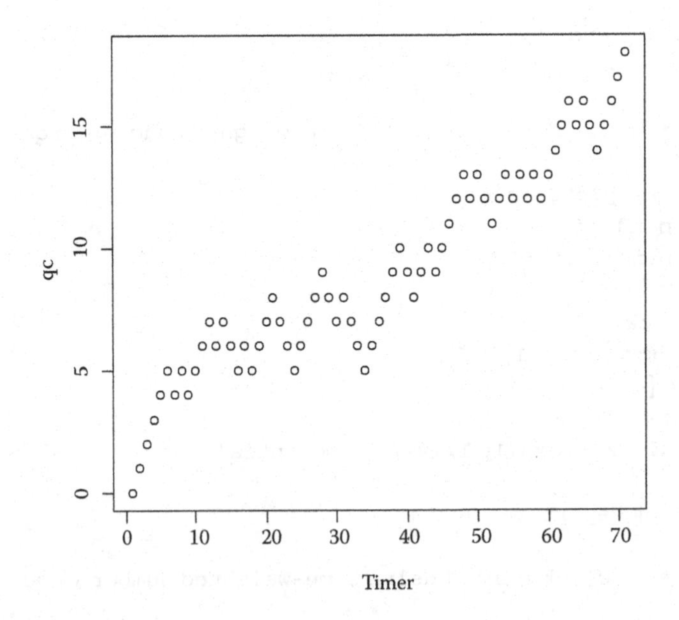

FIGURE 9.1
Queue size versus time.

WinBUGS Code 9.2 is totally Bayesian and based on the data of exponential arrival and service times generated in WinBUGS Code 9.1.

9.2.4 G/M/1 Queue

The G/M/1 model denotes that the arrival times have a general distribution and the service times have an exponential. This is a slight generalization of the M/M/1 queue, where the interarrival times have something other than an exponential distribution. Insua, Ruggeri, and Wiper (pages 180–183)[1] developed Bayesian inferences for this queue by assuming that the interarrival times X are i.i.d. with an Erlang (ν, λ) distribution and density of

$$f(x|\nu,\lambda) = [(\nu\lambda)^{\nu}/\Gamma(\nu)]x^{\nu-1}\exp(-\nu\lambda x), \quad x > 0 \tag{9.13}$$

and mean and variance of

$$\begin{aligned} &E(X) = 1/\lambda \\ &\text{and} \\ &V(X) = 1/\nu\lambda^2, \end{aligned} \tag{9.14}$$

respectively.

As was seen earlier, the traffic intensity is $\rho = \lambda/\mu$, where the service times have an exponential distribution with parameter μ. It is easy to see that the Erlang distribution is a gamma distribution with parameters $\alpha = \nu$ and $\beta = \nu\lambda$; thus, the traffic intensity can be expressed as

$$\rho = \alpha/\mu\beta. \tag{9.15}$$

A Bayesian analysis is presented, assuming that $\alpha = 2$ and $\beta = 4$ for the interarrival times of the gamma distribution or, equivalently, in terms of the Erlang $\nu = 2$ and $\lambda = \beta/\alpha = 2$. Also, the arrival rate is $\beta/\alpha = 1/\lambda = 2$, and finally, the service time exponential distribution is assigned the parameter $\mu = 4$; hence, the traffic intensity is $\rho = \alpha/\mu\beta = 1/2$.

WinBUGS Code 9.3 generates the 32 interarrival times X with a gamma (2,4) distribution, while the 24 service times Y have a distribution that is exponential with parameter of 4.

WinBUGS Code 9.3

```
model;
{
for( i in 1:32){
x[i]~dgamma(2,4)}
for( j in 1:24){
y[j]~dexp(4)
}
list(
x = c(
1.215,0.7093,0.3025,0.2905,0.202,
0.3462,0.4008,0.314,0.3074,0.9409,
0.5384,1.022,0.7333,0.2133,0.7381,
```

```
0.3969,0.1987,0.4238,0.3841,0.311,
0.556,0.3274,0.5975,0.2693,0.1435,
0.4527,0.8288,1.002,0.2473,1.051,
1.672,0.7948),
y = c(
0.4112,0.1581,0.1171,0.2083,0.03417,
0.1366,0.0466,1.242,0.0858,0.1086,
0.03962,0.1377,0.1301,0.1286,0.2874,
0.0092,0.2794,0.3757,0.002431,0.1671,
0.04103,0.6679,0.06998,0.05484))
```

Now based on the observations for the interarrival times and service times generated in WinBUGS Code 9.3, WinBUGS Code 9.4 is executed for the Bayesian analysis with 35,000 observations for the simulation and 5,000 for the burn-in. One is tacitly assuming that the parameters μ and β are unknown. The main goal is to provide Bayesian inferences for α,β,μ,ρ, and $P[\rho < 1|\text{data}]$. Noninformative gamma (.001,.001) prior distributions are assigned to α, β, and μ.

WinBUGS Code 9.4

```
model;
{
alpha~dgamma(.001,.001)
beta~dgamma(.001,.001)
mu~dgamma(.001,.001)
for ( i in 1:32){
x[i]~dgamma(alpha,beta)}
for( j in 1:24){
y[j]~dexp(mu)}
lamda<-alpha/beta
rho<-alpha/(beta*mu)
prob<-step(1-rho)
}
list(
x = c(
1.215,0.7093,0.3025,0.2905,0.202,
0.3462,0.4008,0.314,0.3074,0.9409,
0.5384,1.022,0.7333,0.2133,0.7381,
0.3969,0.1987,0.4238,0.3841,0.311,
0.556,0.3274,0.5975,0.2693,0.1435,
0.4527,0.8288,1.002,0.2473,1.051,
1.672,0.7948),
y = c(
0.4112,0.1581,0.1171,0.2083,0.03417,
0.1366,0.0466,1.242,0.0858,0.1086,
0.03962,0.1377,0.1301,0.1286,0.2874,
0.0092,0.2794,0.3757,0.002431,0.1671,
0.04103,0.6679,0.06998,0.05484))
list(mu=4,alpha=2,beta=4)
```

TABLE 9.2

Posterior Analysis of G/M/1

Parameter	Mean	SD	Error	2 1/2	Median	97 1/2
α	2.846	0.6811	0.008167	1.682	2.79	4.333
β	5.081	1.382	0.01587	2.819	4.97	7.988
λ	.5668	.06182	.000218	.4584	.5621	.7013
μ	4.855	0.9883	0.0035	3.121	4.786	6.972
$P(\rho < 1)$	1	.004851	.000016	1	1	1
ρ	.1218	.02921	.000147	.07668	.1177	.1903
rate = $1/\lambda$	1.786	0.1923	0.00111	1.425	1.78	2.181

The posterior analysis is reported in Table 9.2.

The posterior distributions appear to be symmetric about their posterior means. Note $\alpha = 2$ was used to generate the interarrival times and that the posterior median is 2.79 with a 95% credible interval of (1.682,4.333), and in a similar approach, $4 = \beta$ was used to generate the interarrival times, and its posterior median is 4.97 with 95% credible interval of (2.819,7.988).

Also, note that the rate clients arrive at the queue is estimated as 1.786 with the posterior mean. Lastly, the posterior probability $P[\rho < 1|\text{data}]$ has a posterior mean of 1, indicating a stable queue.

9.2.5 G/G/1 Queue

In this situation, the interarrival and the service times have a general distribution, but for our purposes, both will be assigned gamma distributions. To that end, suppose the interarrival time has a gamma with parameters α_a and β_a, while for the service time distribution, the parameters are α_s and β_s. This implies that the traffic intensity is

$$\rho = (\beta_a/\alpha_a)/(\beta_s/\alpha_s). \tag{9.16}$$

The Bayesian analysis will be based on observations generated from the appropriate gamma distribution, namely, 32 observations from the gamma(2,4) for the interarrival times, and for the 24 service times, the gamma(2,8). Using 35,000 for the simulation and a burn-in of 5,000, WinBUGS Code 9.5 generates the appropriate interarrival X and service Y times.

WinBUGS Code 9.5

```
model;
{
for( i in 1:32){ x[i]~dgamma(2,4)}
for( j in 1:24){y[j]~dgamma(2,8)}
}
list(
x = c(
```

```
0.2072,0.1701,0.8705,0.3998,0.1295,
0.2305,0.3799,0.2855,0.6025,0.5822,
0.1772,0.6122,0.1712,0.3514,0.2978,
0.25,0.4489,0.5986,1.262,0.6811,
0.4534,0.2926,0.3596,1.104,0.201,
0.2574,0.2824,0.7037,0.3016,0.1991,
1.147,0.1383),
y=c(
0.6385,0.4613,0.168,0.2391,0.1547,
0.2135,0.2841,0.3483,0.7737,0.07918,
0.1383,0.2535,0.3224,0.1256,0.2765,
0.23,0.5598,0.1059,0.1343,0.3265,
0.0195,0.1944,0.1163,0.4366))
```

The purpose of the Bayesian analysis is to estimate the unknown parameters $\alpha_a,\beta_a,\alpha_s,\beta_s,\mu$, and λ and the posterior probability $P[\rho < 1|\text{data}]$.

WinBUGS Code 9.6 is executed with 35,000 observations for the simulation and a burn-in of 5,000. Noninformative gamma priors were assigned to $\alpha_a,\beta_a,\alpha_s$, and β_s.

WinBUGS Code 9.6

```
model;
{
alphaa~dgamma(.001,.001)
betaa~dgamma(.001,.001)
alphas~dgamma(.001,.001)
betas~dgamma(.001,.001)
for( i in 1:32){x[i]~dgamma(alphaa,betaa)}

for ( j in 1:24){y[j]~dgamma(alphas,betas)}

rho<-(betaa/alphaa)/(betas/alphas)
prob<-step(1-rho)

lamda<-betaa/alphaa
mu<-betas/alphas

}
list(
x=c(
0.2072,0.1701,0.8705,0.3998,0.1295,
0.2305,0.3799,0.2855,0.6025,0.5822,
0.1772,0.6122,0.1712,0.3514,0.2978,
0.25,0.4489,0.5986,1.262,0.6811,
```

```
0.4534,0.2926,0.3596,1.104,0.201,
0.2574,0.2824,0.7037,0.3016,0.1991,
1.147,0.1383),
y = c(
0.6385,0.4613,0.168,0.2391,0.1547,
0.2135,0.2841,0.3483,0.7737,0.07918,
0.1383,0.2535,0.3224,0.1256,0.2765,
0.23,0.5598,0.1059,0.1343,0.3265,
0.0195,0.1944,0.1163,0.4366))

list( alphaa=2,betaa=4,alphas=2,betas=8)
```

The posterior analysis is displayed by Table 9.3, and the posterior distributions appear to be symmetric about the posterior mean. Consider first the traffic intensity ρ, which has a 95% credible interval of (.4365, .9032), implying informally that the process is stable, which is also implied by the posterior probability $P[\rho < 1|\text{data}]$, which has a posterior mean of .9935.

The values of $\alpha_a = 2$ and $\beta_a = 4$ were used to generate the interarrival times of the clients, and it is noted that the posterior mean of α_a is 2.616 with a 95% credible interval of (1.54,4.02), which indeed includes the value 2.

In fact, the values used to generate the interarrival and service time data do in fact contain those values with the corresponding 95% credible interval. The per unit arrival rate for clients is estimated as 2.262, while the per unit time service rate is 3.638 clients.

This section has presented three examples of queues with one server, what do they have in common? This will be further investigated in the problems at the end of the chapter. For additional information about the Bayesian analysis of the M/G/1 queue, see pages 183–187 of Insua, Ruggeri, and Wiper,[1] who in turn refer to Ausin and Lopes,[4] and for additional information about the Bayesian analysis of G/M/1 queues, see Wiper.[5]

TABLE 9.3

Posterior Analysis for the G/G/1 Queue

Parameter	Mean	SD	Error	2 1/2	Median	97 1/2
α_a	2.616	0.634	0.01081	1.541	2.564	4.023
β_a	5.918	1.575	0.02704	3.241	5.782	9.401
α_s	2.164	0.5873	0.00926	1.17	2.109	3.459
β_s	7.874	2.403	0.03784	3.843	7.647	13.15
λ	2.262	0.2548	0.00139	1.788	2.255	2.794
μ	3.638	0.522	0.00269	2.673	3.619	4.735
$P[\rho < 1\|\text{data}]$	.9935	.08051	.000448	1	1	1
ρ	.635	.1186	.00067	.43656	.6233	.9032

9.3 Time Series

9.3.1 Introduction

Time series are stochastic processes but typically involve a large number of observations and, for some time, have been analyzed by Bayesian methods. See, for example, Chapter 5 of Broemeling[6] and an early work on Bayesian forecasting by Smith.[7] More recent Bayesian books on the subject are Pole, West, and Harrison[8] and Barber, Cemgil, and Chippa.[9] There are several good books on time series using R, including Petris, Petrione, and Campagnoli[10] and Cowpertwait and Metcalfe.[11] Material from the last reference has been used in this section. The usual Bayesian methods for inference will be employed to estimate, test hypotheses, and predict about the parameters of time series models. Using R, the time series will simulate the model with known parameters, and then tacitly assuming that the parameters are unknown, Bayesian approaches using WinBUGS will allow one to provide inferences for those "unknown" parameters.

Time series are studied so that one may understand the past in order to predict the future. This is a necessary activity for managers and policy makers to make an intelligent decision. A time series investigation quantifies the main features of the sample information and the random variation. Improved computer techniques and probability theory have made it possible for time series methods to apply to real-world endeavors in business, commerce, science, and industry.

Several examples of using time series are very interesting and show the practical utility of the subject. For example, Cowpertwait and Metcalfe (pages 1 and 2)[11] mention the Kyoto Protocol on climate change that emphasized the need to reduce the emission of greenhouse gases and that this assertion heavily depended on science, economics, and time series analysis. Another interesting example is in business when Singapore Airlines placed an initial order for 20 Boeing 787–9s and signed an order of intent to buy 29 Airbus, namely, 20 A350s and 9 A380s. This decision relied on a combination of time series analysis of airline passenger trends and a corporation plans for maintaining and increasing the market.

It is often true that time series observations are an essential part of scientific studies, engineering projects, and measures of economic activity. For example, our government's statistic office estimates the annual gross domestic product, and the US Federal Reserve Bank relies on good records and measures of economic activity (expressed as time series) in order to set interest rates.

Our approach will be to use R to generate observations from basic stochastic models including white noise, random walks, autoregressive, moving average, and regression models. Then, Bayesian inferences for estimating, testing, and prediction of the parameters of the model will utilize WinBUGS for the posterior analysis.

The main features of many time series are trends and seasonal variation that can be modeled statistically with function of time. The most important probabilistic properties of a time series are that the observations are correlated, with the characteristic that those observations nearer to each other have a higher correlation than those that are separated more from each other. Thus, statistical methods are used to explore and estimate the correlation between them, including models that allow for correlation in a specific way. Various alternative models are proposed, and there are ways to determine the one with the best fit, and then once one is chosen, the model will predict future observations. Time series occur at various points in time and can be regular using equally spaced points or can vary with time depending on the application.

9.3.2 Fundamentals of Time Series Analysis

The best way to discern trends and seasonal variation is to plot the data; thus, consider the monthly International Air Passenger 1949–1960 data of US booking in the United States and those found on pages 4–6 of Cowpertwait and Metcalfe[11] and are available from the Federal Aviation Administration, which appear as follows:

	Jan	Feb	Mar	Apr	May	Jun	Jul	Aug	Sep	Oct	Nov	Dec
1949	112	118	132	129	121	135	148	148	136	119	104	118
1950	115	126	141	135	125	149	170	170	158	133	114	140
1951	145	150	178	163	172	178	199	199	184	162	146	166
1952	171	180	193	181	183	218	230	242	209	191	172	194
1953	196	196	236	235	229	243	264	272	237	211	180	201
1954	204	188	235	227	234	264	302	293	259	229	203	229
1955	242	233	267	269	270	315	364	347	312	274	237	278
1956	284	277	317	313	318	374	413	405	355	306	271	306
1957	315	301	356	348	355	422	465	467	404	347	305	336
1958	340	318	362	348	363	435	491	505	404	359	310	337
1959	360	342	406	396	420	472	548	559	463	407	362	405
1960	417	391	419	461	472	535	622	606	508	461	390	432

R Code 9.2 downloads the data and plots the time series.

R Code 9.2

```
data(AirPassengers)
AP<-AirPassengers
AP
plot(AP, ylab="Passenger (1000's)")
```

The corresponding graph is portrayed in Figure 9.2.

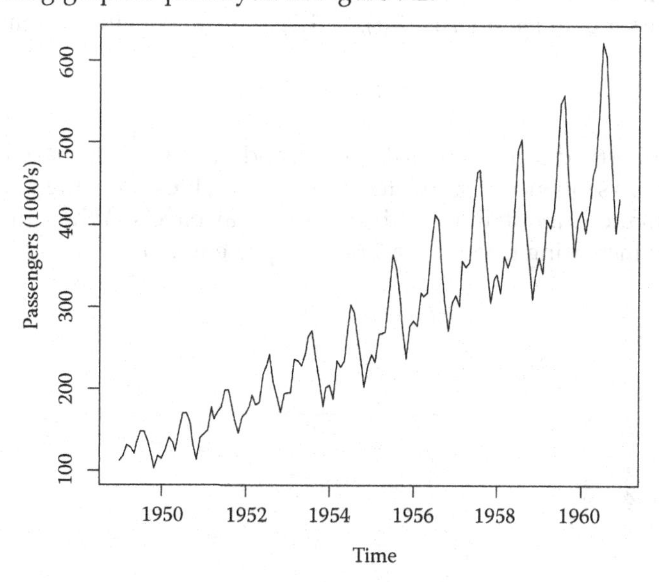

FIGURE 9.2
Monthly international air passenger information.

One can see the overall trend and seasonal (annual variation) variation, where the trend appears linear, and the annual variation has the same pattern from year to year. A statistical analysis would propose mathematical models to account for the trend and seasonal variation.

There are a number of features in the plot that are common to many time series. Generally speaking, a deterministic change in the series that is not periodic is referred to as trend, and the most elementary change is linear. A repeating pattern annually is called seasonal, although the term is also pertinent to a pattern that repeats itself over a fixed period.

The seasonal variation in the air passenger data reveals that bookings were higher during the summer months and lowest in November and February. What is a reasonable explanation for the overall increasing trend? As proposed by Cowpertwait and Metcalfe, the overall increasing prosperity after World War II, the availability of more aircraft, and cheaper tickets because of competition are seen to be reasonable causes. With a better view of the overall trend, the data can be aggregated, which for the present example is executed with R Code 9.3:

R Code 9.3

```
layout(1:2)
plot(aggregate(AP))
```

One sees the absence of seasonal variation and a clear linear increasing trend displayed in Figure 9.3. Additional information can be achieved by aggregating over the years and computing the box plot for each month as displayed in Figure 9.4. The graph in this figure accounts for the overall trend after aggregating over the 12 years and for the variation within the 12 years.

9.3.3 Decomposition of a Time Series

So far so good with plotting the data to detect trend and seasonal variation, now is the time to introduce some notation and the ideas for decomposition models.

Denote the time series of length n as $\{X(t), t = 1,2, \ldots, n\}$, and let the decompose the series as

$$x(t) = m(t) + s(t) + z(t), \tag{9.17}$$

where $x(t)$ is the observed series, $m(t)$ is the trend, and $s(t)$ is the seasonal effect, and $z(t)$ is an error term which is a sequence of correlated random variables. Now the principal problem is extracting information about the trend and seasonal effects. When the seasonal effect increases with an increasing trend, a reasonable representation is

$$x(t) = m(t)s(t) + z(t). \tag{9.18}$$

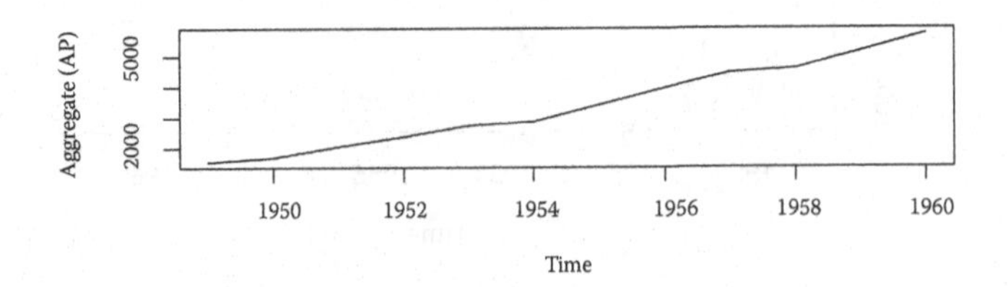

FIGURE 9.3
Aggregated air passenger data.

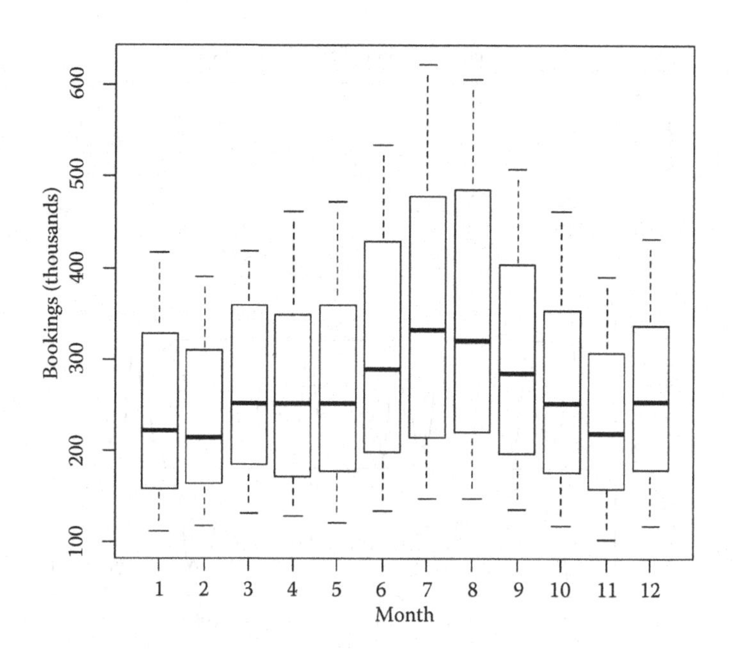

FIGURE 9.4
Box plot of air passenger data.

With a positive variation to account for, a plausible Model is (9.19)

$$\log[x(t)] = m(t)s(t) + z(t). \tag{9.19}$$

Then based on Equation 9.18, the predicted mean of $x(t)$ is

$$\widehat{x(t)} = \exp[m(t)s(t)]\exp[\sigma^2/2], \tag{9.20}$$

and the errors $z(t)$ are a correlated sequence of normally distributed random variables with mean of 0 and variance σ^2.

In R, the function decompose estimates trend and seasonal effects with a moving average method. For example, consider the air passenger data and use the command

```
plot(decompose(AirPassengers))
```

with results depicted in the graph in Figure 9.5.

There are various ways to estimate the trend $m(t)$, and the moving average is a relatively simple method. This approach is an average of the observations over a specified time range around each value of the observed series; thus, if there are 100 observations, there are 100 moving averages. In this way, the seasonal effects are averaged out, leaving the trend unaffected by the seasonal effects. An estimate of the monthly additive effect at time t is found by realizing that seasonal effect is estimated as

$$\widehat{s(t)} = x(t) - \widehat{m(t)}, \tag{9.21}$$

and if the model is multiplicative, see Equation 9.18, the seasonal effect is

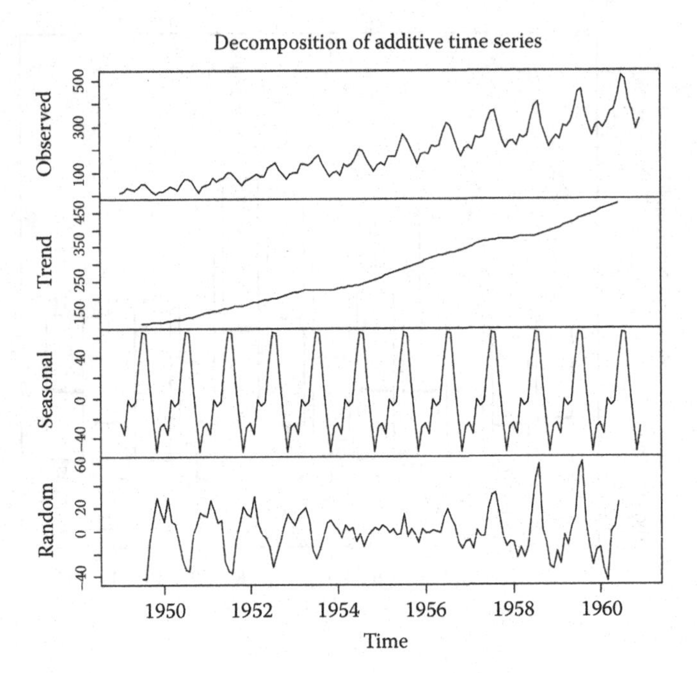

FIGURE 9.5
Decomposition of air passenger information.

$$\widehat{s(t)} = x(t)/\widehat{m(t)}. \tag{9.22}$$

These estimates of trend (via the moving average) and seasonal effects (via Equations 9.20 or 9.21) are reflected in Figure 9.5, the graphical representation of decomposition of the series via R.

9.3.4 Autocorrelation of a Time Series

When building a model for a time series, the trend, seasonal effects, and correlation must be accounted for. This section will define the autocorrelation of a time series and show how it is computed with R.

First the mean and variance of a time series $\{X(t), t \geq 0\}$ is defined as

$$E[X(t)] = \mu(t). \tag{9.23}$$

for the mean value function, and for the variance function,

$$\text{Var}[X(t)] = E[X(t) - \mu(t)]^2. \tag{9.24}$$

Many time series are correlated and are an important parameter to estimate in a statistical analysis and defined in terms of the covariance

$$\text{Cov}[X(s), X(t)] = E[[X(s) - \mu(s)][X(t) - \mu(t)]$$

as

$$\rho[X(s), X(t)] = \text{Cov}[X(s), X(t)]/\sqrt{\text{Var}[X(s)]\text{Var}[X(t)]},$$

where

$$\text{Var}[X(s)] = \text{Cov}[X(s), X(s)]. \tag{9.25}$$

Of course, one is assuming that the second moments exist.

The mean function $\mu(t)$ is an average taken over all possible ensembles of the time series $\{X(t), t \geq 0\}$. That is to say, think of all possible continuous realizations of the time series, then for each t, compute the usual average, then this determines $\mu(t)$, and the process is said to be stationary in the mean if $\mu(t)$ is a constant (the same for all t). Similar considerations apply to the variance function Var$[X(t)]$, which is also interpreted as an ensemble average, and the process is said to be stationary in the variance if it is constant.

Consider a time series that is stationary in the mean and variance, and then the model is second-order stationary if the correlation between variables depends only on the distance between the variables; thus, for such processes, the autocovariance function of lag k is defined as

$$\gamma(k) = \text{Cov}[X(t+k), X(t)]$$

and the corresponding autocorrelation function of lag k is

$$\rho(k) = \gamma(k)/\sigma^2, \tag{9.26}$$

where

$$\sigma^2 = \text{Var}[X(t)] \tag{9.27}$$

is the constant variance function. It is important to know that the autocorrelation function does not depend on t, only on k.

As an example, a minimum of 60 water levels of the Nile River were measured, in millimeters, in a gauge near Cairo. This appears to be a process with a constant mean level, and the first 31 values are plotted in Figure 9.6. There does not appear to be a trend in the observations; thus, the autocorrelation function will be plotted, and the autocorrelation values, computed for a few lags. Can a pattern be discerned in this information about the minimum values of the Nile River for 60 years?

The autocorrelation function is plotted in Figure 9.7 and show a large positive correlation at lag 0, which had the required value 0, while at lag 1, the positive correlation is computed as .08699, and at lag 2, a small negative value of −.1895, etc. For example, use the R function `acf(nile)` for the autocorrelation function and `acf(nile)$acf[2]` for the lag 1 correlation.

As will be seen, the pattern of how the autocorrelations die out is quite useful in the identification of the appropriate model. The acf plot of Figure 9.7 is referred to as the correlogram.

The command `plot(nile[1:59],nile[2:60])` is a plot of the lag 1 values, that is, the pairs $[X(i-1), X(i)]$, $i = 2, 3, \ldots, 60$, where the graph should approach a straight line with

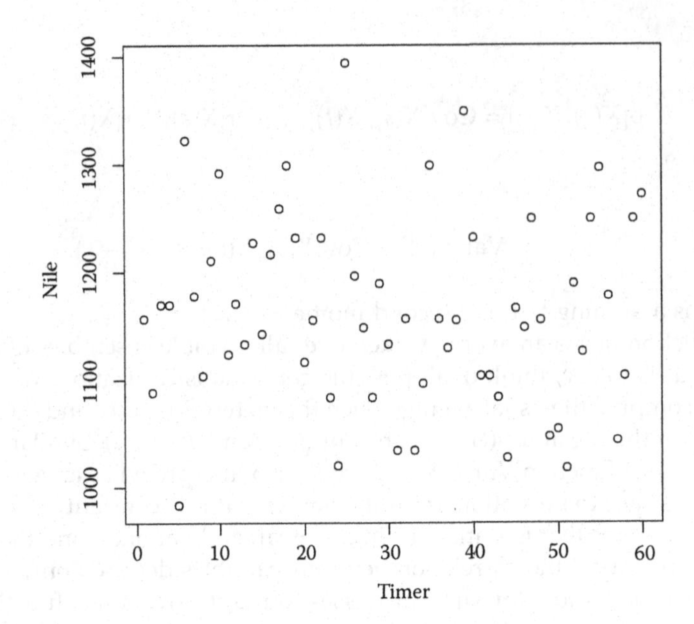

FIGURE 9.6
Nile River data.

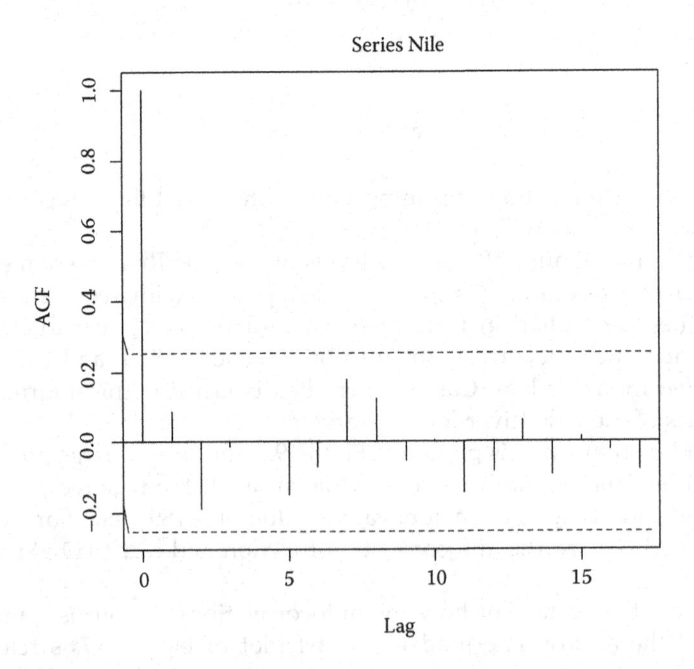

FIGURE 9.7
Autocorrelation function of Nile River levels.

slope 1 when the lag 1 autocorrelations approach 1. The reader will be asked to execute this command as an exercise at the end of this chapter.

9.3.5 Basic Time Series Models

Up to this point, two approaches for modeling time series have been discussed. As previously explained, the first is founded on the belief that there is a fixed seasonal pattern about a trend, such as the air passenger data, and these two features can be delineated with the R decompose command. The second approach takes account of the seasonal pattern and trend to change over time. With mathematical time series, the difference between the fitted value and the observed value allows one to calculate the random error series. If the model includes various aspects of the deterministic characteristic of the series, then the residuals should exhibit as a realization of independent random variables, but this may not be the case. That is to say, the residuals may show some structure such as a trend or positive autocorrelation.

Because a good fit implies that the residuals are independent random models, the models to be presented next will be with a foundation of white noise.

White noise is a time series

$$\{W(t), t = 1, 2, \ldots, n\} \tag{9.28}$$

of variables $W(1), W(2), \ldots, W = (n)$, which are independent and identically distributed with mean of 0, constant variance of σ^2, and, of course, $\mathrm{cor}[W(i), W(j)] = 0,\ i \neq j$. In addition, if $W(i) \sim N(0,\sigma^2)$, the noise is referred to as Gaussian or normal white noise.

R is useful to simulate time series, and this will be done for the basic stochastic models, such as white noise, random walks, and random walks with drift. Consider the scenario where a fitted time series can be used to simulate data. As has been seen throughout this book, simulation is used for a variety of reasons. In R, simulation is a simple operation where most of the well-known distributions are simulated with an R function. For example, for a simulation of normal random variables, rnorm(100) generates 100 standard normal variables. Now consider, the code

R Code 9.4

```
set.seed(1)
w<-rnorm(100)
time<-seq(1,100,1)
plot(time,w)
```

that generates 100 white noise values, and the plot abscissa has a unit of one. One should check to see how well the random number generator simulates white noise. For example, the sample mean(w) = .10887, sample standard deviation sd(w) = .8067, the lag 1 correlation is acf(w)$acf[2] = −.00365, and, finally, the lag 2 is given by acf(w)$acf[3] = −.02707. Of course, one should also use the command `acf(w)` to plot the autocorrelation function of the w series. The command `hist(w)` generates the default histogram of the w-purported white noise series and is valuable in detecting departures from a normal distribution. Such evaluations are necessarily subjective, in that, another individual might detect different deviations from normality.

The next stochastic model to be studied is the random walk series

$$X(t) = X(t-1) + W(t), \quad t = 1,2,\ldots, \tag{9.29}$$

where $\{W(t), t = 1,2,\ldots\}$ is a white noise process. Another representation of the random walk is an infinite series

$$X(t) = W(t) + W(t-1) + W(t-2) + \ldots,$$

and since the series begins at some point, say, $t = 1$, it can be expressed as

$$X(t) = W(1) + W(2) + \ldots + W(t) \tag{9.30}$$

One may show that the first and second moments are

$$\mu(t) = E[X(t)] = 0,$$

$$\gamma_k(t) = \text{Cov}[X(t), X(t+k)] = t\sigma^2 \tag{9.31}$$

and are a function of t; thus, the process is not covariance stationary. Also, the autocorrelation is

$$\begin{aligned}\rho_k(t) &= \text{Cov}[X(t), X(t+k)]/\left(\sqrt{t\sigma^2(t+k)\sigma^2}\right)\\ &= 1/\sqrt{1+k/t};\end{aligned} \tag{9.32}$$

thus, the autocorrelation function is positive and decays from 1 to 0.

Consider a simulation of a random walk with R Code 9.5:

R Code 9.5

```
x<-w<-rnorm(1000)
for ( t in 2:1000) x[t]<-x[t-1]+w[t]
time<-1:1000
plot(time,x)
```

The plot of the random walk values over 1000 time points is shown in Figure 9.8. I computed the following autocorrelations: the lag 2 as acf(x)$acf[3] = .9901 and lag 28 as acf$acf[29] = .87105, which follows from Equation 9.32. The student will be asked to display the autocorrelation function with the code acf(x). The random walk values start at 0, increase to 9, and then rapidly fall to an average value of −25. The average value of the 1000 values is −10.92, which is reasonable after looking at Figure 9.8.

Consider a random walk model with drift, namely,

$$X(t) = X(t-1) + \zeta + W(t), \quad t = 1,2,\ldots,n, \tag{9.33}$$

where $W(t)$ is Gaussian white noise with variance σ^2.

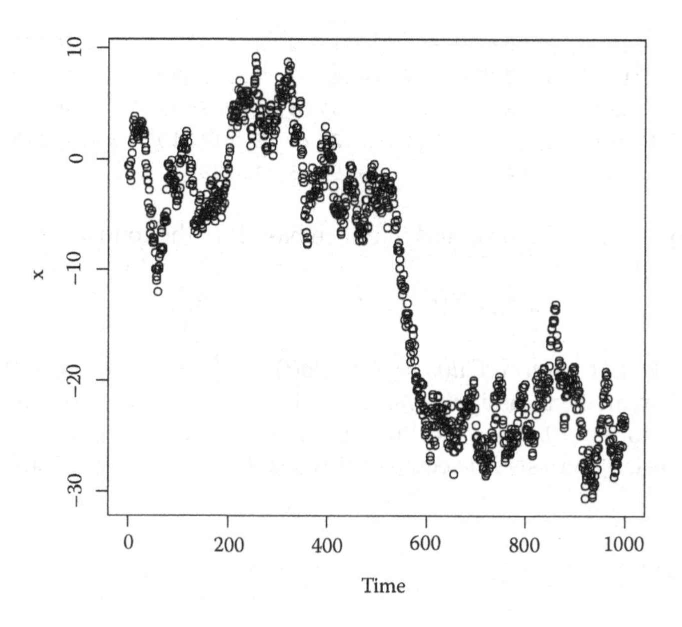

FIGURE 9.8
Random walk from 1 to 1000.

One hundred values for this random walk with drift of $\zeta = 3$ and variance of $\sigma^2 = 1$ is simulated with R Code 9.6:

R Code 9.6

```
delta=3
x<-w<-d<-rnorm(100)
for (t in 2:100) {x[t]<-x[t-1]+ delta+w[t]
d[t]<-x[t]-x[t-1]}
time<-1:100
plot(time,x)
```

The 100 increments $d(t)$ are the components of the vector d:

```
d=c( 0.7140855, 3.5813846, 2.8532761, 4.5069818, 2.7204674,
5.0277387, 1.8042598, 4.3123179, 2.4759925, 3.3542495, 2.9283255,
2.8668557, 2.9119226, 3.9177889, 3.0314393, 4.3589306, 3.1134948,
3.1743065, 2.9465196, 2.5752420, 3.2079116, 3.9929229, 3.8138579,
3.9960056, 4.7327443, 3.3570420, 3.1557263, 4.5527887, 2.9359304,
1.9917812, 2.1071045, 2.3454854, 1.5662172, 4.4181844, 2.4061143,
4.1189718, 5.3840816, 2.5967135, 3.3701086, 2.4247070, 5.9192288,
1.9210852, 4.1784913, 3.8696028, 3.9311663, 4.0246812, 3.6129751,
3.6081901, 2.1071083, 3.4520356, 2.6252877, 5.8401851, 3.7323935,
3.8405189, 0.7803479, 1.6725362, 2.0233393, 2.4129241, 2.4319872,
3.6436410, 2.7626929, 2.7014220, 4.1895626, 0.8910588, 2.9267180,
3.8634029, 4.0747075, 2.7941474, 2.5663862, 2.8177762, 2.5689213,
```

```
2.5753904, 1.6461803, 2.8034051, 3.2341161, 2.4427937, 1.8866747,
1.3693646, 2.6121761, 4.7226320, 2.4414997, 2.3633392, 5.7813871,
1.0427441, 3.1425279, 3.6644988, 2.5213725, 4.1836896, 3.8054373,
3.0533178, 1.8326536, 3.3233502, 2.6792170, 2.1140497, 3.1512782,
3.5793635, 2.9208332, 3.8487676, 0.4889665, 3.7384771).
```

The goal is to estimate the drift and variance based on the following differences:

$$d(t) = X(t) - X(t-1), \quad t = 2,3,\ldots,n, \tag{9.34}$$

where $d(t) = \zeta + W(t)$; therefore $E[d(t)] = 3$, $\mathrm{Var}[d(t)] = \sigma^2$, and $\mathrm{Cov}[d(s), d(t)] = 0$.

For the Bayesian analysis, assign noninformative priors for the mean μ and variance σ^2, namely, for μ, a normal (0,.001) distribution, and for $\tau = 1/\sigma^2$, a gamma (.001,.001) distribution. The Bayesian analysis is executed with 35,000 for the simulation and 5,000 for the burn-in.

WinBUGS Code 9.7

```
model;
{
# the prior distributions
mu~dnorm(0,.001)
tau~dgamma(.001,.001)
for ( i in 1:100){
d[i]~dnorm(mu, tau)}
sigma<-1/tau
}

list( d=c( 0.7140855, 3.5813846, 2.8532761, 4.5069818, 2.7204674,
5.0277387, 1.8042598, 4.3123179, 2.4759925, 3.3542495, 2.9283255,
2.8668557, 2.9119226, 3.9177889, 3.0314393, 4.3589306, 3.1134948,
3.1743065, 2.9465196, 2.5752420, 3.2079116, 3.9929229, 3.8138579,
3.9960056, 4.7327443, 3.3570420, 3.1557263, 4.5527887, 2.9359304,
1.9917812, 2.1071045, 2.3454854, 1.5662172, 4.4181844, 2.4061143,
4.1189718, 5.3840816, 2.5967135, 3.3701086, 2.4247070, 5.9192288,
1.9210852, 4.1784913, 3.8696028, 3.9311663, 4.0246812, 3.6129751,
3.6081901, 2.1071083, 3.4520356, 2.6252877, 5.8401851, 3.7323935,
3.8405189, 0.7803479, 1.6725362, 2.0233393, 2.4129241, 2.4319872,
3.6436410, 2.7626929, 2.7014220, 4.1895626, 0.8910588, 2.9267180,
3.8634029, 4.0747075, 2.7941474, 2.5663862, 2.8177762, 2.5689213,
2.5753904, 1.6461803, 2.8034051, 3.2341161, 2.4427937, 1.8866747,
1.3693646, 2.6121761, 4.7226320, 2.4414997, 2.3633392, 5.7813871,
1.0427441, 3.1425279, 3.6644988, 2.5213725, 4.1836896, 3.8054373,
3.0533178, 1.8326536, 3.3233502, 2.6792170, 2.1140497, 3.1512782,
3.5793635, 2.9208332, 3.8487676, 0.4889665, 3.7384771))
list(mu=0,tau=1)
```

The posterior analysis is reported in Table 9.4.

TABLE 9.4

Posterior Analysis for Random Walk with Drift

Parameter	Mean	SD	Error	2 1/2	Median	97 1/2
μ	3.104	0.1103	0.000644	2.887	3.104	3.321
σ^2	1.204	0.1748	0.000948	0.91	1.189	1.597
τ	0.8474	0.1203	0.000652	0.6263	0.841	1.099

Recall that the value of μ used to simulate the differences $d(t)$ is 3 (the drift) and that for τ is 1. Thus, according to Table 9.4, it seems that the simulated values are believable. For example, the posterior mean of μ is 3.104 with a 95% credible interval of (2.887,3.321), which includes 3. In a similar way, the posterior mean of σ^2 is 1.204 with a 95% credible interval of (0.91,1.597), which includes the value 1!

The next class of basic time series models is the autoregressive process AR(p) defined as

$$Y(t) = \sum_{i=1}^{i=p} \theta_i Y(t-i) + W(t) \tag{9.35}$$

or in terms of the backshift operators,

$$\Phi_p(B)Y(t) = \left(1 - \theta_1 B - \theta_2 B^2 - \dots - \theta_p B^p\right)Y(t) + W(t),$$

where B is the backshift operator defined as

$$B^m Y(t) = Y(t-m).$$

Also, $\{W(t), t > 0\}$ is Gaussian white noise with variance σ^2 and θ_i, $i = 1, 2, \dots, p$, is a sequence of unknown autoregressive parameters.

It should be observed that the random walk is a special case of the AR(1) process with $\theta_1 = 1$, and the class is called autoregressive because $Y(t)$ is regressed on past terms of the same process.

AR processes can be stationary or not, depending on the roots of the characteristic equation $\Phi_p(B) = 0$. Treat the operator B as a real or complex number, and then if the roots of the characteristic equation all lie outside the unit circle, then the process is stationary.

Note that the characteristic equation

$$\Phi_p(B) = 0 \tag{9.36}$$

$$\Phi_p(B) = \left(1 - \theta_1 B - \theta_2 B^2 - \dots - \theta_p B^p\right)$$

is a pth-order polynomial in B.

The following three examples show how to determine if an AR process is stationary:

1. The AR(1) process $Y(t) = (1/2)Y(t-1) + W(t)$ is stationary because the root of $1 - B/2 = 0$ is $B = 2$, which is greater than 1! And more generally, the AR(1) process is stationary if $|\theta_1| < 1$.
2. Also, the AR(2) process $Y(t) = Y(t-1) - (1/4)Y(t-2) + W(t)$ is stationary because the roots of the characteristic equation $(1/4)(B-2)^2 = 0$ has two roots, which are both $B = 2$.
3. On the other hand, the AR(2) process $Y(t) = (1/2)Y(t-1) + (1/2)Y(t-2) + W(t)$ is not stationary. This is left as an exercise for the reader.

For the AR(1) model, $Y(t) = \theta Y(t-1) + W(t)$, the first two moments are $\mu(t) = 0$ and covariance function with lag $k = 1, 2, \ldots,$

$$\gamma_k = \theta^k \sigma^2 / (1 - \theta^2), \tag{9.37}$$

and with autocorrelation function

$$\rho_k = \theta^k, \quad |\theta| < 1. \tag{9.38}$$

It is obvious from Equation 9.38 that the autocorrelations are nonzero and decay exponentially with k. Another important second-order property of the AR(p) process is the partial correlation function at lag k, which is the correlation that results after removing the effects of correlations of terms with lags less than k.

R Code 9.7 is given on page 81 of Cowpertwait and Metcalfe,[11] which executes 100 values from the AR(1) process with autoregression coefficient $\theta = .6$. The autocorrelation function is acf while that for the partial autocorrelation is pacf.

R Code 9.7

```
set.seed(1)
y<-w<-rnorm(100)
for( t in 2:100) y[t]<-.6*y[t-1]+w[t]
time <- 1:100
plot(time,y)
acf(y)
pacf(y)
```

The first 50 values are labeled by the vector y = (−0.62645381, −0.19222896, −0.95096599, 1.02470121, 0.94432850,

```
-0.25387129, 0.33510628, 0.93938847, 1.13941444, 0.37826027,
1.73873733, 1.43308564, 0.23861080, -2.07153341, -0.11798913,
-0.11572708, -0.08562651, 0.89246030, 1.35669738, 1.40791975,
1.76372922, 1.84037383, 1.17878928, -1.28207813, -0.14942113,
-0.14578142, -0.24326436, -1.61671100, -1.44817665, -0.45096443,
1.08810089, 0.55007281, 0.71771530, 0.37682414, -1.15096507,
-1.10557361, -1.05763412, -0.69389387, 0.68368905, 1.17338918,
0.53950991, 0.07034427, 0.73916994, 1.00016516, -0.08865660,
-0.76068912, -0.09183151, 0.71343402, 0.31571420, 1.07053625).
```

The posterior density of θ is derived as follows.

The likelihood function is

$$l(\theta|y) = (\tau/2\pi)^{n/2} \exp -(\tau/2) \sum_{t=2}^{t=n} [y(t) - \theta y(t-1)]^2, \tag{9.39}$$

where y is the vector of observations $y = (y(1), y(2), \ldots, y(n))$ and $|\theta| < 1$. Suppose the prior density of θ is uniform

$$\zeta(\theta) = 1,$$

and for τ,

$$\varsigma(\tau) = 1/\tau, \quad \tau > 0.$$

It can be shown that the posterior density of θ is

$$\propto f(\theta|\text{data}) \propto \{1 + \lambda[\theta - \nu]^2\}^{-n/2}, \tag{9.40}$$

where $\nu = \left[\sum_{t=2}^{t=n} y(t)y(t-1)\right] \Big/ \sum_{t=2}^{t=n} y^2(t)$. This is the density of a t distribution with $n-1$ degrees of freedom, mean of ν, and precision of $\lambda = (n-1)/c$, where

$$c = \sum_{t=1}^{t=n} y^2(t) - \left[\sum_{t=2}^{t=n} y(t)y(t-1)\right]^2 \Big/ \sum_{t=2}^{t=n} y^2(t-1).$$

I computed the following:

$$\sum_{t=1}^{t=n} y^2(t) = 46.617288,$$

$$\sum_{t=2}^{t=n} y^2(t-1) = 46.22491,$$

and

$$\sum_{t=2}^{t=n} y(t)y(t-1) = 28.87423;$$

thus, $c = 28.58109$ and $\lambda = 1.71442$. The posterior distribution of θ is a univariate t with 49 degrees of freedom, mean of $\nu = .62464$, and precision of $\lambda = 49/28.58109 = 1.7144$. The posterior mean of θ is .62464 and compares favorably with the value $\theta = .6$ used to generate the data vector y generated by R Code 9.7.

Note that the AR(1) process is represented as

$$Y(t) = \theta Y(t-1) + W(t), \tag{9.41}$$

where $\{W(t), t > 0\}$ is Gaussian white noise with variance σ^2, and then its mean is 0 and the autocovariance function is

$$\gamma(k) = \theta^k \sigma^2 / (1 - \theta^2), \tag{9.42}$$

and the variance is

$$V[Y(t)] = \sigma^2 / (1 - \theta^2) \tag{9.43}$$

Based on the representation (Equation 9.41) of an autoregressive process, the Bayesian analysis is repeated with WinBUGS Code 9.8 executed with 35,000 observations for the simulation and a burn-in of 5,000. Note that the prior for θ is beta (6,4) and that for the variance is expressed as a gamma(.001,.001) distribution. This induces a prior for the variance–covariance matrix of the 50 observations Y. The code is based on the fact that the 1×50 vector Y has a multivariate normal distribution with mean vector of 0 and covariance matrix given by Equation 9.42.

WinBUGS Code 9.8

```
model;
{

v~dgamma(.01,.01)
theta~dbeta(6,4)

Y[1,1:50] ~ dmnorm(mu[], tau[ , ]
for( i in 1:50){mu[i]<-0}

tau[1:50,1:50]<-inverse(Sigma[,])

for(i in 1:50){Sigma[i,i]<-v/(1-theta*theta)}
for(i in 1:50){for(j in i+1:50){Sigma[i,j]<-
v*pow(theta,j)*1/(1-theta*theta)}}
for( i in 2:50){ for(j in 1: i-1){Sigma[i,j]<-v*pow(theta,i-
1)*1/(1-theta*theta)}}
}

list(Y= structure(.Data=c( -0.62645381, -0.19222896,
-0.95096599, 1.02470121, 0.94432850,
-0.25387129, 0.33510628, 0.93938847, 1.13941444, 0.37826027,
 1.73873733, 1.43308564, 0.23861080, -2.07153341, -0.11798913,
-0.11572708, -0.08562651, 0.89246030, 1.35669738, 1.40791975,
1.76372922, 1.84037383, 1.17878928, -1.28207813, -0.14942113,
-0.14578142, -0.24326436, -1.61671100, -1.44817665, -0.45096443,
1.08810089, 0.55007281, 0.71771530, 0.37682414, -1.15096507,
-1.10557361, -1.05763412,-0.69389387, 0.68368905, 1.17338918,
 0.53950991, 0.07034427, 0.73916994, 1.00016516, -0.08865660,
-0.76068912, -0.09183151, 0.71343402, 0.31571420,
 1.07053625),.Dim=c(1,50)))

list(theta=.6,v=1)
```

TABLE 9.5

Posterior Analysis for AR(1) Process

Parameter	Mean	SD	Error	2 /12	Median	97 1/2
θ	.6258	.1509	.002015	.3096	.6383	.8683
σ^2	0.5724	0.2138	0.002894	0.228	0.552	1.043

Table 9.5 reports the posterior analysis.

The θ parameter is the main parameter of interest and is the correlation between observations spaced one unit apart and is estimated as 0.6258 with the posterior mean and 95% credible interval of (.3096,.8683), which contains the value $\theta = .6$ used to generate the data. Also, remember that $\sigma^2 = 1$ is the value used to generate the data, which compares to a posterior mean of .5724 and 95% credible interval of (0.228,1.043).

R can perform the analysis of an AR process with maximum likelihood estimation. Consider the 100 observations of an AR(1) process generated by R Code 9.7; then the following commands invoke the AR function followed by the code for estimating the order, which is 1:

```
> y.ar<-ar(y,method="mle")
> y.ar$order
[1] 1
```

The following command estimates the coefficient θ via maximum likelihood as .52311:

```
> y.ar$ar
[1] 0.5231187
```

This command generates a 95% confidence interval for θ:

```
> y.ar$ar+c(-2,2)*sqrt(y.ar$asy.var)
( 0.3521863, 0.6940510).
```

Note the difference in the MLE of .523 compared to the posterior mean of .5993. Why the difference? This is left as an exercise at the end of this chapter.

9.3.6 Bayesian Inference of Regression Models with Correlated Errors

In this section, various regression models are introduced. First to be considered is the estimation of the trend in linear models with autocorrelated errors, then later to provide inferences for the seasonal effects using harmonic and latent variables. Time series regression models are different from the usual regression models in that the errors are autocorrelated. One of the first linear models to be studied is simple linear regression following an AR(1) process. What is the effect of autocorrelation on the usual estimates of the regression coefficients? If the correlation is positive, the estimated standard errors of the estimates tend to be less than the estimated standard errors of the estimates assuming no correlation. Of course, a corresponding scenario holds for the Bayesian estimates of the regression coefficients.

A model is linear if

$$Y(t) = \beta_0 + \beta_1 X_1(t) + \beta_2 X_2(t) + \ldots + \beta_m X_m(t) + Z(t), \tag{9.44}$$

where $Y(t)$ is the observation of the dependent variable at time t; $X_i(t)$ is the observation of the ith independent variable at time t; and, finally, $Z(t)$ is the error term at time t. The errors $Z(t)$, $t = 1, 2, \ldots, n$, are assumed to have mean of 0, have a constant variance, and are autocorrelated. Our goal is to compute Bayesian inferences for the m regression coefficients β_i and the unknown parameters of the error process.

As a first example, consider the linear regression model

$$Y(t) = 50 + 3t + Z(t), \quad t = 1, 2, \ldots, 100, \tag{9.45}$$

where the $Z(t)$ follow an AR(1) process with autocorrelation θ.

Consider R Code 9.8, which will generate 100 observations from the linear regression model. The model is assumed to have a standard deviation of 10 for the Gaussian white noise of the AR(1) process, and the autocorrelation coefficient is assigned the value $\theta = .6$, while the regression coefficients are assumed to be 3 for the slope and 50 for the intercept. The first 50 values generated by R Code 9.8 appear below the code. The 50 values of the dependent variable are the components of the vector 7.

R Code 9.8

```
> set.seed(1)
> u<-w<-rnorm(100,sd=10)
> for ( t in 2:100) u[t]<-.6*u[t-1]+w[t]
> time<-1:100
> y<-50+3*time+u
> plot(time, y)
Y=( 46.73546, 54.07771, 49.49034, 72.24701, 74.44328, 65.46129,
74.35106, 83.39388, 88.39414, 83.78260, 100.38737, 100.33086,
91.38611, 71.28467, 93.82011, 96.84273, 100.14373, 112.92460,
120.56697, 124.07920, 130.63729, 134.40374, 130.78789, 109.17922,
123.50579, 126.54219, 128.56736, 117.83289, 122.51823, 135.49036,
153.88101, 151.50073, 156.17715, 155.76824, 143.49035, 146.94426,
150.42366, 157.06106, 173.83689, 181.73389, 178.39510, 176.70344,
186.39170, 192.00165, 184.11343, 180.39311, 190.08168, 201.13434,
200.15714, 210.70536)
```

A plot of the simple linear regression-dependent variables versus time over 50 days appears in Figure 9.9, and the linear trend is obvious. What is the posterior mean of the intercept, slope, and autocorrelation?

It appears that the observation begins at 50 starting at time 0.

A goal of the Bayesian analysis is to estimate the regression coefficient and the autocorrelation θ. In the statements of WinBUGS Code 9.9, beta0 is the intercept and beta1 is the slope of the regression model with autocorrelation theta. The 20 observations is a vector with a multivariate normal distribution with mean vector consisting of 20 values of $50 + 3t$, $t = 1, 2, \ldots, 20$, and a 20×20 precision matrix, which is the inverse of the variance–covariance matrix with components specified by Equation 9.42, the variance–covariance matrix of an AR(1) error process.

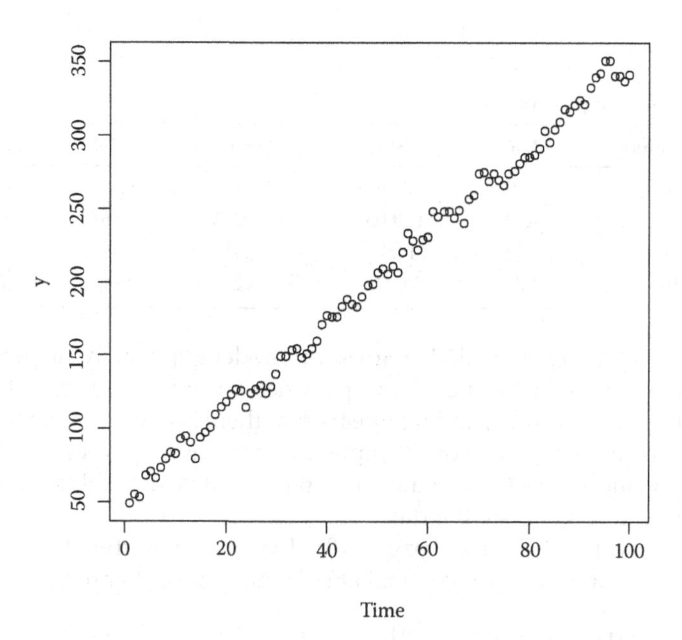

FIGURE 9.9
Simple linear regression with autocorrelated errors.

WinBUGS Code 9.9

```
model;
{
v~dgamma(.1,.01)
theta~dbeta(6,4)
beta0~dnorm(0,.001)
beta1~dnorm(0,.001)
   Y[1,1:20] ~ dmnorm(mu[], tau[ , ])

      for( i in 1:20){mu[i]<-beta0+beta1*i}

      tau[1:20,1:20]<-inverse(Sigma[,])
      for(i in 1:20){Sigma[i,i]<-v/(1-theta*theta)}
      for(i in 1:20){for(jin i+1:20){Sigma[i,j]<-
v*pow(theta,j)*1/(1-theta*theta)}}
      for(i in 2:20){ for(j in 1: i-1){Sigma[i,j]<-v*pow(theta,i-
1)*1/(1-theta*theta)}}
}

list(Y= structure(.Data=c( 46.73546, 54.07771, 49.49034,
72.24701, 74.44328, 65.46129, 74.35106, 83.39388, 88.39414,
83.78260, 100.38737, 100.33086, 91.38611, 71.28467, 93.82011,
96.84273, 100.14373, 112.92460, 120.56697,
124.07920),.Dim=c(1,20)))
list(theta=.6,v=100, beta0=50, beta1=3)
```

TABLE 9.6

Posterior Analysis for Regression Model

Parameter	Value	Mean	SD	Error	2 1/2	Median	97 1/2
β_0	50	47.99	6.828	0.1323	32.9	48.35	60.57
β_1	3	3.483	0.4805	0.009213	2.567	3.47	4.499
θ	.6	.6509	.1487	.006511	.3352	.6653	.8933
σ^2	100	54.51	24.33	0.2522	20.92	50.01	114

The posterior analysis for the AR(1) regression model appears in Table 9.6.

The second column is the value of the parameter used to generate the independent variable of the regression model, and it appears that there is good agreement between those values and the posterior means. For example, the posterior mean of the slope β_1 is 3.483 compared to the value 3 used to generate the data, and all 95% credible intervals do include the values employed to generate the data.

Our next example is taken from pages 101–105 of Cowpertwait and Metcalfe[11] and concerns a model with trend and seasonal effect where the data are based on the model

$$\begin{aligned} Y(t) &= 0.1 + .005t + .001t^2 + \sin(2\pi t/12) + .2\sin(4\pi t/12) \\ &\quad +.1\sin(8\pi t/12) + .1\cos(8\pi t/12) + W(t), \end{aligned} \tag{9.46}$$

where $W(t)$ are autocorrelated with coefficients $\theta = .6$ and $\sigma^2 = .25$, and $n = 120$. Note that the model is a linear model with a quadratic trend and seasonal effects represented by sinusoidal waves with very small amplitudes and frequencies of 1, 2, and 3 cycles per unit time. R Code 9.9 generates 120 values for the dependent variable given by Equation 9.46, and these are in the list statement of WinBUGS Code 9.10.

R Code 9.9

```
set.seed(1)
time<- 1:(10*12)
w<-rnorm(10*12, sd=.5)
Trend<- 0.1+.005*time+.001*time^2
Seasonal<- sin(2*pi*time/12)+0.2*sin(2*pi*2*time/12)+
0.1*sin(2*pi*4*time/12)+0.1*cos(2*pi*4*time/12)
x<-Trend+Seasonal+w
```

The general version of the specific model (Equation 9.46) is defined as

$$\begin{aligned} Y(t) &= \beta_0 + \beta_1 t + \beta_2 t^2 + \beta_3\sin(2\pi t/12) + \beta_4\sin(4\pi t/12) \\ &\quad +\beta_5\sin(8\pi t/12) + \beta_6\cos(8\pi t/12) + W(t), \end{aligned} \tag{9.47}$$

where β_i, $i = 0, 1, \ldots, 6$, are unknown regression coefficients and $W(t)$ is a sequence of errors which are correlated. The Bayesian analysis is based on the 30 values of the dependent variable of the model in Equation 9.46, and the goal is to estimate the β_i, $i = 0, 1, \ldots, 6$, the autocorrelation coefficient θ, and variance σ^2. WinBUGS Code 9.10 is executed with 45,000 observations for the simulation and 5,000 for the burn-in. Note that the vector of 30 observations is specified as a multivariate normal distribution with 30×1 mean vector

(Equation 9.46) and precision matrix τ, which is the inverse of the variance–covariance matrix of an AR(1) process with correlation θ. The parameters of the model are given noninformative prior distributions. See WinBUGS Code 9.10 for the specifications.

WinBUGS Code 9.10

```
model;
{
beta0~dnorm(0,.001)
beta1~dnorm(0,.001)
beta2~dnorm(0,.001)
beta3~dnorm(0,.001)
beta4~dnorm(0,.001)
beta5~dnorm(0,.001)
beta6~dnorm(0,.001)

theta~dbeta(6,4)
v~dgamma(.001,.001)

Y[1,1:30]~dmnorm(mu[],tau[,])

for( t in 1:30){
mu[t]<-beta0+beta1*t+beta2*t*t+beta3*sin(2*3.1416*t/12)
+beta4*sin(4*3.1416*t/12)+beta5*sin(8*3.1416*t/12)+
beta6*cos(8*3.1416*t/12)}
#Sigma is the variance covariance matrix of an AR(1)

   tau[1:30,1:30]<-inverse(Sigma[,])
   for(i in 1:30){Sigma[i,i]<-v/(1-theta*theta)}
   for(i in 1:30){for(j in i+1:30){Sigma[i,j]<-
v*pow(theta,j)*1/(1-theta*theta)}}
   for( i in 2:30){ for(j in 1: i-1){Sigma[i,j]<-v*pow(theta,i-
1)*1/(1-theta*theta)}}
}
list(Y =structure(.Data=c( 0.50258072, 1.10844961, 0.80618569,
1.66306326, 0.50494626, -0.14423419, 0.13752215, -0.25626051,
-0.38610932, -0.90532214, 0.22208296, 0.59892162, 0.73318733,
0.16127800, 2.06246546, 1.14295606, 0.65609725, 1.08591811,
0.67641822, 0.06752780, 0.20548869, 0.08244021, -0.02852513,
-0.09867585, 1.86972050, 1.78056357,
1.98610225,.01804667,1.03711735,1.45897078),.Dim=c(1,30)))
list(
theta=.6,v=.25,beta0=.1,beta1=.1,beta2=.2,beta3=.2,beta4=.2,beta5=.2,
beta6=.1)
```

The posterior analysis for the model with quadratic trend and seasonal effects is reported in Table 9.7.

The values used to generate the data are listed in the second column and should be compared to their corresponding posterior means. For example, the value of β_3 used for the

TABLE 9.7
Posterior Analysis for Seasonal Effects

Parameter	Value	Mean	SD	Error	2 1/2	Median	97 1/2
β_0	1	.086	.459	.02289	−.8377	.0971	.9962
β_1	.005	.0345	.06122	.00341	−.0875	.0321	.159
β_2	.001	.0003	.0018	.00010	−.004	−.0002	.00332
β_3	1	.7606	.1669	.00295	.4304	.7601	1.09
β_4	.2	.1711	.1525	.00094	−.1315	.1722	.4722
β_5	.1	.0453	.1468	.00092	−.2424	.0454	.337
β_6	.1	.1092	.15	.00099	−.186	.1084	.4041
θ	.6	.6012	.1461	.00189	.3023	.6077	.8599
σ^2	.25	.2139	.0901	.00100	.0849	.1991	.428

simulation is 1 compared to its posterior mean of 0.7606 and 95% credible interval of (.4304,1.09), which indeed includes 1! The posterior mean for σ^2 is very close to its nominal value of 0.25 used to generate the data, and its 95% credible interval does indeed include .25!

9.3.7 Bayesian Inference for a Nonlinear Trend in Time Series

As we have seen, linear models have a wide range of applications, but so do time series that have a nonlinear trend and autocorrelated errors. Consider the process

$$Y(t) = \exp(\beta_0 + \beta_1 t) + Z(t), \tag{9.48}$$

where, β_0 and β_1 are unknown parameters, and the residuals $Z(t)$ on the log scale form an AR(1) process with autocorrelation θ. R Code 9.10 is taken from pages 113 and 114 of Cowpertwait and Metcalfe[11] and generates 100 observations from the nonlinear model with $\beta_0 = 1$, $\beta_1 = .05$, $\theta = 0.6$, and $\sigma = 2$.

R Code 9.10

```
set.seed(1)
w<-rnorm(100,sd =2)
z<-rep(0,100)
for ( t in 2:100) z[t]<-0.6*z[t-1]+w[t]
Time<-1:100
f<- function(x) exp(1+0.05*x)
x<-f(Time)+z
```

The first 50 values of the simulation R Code 9.10 appear as components of the vector y:

```
y= ( 2.857651, 3.371453, 1.707308, 5.640147, 5.541377, 3.258980, 4.586094,
5.969050, 6.562987, 5.250836, 8.196521, 7.823749, 5.686929, 1.332517,
5.519606, 5.818782, 6.188920, 8.471027, 9.742210, 10.204972, 11.295405,
11.846945, 10.942453, 6.460867, 9.188900, 9.682623, 9.999043, 7.789756,
8.691994, 11.280566, 14.983306, 14.563884, 15.589469, 15.633380, 13.340702,
```

```
14.233500, 15.172514, 16.786358, 20.473332, 22.432315, 22.194364, 22.338640,
24.814404, 26.532861, 25.613027, 25.591261, 28.319071, 31.390968, 32.131821,
35.256524)
```

The Bayesian analysis is based on 30 observations generated by R Code 9.10 and will focus on the estimation of the unknown parameters β_0, β_1, θ, and σ^2 and is executed with WinBUGS Code 9.11 using 35,000 observations for the simulation and 5,000 for the burn-in.

The 30 observations are the components of a vector Y, which has a multivariate normal distribution with mean vector given by Equation 9.48 and precision matrix, which is the inverse of the variance–covariance matrix of an AR(1) process with correlation θ.

WinBUGS Code 9.11

```
model;
{
beta0~dnorm(0,.001)
beta1~dnorm(0,.001)
theta~dbeta(6,4)
v~dgamma(.001,.001)
for( t in 1:30){mu[t]<-exp(beta0+beta1*t)}
Y[1,1:30]~dmnorm(mu[],tau[,])
   tau[1:30,1:30]<-inverse(Sigma[,])
   for(i in 1:30){Sigma[i,i]<-v/(1-theta*theta)}
for(i in 1:30){for(j in i+1:30){Sigma[i,j]<-
v*pow(theta,j)*1/(1-theta*theta)}}
   for( i in 2:30){ for(j in 1: i-1){Sigma[i,j]<-v*pow(theta,i-
1)*1/(1-theta*theta)}}
}
list(Y=structure(.Data=c( 2.857651, 3.371453, 1.707308, 5.640147,
5.541377, 3.258980, 4.586094, 5.969050, 6.562987, 5.250836,
8.196521, 7.823749, 5.686929, 1.332517, 5.519606, 5.818782,
6.188920, 8.471027, 9.742210, 10.204972, 11.295405, 11.846945,
10.942453, 6.460867, 9.188900, 9.682623, 9.999043, 7.789756,
8.691994, 11.280566),.Dim=c(1,30)))
list(beta0=1,beta1=.05,theta=.6, v=4)
```

The posterior analysis is reported in Table 9.8.

The posterior analysis reveals that the posterior mean of .6349 for θ is very close to the value used to generate the observations, and the same holds for the other parameters, as for

TABLE 9.8

Posterior Analysis for Nonlinear Model

Parameter	Value	Mean	SD	Error	2 1/2	Median	97 1/2
β_0	1	1.407	.1794	.00945	1.021	1.41	1.757
β_1	.05	.0327	.0077	.00040	.0170	.0328	.0484
θ	.6	.6349	.1513	.00269	.3218	.6455	.891
σ^2	4	2.449	1.056	0.01645	0.8932	2.303	4.932

example σ^2, with a posterior mean of 2.449, which is also very close to its nominal value of 4. What about the other parameters?

9.3.8 Stationary Models

As presented earlier, time series will often have a well-defined trend and seasonal components, and a good fitting model will account for these components in such a way that the residuals will tend to have means of 0, constant variance, and, of course, autocorrelation. The adjacent observations have correlations that can be positive or negative, as for example, the airline passenger data where higher values are followed by higher values. On the other hand, adjacent observations can be negatively correlated such as when higher sales values are followed by lower numbers. The models to be discussed will serve as the residuals with complex autocorrelation patterns for time series regression model.

The first class of models to be studied is the moving average, which is a special case of a stationary stochastic process. A stationary process $\{Y(t), t \geq 0\}$ is defined as one that satisfies the time invariance property:

$$[Y(t_1), Y(t_2), ..., Y(t_n)] \sim [Y(t_1 + h), Y(t_2 + h), ..., Y(t_n + h)] \tag{9.49}$$

for all $h > 0$.

There are processes that are stationary in the mean and covariance stationary, but not strictly stationary in the sense of Equation 9.49. However, if a Gaussian process is covariance stationary, it is also strictly stationary.

A moving average process MA(q) is defined as

$$Y(t) = W(t) + \beta_1 W(t-1) + .. + \beta_q W(t-q), \tag{9.50}$$

where $W(t), W(t-1), ..., W(t-q)$ is a sequence of independent white noise random variables with variance σ^2; and the β_i, $i = 1, 2, ..., q$, are unknown real parameters. It is obvious that the mean value function of the process is 0, the variance is

$$\text{Var}[Y(t)] = \sigma^2 \left(1 + \sum_{i=1}^{i=q} \beta_i^2\right), \tag{9.51}$$

and the autocorrelation of lag k is

$$\rho(k) = \sum_{i=0}^{i=q-k} \beta_i \beta_{i+k} / \sum_{i=0}^{i=q} \beta_i^2. \tag{9.52}$$

As an example, consider the MA(1) process

$$Y(t) = W(t) + \beta_1 W(t-1), \quad t = 2, 3, 4, ..., 100, \tag{9.53}$$

where $\beta_1 = .8$ and $\sigma^2 = 1$.

R Code 9.11 generates 1000 observations from the MA(1) process (Equation 9.53):

R Code 9.11

```
set.seed(1)
b<-c(.8)
x<-w<-rnorm(1000)
for ( t in 2:1000){
for ( j in 1:1) x[t]<-w[t]+b[j]*w[t-j]}
```

The first 20 values generated appear as components of the vector y.

```
Y=( -0.626453811, -0.317519724, -0.688713953, 0.926777912,
1.605732414, -0.556862167, -0.168945655, 1.128267947,
1.166441116, 0.155236694, 1.267470459, 1.599268171,
-0.309365991, -2.711692352, -0.646828992, 0.855011125,
-0.052137150, 0.930884000, 1.576290164, 1.250878277).
```

Based on these 20 values, the Bayesian analysis will estimate the parameters of the MA(1) process (Equation 9.53), and the posterior analysis is executed with WinBUGS Code 9.12 with 35,000 observations for the simulation and a burn-in of 5,000. Noninformative prior distributions are used for the following parameters: β_1 is normal(.8,.01), and that for σ^2 it is gamma(.001,.001). Note that it is assumed that the data vector of 20 observations has a multivariate normal distribution with mean vector of 0, and a variance–covariance matrix is given by Equations 9.51 and 9.52, respectively.

WinBUGS Code 9.12

```
model;
{
beta~dnorm(.8,1)
v~dgamma(.1,.1)

for( t in 1:20){mu[t]<-0}
Y[1,1:20]~dmnorm(mu[],tau[,])

for( i in 1:20){Sigma[i,i]<-v*(1+pow(beta,2))}
for ( i in 1:19){Sigma[i,i+1]<-v*beta}
for ( i in 1:18){for ( j in i+2: 20) {Sigma[i,j]<-0}}
for ( i in 2:20){Sigma[i,i-1]<-v*beta}
 for( i in 3:20){ for ( j in 1:i-2 ){Sigma[i,j]<-0}}
   tau[1:20,1:20]<-inverse(Sigma[,])
}
list(Y=structure(.Data=c(-0.626453811, -0.317519724, -0.688713953,
0.926777912, 1.605732414, -0.556862167, -0.168945655,
1.128267947, 1.166441116, 0.155236694, 1.267470459,
1.599268171, -0.309365991, -2.711692352, -0.646828992,
0.855011125, -0.052137150, 0.930884000, 1.576290164,
1.250878277),.Dim=c(1,20)))
list(v=1, beta=.8)
```

TABLE 9.9

Posterior Analysis for MA(1) Process

Parameter	Value	Mean	SD	Error	2 1/2	Median	97 1/2
β_1	0.8	1.189	0.5342	0.01519	0.3447	1.196	2.366
σ^2	1	0.658	0.3811	0.00869	0.1552	0.592	1.571

The posterior analysis for the MA(1) process is reported in Table 9.9.

Note that the posterior 95% credible intervals for the parameters include the values of the parameters used to generate the 20 observations used for the data and in the list statement of WinBUGS Code 9.12. Additional details of the posterior analysis for the MA(1) process is left as several exercises for the student.

Before leaving the MA model, the MA(1) process will serve as errors for a regression model

$$Y(t) = \gamma_0 + \gamma_1 t + \gamma_2 t^2 + Z(t), \quad t = 1, 2, \ldots, n, \tag{9.54}$$

where the errors follow the MA(1) process

$$Z(t) = W(t) + \beta W(t-1), \tag{9.55}$$

with $\gamma_0 = 1$, $\gamma_1 = 2$, $\gamma_2 = 3$, and $\beta = .8$, and the variance of white noise is $\sigma^2 = 1$.

R Code 9.12 generates 20 observations from the regression process (Equation 9.54) with the known parameters values as given earlier.

R Code 9.12

```
set.seed(1)
b<-c(.8)
y<-x<-w<-rnorm(20)
for ( t in 2:20)
{for ( j in1:1) x[t]<-w[t]+b[j]*w[t-j]}
{for( t in1:20)y[t]<-1+2*t+3*t^2+x[t]}
Y=( 5.373546, 16.682480, 33.311286, 57.926778, 87.605732,
120.443138, 161.831054, 210.128268, 263.166441, 321.155237,
387.267470, 458.599268, 533.690634, 614.288308, 705.353171,
801.855011, 901.947863, 1009.930884), 1123.576290, 1242.250878)
```

Of course, the goal of the Bayesian analysis is to estimate the unknown regression parameters γ_i, $i = 1, 2, 3$, the moving average coefficient β, and Gaussian noise variance σ^2. Using the above 20 values generated by R Code 9.12 according to the model in Equations 9.54 and 9.55, WinBUGS Code 9.13 is executed with 35,000 observations for the simulation and 5,000 for the burn-in. The vector Y of 20 observations are normally distributed with mean vector given by Equation 9.54 and variance–covariance matrix appropriate to the MA (1) errors with moving average parameter β.

WinBUGS Code 9.13

```
model;{
beta~dnorm(.8,1)
v~dgamma(.1,.1)
```

```
g0~dnorm(1,1)
g1~dnorm(2,1)
g2~dnorm(3,1)
for( t in 1:20){mu[t]<-g0+g1*t+g2*t*t}
Y[1,1:20]~dmnorm(mu[],tau[,])
for( i in 1:20){Sigma[i,i]<-v*(1+pow(beta,2))}
for ( i in 1:19){Sigma[i,i+1]<-v*beta}
for ( i in 1:18){for ( j in i+2: 20) {Sigma[i,j]<-0}}
for ( i in 2:20){Sigma[i,i-1]<-v*beta}
for( i in 3:20){ for ( j in 1:i-2 ){Sigma[i,j]<-0}} tau[1:20,1:20]<-
inverse(Sigma[,])
}
list(Y=structure(.Data=c(5.373546, 16.682480, 33.311286,
 57.926778, 87.605732, 120.443138,161.831054, 210.128268,
263.166441, 321.155237, 387.267470, 458.599268, 533.690634,
614.288308, 705.353171, 801.855011, 901.947863, 1009.930884,
1123.576290, 1242.250),.Dim=c(1,20)))
list(v=1, g0=1,g1=2,g2=3, beta=.8)
```

The posterior analysis is reported in Table 9.10.

Bayesian posterior means appear to be very close to the values of the parameters used to generate the data, and this is most obvious for γ_2 with a posterior mean of 3.001 compared to its nominal value of 3!

The next example is a regression model with MA(1) errors but with harmonic seasonal effects, which is specified as

$$\begin{aligned} Y(t) &= \gamma_0 + \gamma_1 \sin(2\pi t/12) + \gamma_2 \sin(4\pi t/12) \\ &\quad + \gamma_3 \sin(8\pi t/12) + \gamma_4 \cos(8\pi t/12) + Z(t), \end{aligned} \tag{9.56}$$

where

$$Z(t) = W(t) + \beta W(t-1) \tag{9.57}$$

is an MA(1) process with parameter β.

The Bayesian analysis will consist of estimating the parameters γ_i, $i = 0, 1, 2, 3, 4$; β; and σ^2, the variance of the Gaussian noise process. R Code 9.13 generates 100 observations from the harmonic seasonal regression model (Equation 9.56) with parameter values $\beta = .8$, $\gamma_0 = 1 = \gamma_1, \gamma_2 = .2$, and $\gamma_3 = \gamma_4 = .1$.

TABLE 9.10

Posterior Analysis for Regression Model with MA(1) Errors

Parameter	Value	Mean	SD	Error	2 1/2	Median	97 1/2
β	0.8	1.184	0.5182	0.01273	0.375	1.182	2.318
γ_0	1	0.9514	0.733	0.02000	−0.4733	0.9477	2.393
γ_1	2	2.026	0.1849	0.00681	1.622	2.029	2.381
γ_2	3	3.001	0.0093	0.00033	2.982	3.001	3.018
σ^2	1	0.7069	0.411	0.00804	0.1643	0.637	1.679

R Code 9.13

```
set.seed(1)
b<-c(.8)
y<-x<-w<-rnorm(100,0,.1)
for ( t in 2:100){
for ( j in 1:1) x[t]<-w[t]+b[j]*w[t-j]}
{for( t in 1:100)y[t]<-1+sin(2*pi*t/12)+0.2*sin(4*pi*t/12)+0.1*sin(8*pi*t
/12)+0.1*cos(8*pi*t/12)+x[t]}
```

The 100 values generated from the harmonic seasonal regression model are contained in the vector Y:

Y = (1.647162240, 1.870875972, 2.031128605, 1.822100655,
1.350765620, 1.044313783, 0.692913056, 0.283403931, 0.216644112,
0.012895725, 0.316939425, 1.259926817, 1.678871022, 1.631458709,
2.035317101, 1.814923976, 1.184978664, 1.193088400, 0.867436637,
0.295664964, 0.239409843, 0.149103876, 0.260219781, 0.907030029,
1.612642060, 1.946601130, 2.079930150, 1.569883984, 1.024717183,
1.103542152,
0.879110901, 0.268992728, 0.130544143, 0.023005281, 0.048182020,
0.948335779, 1.637179061, 1.865153408, 2.205257465, 1.893742468,
1.234794079, 1.061501944, 0.759235024, 0.282000526, 0.075657486,
-0.128477915, 0.170050963, 1.206019849, 1.760055634, 1.981751020,
2.210299206, 1.700068695, 1.175342237, 1.014353266, 0.762760944,
0.483259023, 0.221709844, -0.136419125, 0.163633572, 1.132072110,
1.939165029, 2.090833365, 2.165834736, 1.787402228, 1.118105231,
1.059417373, 0.544415142, 0.172735932, 0.232569723, 0.226893490,
0.411552265, 1.067046119, 1.714084542, 1.858076289, 1.899908849,
1.658276815, 1.169178890, 1.064647185, 0.717330182, 0.117572348,
-0.004028549, -0.061639304, 0.297186789, 1.041890280, 1.647316896,
1.983438676, 2.232946013, 1.784052458, 1.202859546, 1.156311384,
0.676923521, 0.247962315, 0.312669686, 0.160225630, 0.404892816,
1.282795319, 1.626827314, 1.743174026, 1.931677505, 1.584113791)

Based on the first 20 values generated from the regression model with harmonic seasonal effect, and using noninformative priors for the regression coefficients γ_i, $i = 0, 1, 2, 3, 4$, moving average parameter β, and variance of the Gaussian noise σ^2, the Bayesian analysis is executed with WinBUGS Code 9.14 using 35,000 observations for the simulation and 5,000 for the burn-in.

WinBUGS Code 9.14

```
model;
{
g0~dnorm(0,.01)
```

```
g1~dnorm(0,.01)
g2~dnorm(0,.01)
g3~dnorm(0,.01)
g4~dnorm(0,.01)

v~dgamma(.01,.01)
beta~dbeta(8,2)

for( t in 1:20){mu[t]<-g0+g1*sin(2*3.1416*t/12)+g2*sin(4*3.1416*t/
12)+g3*sin(8*3.1416*t/12)+g4*cos(8*3.1416*t/12)}
Y[1,1:20]~dmnorm(mu[],tau[,])

for( i in 1:20){Sigma[i,i]<-v*(1+pow(beta,2))}
for ( i in 1:19){Sigma[i,i+1]<-v*beta}
for ( i in 1:18){for ( j in i+2: 20) {Sigma[i,j]<-0}}
for ( i in 2:20){Sigma[i,i-1]<-v*beta}
for( i in 3:20){ for ( j in 1:i-2 ){Sigma[i,j]<-0}}tau[1:20,1:20]<-
inverse(Sigma[,]
}
list(Y=structure(.Data=c(1.647162240, 1.870875972, 2.031128605,
1.822100655, 1.350765620, 1.044313783, 0.692913056, 0.283403931,
0.216644112, 0.012895725, 0.316939425, 1.259926817, 1.678871022,
1.631458709, 2.035317101, 1.814923976, 1.184978664, 1.193088400,
0.867436637, 0.295664964),.Dim=c(1,20)))
list( g0=1,g1=1,g2=.2,g3=.1,g4=.1,beta=.8,v=.01))
```

Table 9.11 reports the results of the posterior analysis for the regression model with moving average errors.

Comparing the actual values of the parameters to their corresponding posterior means reveals that the estimates are very close. In fact, the 95% credible intervals contain the actual values of the parameters used to generate the data in the list statement of WinBUGS Code 9.14, the sample information used for the Bayesian analysis.

One way to generalize the moving average and autoregressive processes is to combine the two into the ARMA(p,q) process defined as

$$Y(t) = \sum_{i=1}^{i=p} \alpha_i Y(t-i) + W(t) + \sum_{j=1}^{j=q} \beta_j W(t-j), \tag{9.58}$$

TABLE 9.11

Posterior Analysis for Harmonic Seasonal Effects with MA(1) Errors

Parameter	Value	Mean	SD	Error	2 1/2	Median	97 1/2
β	.8	.7756	.1064	.00047	.5367	.7862	.9542
γ_0	1	1.046	0.039	0.00016	0.9682	1.046	1.124
γ_1	1	0.9134	0.0533	0.00021	0.8077	0.9132	1.019
γ_2	.2	.1414	.0475	.00018	.0475	.1413	.2363
γ_3	.1	.1043	.0273	.00010	.0500	.1043	.1589
γ_4	.1	.0946	.0284	.00011	.0382	.094	.1511
σ^2	.01	.0094	.0040	.00002	.0044	.0085	.0196

where $\{W(t), t > 0\}$ is white noise with variance σ^2; α_i, $i = 1,2,...,p$, is a sequence of unknown autoregressive parameters; and β_j, $j = 1,2,...,q$, is a sequence of moving average parameters. As expected, the autocorrelation patterns are quite involved and represent complex correlation structure.

Consider an ARMA(1,1) process

$$
\begin{aligned}
Y(t) &= \theta Y(t-1) + W(t) + \beta W(t-1), \quad t = 2,3,..,n \\
&= W(t) + (\theta + \beta) \sum_{i=1}^{i=\infty} \theta^{i-1} W(t-i),
\end{aligned} \tag{9.59}
$$

and it follows that

$$
\mathrm{Var}[Y(t)] = \sigma^2 + \sigma^2(\theta + \beta)^2 / (1 - \theta^2), \tag{9.60}
$$

and the autocovariance is

$$
\mathrm{cov}[Y(t), Y(t+k)] = (\theta + \beta)\theta^{k-1}\sigma^2 + (\theta + \beta)^2 \sigma^2 \theta^k / (1 - \theta^2). \tag{9.61}
$$

This is sufficient information for a Bayesian analysis in that the variance–covariance matrix of the vector of observations Y can be coded in WinBUGS. R Code 9.14 generates observations from an ARMA(1,1) process with $\theta = .5$, $\beta = .5$, and $\sigma^2 = 1$.

R Code 9.14

```
set.seed(1)
x<-arima.sim(n=10000, list(ar=.5,ma=0.5))
coef(arima(x, order =c(1,0,1)))
```

The first 24 observations from the ARMA (1,1) process are the components of the vector Y:

```
Y=(2.00439112, 0.57587660, -2.23738188, -1.10110996, -0.03302313,
-0.05516863, 0.90815676, 1.74721768, 1.87812076, 2.15498841,
2.31911919, 1.62519273, -1.13947284, -0.94458652, -0.21850913,
-0.29311444, -1.69520736, -2.06112993, -0.85169843, 1.14180112,
1.14745261, 0.91000405, 0.59503279, -1.10644568, -1.65674718)
```

Based on the preceding sample information and prior information for the unknown parameters β, θ, and σ^2, the Bayesian analysis is executed with 35,000 observations for the simulation and a burn-in of 5,000.

WinBUGS Code 9.15

```
model;
{
theta~dbeta(5,5)
beta~dbeta(5,5)
v~dgamma(.01,.01)
for ( t in 1:25){ mu[t]<-0}
Y[1,1:25]~dmnorm(mu[],tau[,])
for( i in 1:25){Sigma[i,i]<-v+v*pow(theta+beta,2)/(1-theta*theta)}
for( i in 1:24){for(j in i+1:25){Sigma[i,j]<-(theta+beta)*pow(theta,j-1)*v+
pow(theta+beta,2)*v*pow(theta,j)/(1-theta*theta)}}
for( i in 2:25){ for ( j in 1:i-1){Sigma[i,j]<-(theta+beta)*pow(theta,j-1)*v+
pow(theta+beta,2)*v*pow(theta,j)/(1-theta*theta)}}
tau[1:25,1:25]<-inverse(Sigma[,])
}
list(Y=structure(.Data=c( 2.00439112, 0.57587660, -2.23738188, -1.10110996,
-0.03302313,-0.05516863, 0.90815676, 1.74721768, 1.87812076, 2.15498841,
2.31911919, 1.62519273, -1.13947284, -0.94458652, -0.21850913, -0.29311444,
-1.69520736, -2.06112993, -0.85169843, 1.14180112, 1.14745261, 0.91000405,
0.59503279, -1.10644568, -1.65674718
),.Dim=c(1,25)))
list(theta=.5,v=1,beta=.5)
```

Bayesian inferences for the parameters of the ARMA(1,1) model are reported in Table 9.12. Comparing the posterior means to the actual value of the parameter reveals that the Bayesian analysis is providing sound inferences. For example, consider β with actual value of .5 and posterior mean of .4979, then one would conclude that indeed, the Bayesian analysis is quite accurate, at least in this case.

TABLE 9.12
Posterior Distribution for ARMA(1,1)

Parameter	Value	Mean	SD	Error	2 1/2	Median	97 1/2
β	.5	.4979	.1512	.00071	.2094	.4978	.7857
θ	.5	.4604	.1407	.00081	.1956	.459	.7339
σ^2	1	1.01	0.4248	0.00386	0.4092	0.9369	2.05

9.4 Comments and Conclusions

This chapter presents a Bayesian analysis for two general classes of stochastic processes: (1) queues and (2) time series.

The chapter begins with describing the fundamental properties of queues, including the interarrival time of customers, the service times of the servers for the customers, and the traffic intensity. Queue studies begin with the M/M/1 and then proceed to the G/M/1 and M/G/1 and, finally, to the G/G/1 queue.

For the M/M/1 queue, WinBUGS Code 9.1 generates observations for the interarrival times between customers and the service time provided by a single server to the customer, and then WinBUGS Code 9.2 provides the Bayesian analysis for the two parameters, plus the traffic intensity parameter. This scenario is essentially repeated for the G/M/1, M/G/1, and G/G/1 queues, but where the general distribution is an Erlang distribution.

The section for time series begins with a discussion of the fundamentals including the ideas of the trend, the seasonal effects, and the noise (errors) of the series. Airline passenger data illustrate these concepts, and the R Code decompose delineates the trend, seasonal effects, and noise in one graph portrayed in Figure 9.2. The autocorrelation of a time series is explained and then illustrated with Nile River level data set portrayed in Figure 9.7.

The R Code command acf(x) computes the autocorrelation function of a time series denoted by x and is computed for the Nile River level data. Also computed are the autocorrelations of lag $k = 1, 2, \ldots$, etc.

Random walk and random walk with drift are introduced in Section 9.3.5. R Code 9.5 generates observations for the random walk with drift, and WinBUGS Code 9.7 provides the Bayesian analysis for the parameters, and the posterior analysis is reported in Table 9.4.

Chapter 9 proceeds by introducing the Bayesian analysis of more sophisticated time series such as the autoregressive and regression models with errors that follow an autoregressive process. For example, for the latter, R Code 9.8 generates observations for a regression model with linear trend and AR(1) errors.

WinBUGS executes the Bayesian analysis where the vector of observations follows a multivariate normal distribution; thus, the mean and variance–covariance matrix need to be specified in the code. The Bayesian analysis for a more complex regression model with quadratic trend plus harmonic seasonal effects is described, and the analysis is reported in Table 9.6.

Stationary time series such as the moving average process is introduced in Section 9.3.8. First, a general MA(q) process is defined by Equation 9.50, and an MA(1) process is used to demonstrate the Bayesian approach to inference. For example, R Code 9.11 generates observations for the first-order process with two parameters: the moving average parameter $\beta = .8$ and variance $\sigma^2 = 1$ of the Gaussian noise, and then based on those observations, the Bayesian analysis is executed with WinBUGS Code 9.12 and is reported in Table 9.9.

This chapter concludes with the Bayesian analysis for a regression model with MA(1) errors and, lastly, with the autoregressive moving average model ARMA(1,1).

9.5 Exercises

1. Write an essay about the contents of Chapter 9.
2. For a queue, define the following:
   a. The number of customers who arrive at time t
   b. The number of busy servers at time t
   c. The number in the system at time t
   d. The time at which the person arrives at the queue
   e. The time in the system by a customer arriving at time t
3. a. Define the traffic intensity of a queue.
   b. When is a queue unstable? Explain carefully.
4. For an M/M/1 queue show that the number of people in the queue has a limiting distribution, which is geometric given by Equation 9.60. Assume that the queue is stable.
5. Refer to WinBUGS Code 9.1.
   a. Generate the 32 interarrival times with an exp(2) distribution.
   b. Generate the 24 service times with an exp(4) distribution.
   c. What is the average interarrival time?
   d. What is the average service time?
6. Refer to WinBUGS Code 9.2. Based on the 32 interarrival times X and 24 service times Y generated by WinBUGS Code 9.1,
   a. What is the posterior distribution of λ?
   b. What is the posterior distribution of μ?
   c. What is the 95% credible interval for the traffic intensity ρ?
   d. Is the queue stable?
7. Refer to the G/G/1 queue of Equation 9.25.
   a. Define the traffic intensity ρ for this queue. See Equation 9.16.
   b. Refer to WinBUGS Code 9.5 and generate the 32 interarrival times with a gamma(2,4) distribution.
   c. Refer to WinBUGS Code 9.6 and generate the 24 service times with a gamma (2,8) distribution.
   d. Refer to WinBUGS Code 9.6 and find the posterior distribution α_a, β_a, the parameters of the gamma posterior distribution of the interarrival times. See Table 9.3.
   e. Using WinBUGS Code 9.6, what are the posterior distribution of α_s, β_s, the parameters of the gamma posterior distribution of the service times of the queue? See Table 9.3.

8. Refer to Section 9.3.1 and write a two-page essay on the Bayesian inferences used in time series analysis.
9. Using R Code 9.2
   a. Duplicate Figure 9.2, a plot of the monthly airline passenger data for 12 years.
   b. Using R Code 9.3., duplicate Figure 9.3, a plot of the aggregation of the monthly airline passenger data.
10. Refer to Section 9.3.3.
    a. Describe the trend, seasonality, and errors of a time series.
    b. Using the R Code command plot(decompose(AirPassengers)), duplicate Figure 9.5, a plot of the decomposition of the airline passenger data.
11. Given a time series, define the mean value function, the variance function, and the autocorrelation function. See Equations 9.23 and 9.24.
12. a. Duplicate the plot of the Nile River level information in Figure 9.6.
    b. Verify the acf plot of the Nile River data portrayed in Figure 9.7.
    c. What is the lag 1 and lag 2 autocorrelations?
13. a. Define the AR(p) model.
    b. For an AR(1) model, what is the autocorrelation function?
    c. Given a uniform prior density for θ, $|\theta| < 1$, derive the posterior distribution θ given by Equation 9.40.
    d. Using WinBUGS Code 9.8, execute the Bayesian analysis for the AR(1) model with 35,000 for the simulation and 5,000 for the burn-in.
    e. Verify the posterior analysis of the AR(1) model reported in Table 9.8. What is the posterior median of θ? Does the 95% credible interval for θ include the value .8?
14. Refer to the regression model with autocorrelated errors defined by Equation 9.44 and consider the simple linear regression model with AR(1) errors of Equation 9.45.
    a. Use R Code 9.8 to generate 100 observations from the regression model (Equation 9.45), where $\theta = .5$ and $\sigma = 10$.
    b. Based on these 100 observations generated with R Code 9.8, execute a Bayesian analysis using WinBUGS Code 9.8 with 45,000 observations for the simulation and 5,000 for the burn-in.
    c. Verify the posterior analysis of Table 9.6.
    d. What is the 95% credible interval for β_1?

15. Refer to Equation 9.46, a regression model with quadratic effects for trend and including harmonic seasonal effects with $\theta = .6$ and $\sigma = 0.5$.
    a. Use R Code 9.9 to generate 120 observations from Equation 9.46.
    b. Using WinBUGS Code 9.10, perform a Bayesian analysis to estimate the parameters of the model in Equation 9.47.
    c. What prior distributions are used for the parameters of this regression model (Equation 9.47)?
    d. What is the mean vector and covariance matrix of the 30×1 vector of observations?
    e. Verify the posterior analysis reported in Table 9.7. What is the posterior median of β_3?
16. a. Refer to Equation 9.50 and define an MA(q) process.
    b. Derive the autocorrelation function (Equation 9.52) of an MA(q) process.
    c. Use R Code 9.11 to generate 1000 observations from an MA(1) process with parameters $\beta = .8$ and $\sigma^2 = 1$.
    d. Using the first 20 observations generated by R Code 9.11, execute the Bayesian analysis given by WinBUGS Code 9.12 with 35,000 observations for the simulation and 5,000 for the burn-in.
    e. Verify the Bayesian analysis of Table 9.9.
    f. What is the posterior mean of β and is it reasonable?
17. a. Define an ARMA(p,q) time series.
    b. For an ARMA(1,1) model with parameters θ, β, and σ^2, derive the autocorrelation function in Equation 9.61.
    c. Use R Code 9.14 to generate 10,000 observations from an ARMA(1,1) process with $\theta = .5$, $\beta = .5$, and $\sigma^2 = 1$.
    d. Based on the first 24 observations generated by R Code 9.14, use WinBUGS Code 9.15 to execute a Bayesian analysis with 45,000 observations for the simulation and burn-in of 5,000.
    e. What prior distributions are used for the Bayesian analysis?
    f. What is the posterior mean of β?
    g. What is the 95% credible interval for θ?

References

1. Insua, D. R., Ruggeri, F., and Wiper, M. P. 2012. *Bayesian Analysis for Stochastic Process Models.* New York: John Wiley & Sons.
2. Little, J. D. C. 1961. A proof of the queuing formula $L = \lambda W$. *Operations Research* 9:383–387.
3. Gross, D., Shortle, J. F., Thompson, J. M., and Harris, C. M. 2008. *Fundamentals of Queuing Theory, Third Edition*. New York: John Wiley & Sons.
4. Ausin, M. C., and Lopes, H. 2007. Bayesian estimation of ruin probabilities with heterogeneous and heavy-tailed insurance claim size distribution, *Australian and New Zealand Journal of Statistics* 49:415–452.
5. Wiper, M. P. 1998. Bayesian analysis of Er/M/1 and Er/M/c queues, *Journal of Statistical Planning and Inference* 69:65–79.
6. Broemeling, L. D. 1984. *Bayesian Analysis of Linear Models*. New York: Marcel Dekker.
7. Smith, J. O. 1987. A generalization of Bayesian steady forecasting model, *Journal of the Royal Statistical Society, Series B* 41(3):375–387.
8. Pole, A., West, M., and Harrison, J. 1994. *Applied Bayesian Forecasting and Time Series Analysis*. Boca Raton, FL: Chapman and Hall.
9. Barber, D., Cemgil, A. T., and Chippa, S. (Editors). 2011. *Bayesian Time Series Models*. Cambridge, UK: Cambridge University Press.
10. Petris, G., Pertrone, S., and Campagnoli, P. 2009. *Dynamic Linear Models*. New York: Springer-Verlag.
11. Cowpertwait, P., and Metcalfe, A. V. 2008. *Introduction to Time Series Analysis with R*. New York: Springer-Verlag.

Index

Page numbers followed by f and t indicate figures and tables, respectively.

M

N

Y

Z

Printed in the United States
by Baker & Taylor Publisher Services